FOOD BIOPRESERVATIVES *of* MICROBIAL ORIGIN

Bibek Ray, Ph.D.
Professor
Department of Animal Science
University of Wyoming
Laramie, Wyoming

Mark Daeschel, Ph.D.
Associate Professor
Department of Food Science
and Technology
Oregon State University
Corvallis, Oregon

CRC Press
Boca Raton Ann Arbor London Tokyo

Library of Congress Cataloging-in-Publication Data

Food biopreservatives of microbial origin / editors, Bibek Ray, Mark Daeschel.
p. cm.
Includes bibliographical references and index.
ISBN 0-8493-4943-5
1. Food—Microbiology. 2. Food—Preservation. I. Ray, Bibek. II. Daeschel, Mark.
QR115.F634 1992
664′.0287—dc20

92-3191
CIP

Direct all inquiries to CRC Press, Inc., 2000 Corporate Blvd., N. W., Boca Raton, Florida, 33431.

International Standard Book Number 0-8493-4943-5

Library of Congress Card Number 92-3191
Printed in Mexico 1 2 3 4 5 6 7 8 9 0
Printed on acid-free paper

DEDICATION

To Purnima, Purbita, and Ranjan for their patience

Bibek

EDITORS

Bibek Ray, Ph.D., is professor of Food Microbiology in the Department of Animal Science, University of Wyoming, Laramie. Professor Ray obtained a B.S. and M.S. degree in Veterinary Science from the University of Calcutta and University of Madras, in India, respectively. He received his Ph.D. degree in Food Science from the University of Minnesota in 1970 and joined the faculty in the Department of Food Science, North Carolina State University, Raleigh, in 1971. He joined the University of Wyoming in 1981 as Associate Professor of Food Microbiology.

While at the University of Minnesota and at North Carolina State University, he conducted research in the area of sublethal microbial injury and its influence on their detection from foods. Since joining the University of Wyoming, he expanded his research interest with sublethal microbial injury in the areas of biopreservation of foods and beneficial intestinal microbes. Currently, most of his research is in the area of food biopreservation using antimicrobial metabolites of starter culture microorganisms. His laboratory has conducted extensive studies on various aspects of a particular bacteriocin, pediocin AcH from *Pediococcus acidilactici* H. Except nisin, no other bacteriocin has been studied so thoroughly. In addition to this, his laboratory is conducting suitability of other bacteriocins from *Leuconostoc* spp., *Lactococcus* ssp., *Lactobacillus* spp., and *Propionibacterium* spp. In the last 4 years he has received over $700,000 in grants from different agencies to support these studies, notably, National Live Stock and Meat Board, Binational Agriculture Research Development Fund (with Israel), North Atlantic Treaty Organization (with Turkey), and Wyoming Economic Development Funds.

Professor Ray has authored more than 70 research articles, reviews, and book chapters. He has also edited the book titled *Detection of Injured Index and Pathogenic Bacteria* published by CRC Press in 1989. He is a member of the American Society for Microbiology, Institute of Food Technologists, International Association of Milk, Food, and Environmental Sanitarians. He is also a Fellow of the American Academy of Microbiology and a member of the Editorial Board of the Journal of Food Protection.

Mark A. Daeschel, Ph.D., is Associate Professor of Food Science and Technology at Oregon State University, Corvallis.

Dr. Daeschel obtained his undergraduate training in microbiology at The State University of New York at Plattsburgh, receiving the B.A. degree in 1977. He studied under the late Orvin J. Mundt at the University of Tennessee, receiving the M.S. degree in Microbiology in 1979. The Ph.D. degree was awarded in Food Science in 1982 at North Carolina State University under the tutelage of Dr. Henry P. Fleming. In 1982, Dr. Daeschel was appointed as a USDA-ARS Research Microbiologist and Assistant Professor in the Department of Food Science at North Carolina State University. It was in 1988 that he assumed his current position at Oregon State University.

Dr. Daeschel is a member of the American Society of Microbiology, Institute of Food Technologists, Society of Industrial Microbiology, and the American Society of Enology and Viticulture. He is the author or co-author of more than 40 papers and is the senior inventor on 2 patents relating to novel industrial strains of lactic acid bacteria. His current major research interests relate to basic and applied investigations on bacteriocins of lactic acid bacteria.

PREFACE

Since the 1970s health conscious consumers have started showing dissatisfaction with foods that are "harshly processed" and "chemically preserved" but an interest for foods that are healthy and "natural" or "close-to-natural." To capture this niche, food processors raced to market foods that are minimally processed, without preservatives, and stored under refrigeration to maintain their freshness appeal and to prolong shelf life; many are also packaged with modified atmosphere, such as vacuum packaging to extend the shelf life. From the standpoint of microbiological quality, many of these foods can harbor different types of spoilage and pathogenic bacteria that are capable of multiplication under refrigeration and in the absence of air. Since these products have a relatively long storage life, even a low initial microbial population can reach very high levels which in turn may affect the safety and acceptance quality of the foods. Any deviation from ideal storage temperature can accelerate the growth rate of many bacteria present in these foods. The regulatory agencies and advisory groups have recognized the potential problems and advocated the use of suitable preservatives.

What are such suitable preservatives? Perhaps, the antimicrobial metabolites of fermentation microorganisms, which have been the basis of food biopreservation since ancient times could fill the need. In recent years, while consumers have shown concern about many processed and preserved foods, their consumption of fermented foods have increased. The consumer probably perceives food containing starter culture microorganisms and metabolites as natural, healthy, and beneficial. Thus, if the antimicrobial metabolites of the starter cultures are found to be effective in controlling spoilage and pathogenic bacteria with minimally processed refrigerated foods, their use will likely be acceptable to consumers and regulatory agencies. Some of these are already in use in many foods for different reasons. Among the antimicrobial metabolites which include certain proteins called bacteriocins, several organic acids, diacetyl, ethanol, hydrogen peroxide, and reuterin, the bacteriocins have generated considerable interest in recent years. Many bacteriocin-producing starter culture bacteria have been identified. However, at present, very limited studies have been conducted both on the characteristics of bacteriocins that are necessary for their use in food systems and their antibacterial effectiveness in food systems.

This book is intended to provide information both on the current need of food biopreservatives (Chapters 1 and 2) and the different antimicrobial metabolites of starter culture microorganisms. Basic and applied up-to-date information about antimicrobial effectiveness of organic acid (Chapter 5) and diacetyl (Chapter 6), hydrogen peroxide and reuterin (Chapter 7), bacteriocins (Chapters 8 to 11), and bacterial cells (Chapter 4) have been provided. The reasons for presenting bacteriocins in four chapters are first to give an overview of the occurrence of bacteriocins among starter culture bacteria and the wide differences that exist in the study methods (Chapter 8), to provide as complete information as possible on two bacteriocins, nisin (Chapter 9) and pediocin(s) of *Pediococcus acidilactici* (Chapter 10) that have probably been most thoroughly studied, and a final chapter on some other bacteriocins that have been studied less extensively, especially for the characteristics that are important for their use in food systems (Chapter 11). Also, a chapter has been devoted on methods that could be useful in differentiating these antimicrobial substances (Chapter 3). Finally, a chapter on antimicrobial metabolites from yeast have been included (Chapter 12).

Our intention is to provide both basic and applied information on the antimicrobial metabolites of fermentation starter culture microorganisms that have the potential for use as biopreservatives in many foods that otherwise can be spoiled or involved in foodborne diseases from the growth of undesirable microorganisms. We hope this book will be helpful to researchers working in food preservation, to regulatory agencies, advisory groups, food industries, and to academicians interested in teaching a course in food biopreservation. Finally, we hope it will be of interest to food advocates, health conscious consumers, and individuals in the news media, especially those who are looking for information based on science.

TABLE OF CONTENTS

Chapter 1

THE NEED FOR FOOD BIOPRESERVATION

Bibek Ray

TABLE OF CONTENTS

Civilization is based upon the food supply. The primitive hunters and gatherers made very little or no provision for the unproductive days; they either gorged themselves or starved. They began to be human when they settled down to till the soil and domesticate the animals and learned to use milk as a source of food. Nature taught humans the art of provision, the virtue of prudence, and the concept of time. Watching woodpeckers storing acorns in the trees, and bees storing honey in the hives, they conceived the notion of saving up food for the future. They found ways to preserve meat by smoking it, salting it, and freezing it; better still, they built granaries to protect the grains from rain and dampness, vermin and thieves, and also to provide for the leaner months of the year. Life became more stable and orderly, and the humans found time and reason to be civilized.[1]

I. HISTORY OF FOOD PRESERVATION

A. The Fourth Ice Age

Human progress is an interlude between the Ice Ages. During each glaciation the ice and stone covered the creation of humans and reduced life to a small segment of earth.[2] The humans who lived in Africa, the Middle East, India, China, Europe, the U.S., Canada, Mexico, and Greenland, prior to the fourth and latest Ice Age that lasted between 50,000 and 25,000 BC, were in the advanced stages of Paleolithic or Old Stone Age. The making of the first hunting tool, "coup-de-poing," or the "blow-of-the-fist" stone that was sharp at one end and round at the other so as to fit the palms of these primitive people, marked the beginning of the Old Stone Age about 1 million years ago.[2a,3] Before the beginning of the latest Ice Age, they were making arrowheads, spearheads, knives, hammers, and other tools and gained superiority in hunting. They recognized the advantages of fire (obtained probably from volcanic lava or forest fire), which included overcoming the fear of darkness, warding off ferocious animals, and keeping themselves warm. They were basically meat eaters and ate the raw meat of animals either hunted by them or killed by other animals, or that died of some other causes. There is no strong evidence to suggest that they intentionally preserved the excess meat for future use.[2]

With the beginning of the fourth Ice Age (about 50,000 BC and lasted until about 25,000 BC), the cooling of the earth forced the Neanderthal man to move toward warmer areas and turned him into a cave dweller.[2–4] There in the cave these people developed the techniques for making and maintaining fires and acquired the knowledge for cooking meat and more importantly, for cooking tubers and other unpalatable and indigestible foods so as to make them palatable and digestible. But the food supply was so meager at that time that there was very little need for preserving it for use during leaner days.

B. The Post-Glacial Era

As the Ice Age was followed by the Post-Glacial period, the Cro-Magnon man and other *Homo sapiens sapiens* appeared in the scene. These ancestors of the modern humans, with their magnificent vigor and physique, acquired and developed the skill of making diverse and efficient hunting weapons, fish hooks, harpoons, small utensils, and many other useful and necessary objects, not only from stones, but also from bones, horns, and ivory. More importantly, they acquired the ability to communicate with each other with the aid of a language. With these superiorities they fought with the Neanderthal man and made him extinct over a period of several millennia. In the early years of the Post-Glacial era, the earth was still very cold and, hence, the *Homo sapiens sapiens* lived in caves. But later, as the earth started warming up, they built temporary shelters or camps out on the ground.[2,3] They were hunters and gatherers and lead a nomadic life in small groups. In each group the men were engaged in hunting and fishing while the women gathered the grains, tubers, fruits, and vegetables. The life of a group pivoted around the migrational pattern of animals, the seasonal availability of fish and shellfish, the seasonal ripening of cereal grains and fruits, and the availability of vegetables at different times and in different places within a fixed area. As a result, the hunter-gatherers traveled to different campsites in a specific sequence throughout the year. Whenever a group increased in size to the extent that a fixed area could not supply enough food, some of the members would move out to another area. These ancestors of ours, during the period between 15,000 and 10,000 BC, not only had many new weapons, but also devised new strategies to kill large animals, such as mammoths, reindeer, and ibexes, and in greater numbers, during their migration. They also recognized the migratory pattern of fish and used nets to catch them in large numbers. The excess meat was either frozen in pits or stored in the colder parts of the caves; the excess fish was gutted and dried in the wind and sun, and so were the excess fruits.[2,3] Although the meat and fish were mostly eaten raw with much delight, cereal grains and tubers were cooked to make them palatable. Later, the cereal grains were ground with stone into flour, and the gruel was baked over a stone plate to make nonleavened breads or cakes that could either be eaten immediately or preserved for a short time.[3] However, as the groups were on the move, following the seasonal availability of foods, preservation of foods was limited and was enough only to provide a sufficient supply for their short stay at a site, and they carried on their backs what little they could in the absence of efficient containers and carriers.

1. Between 10,000 BC and 6000 BC

Between 10,000 and 8000 BC (Table 1) as the ice shrank back to the poles, the world started showing abundance in life. The vegetation spread, the animals multiplied, and the rivers and lakes were crowded with fish, especially in the tropical areas. In an area known as the "Fertile Crescent," which currently covers parts of Turkey, Lebanon, Israel, Egypt, Jordan, Syria, Iraq, and Iran and includes the water resources of the rivers Nile, Tigris, and Euphrates as well as the perennial spring at Jericho, the warm climate and the plentiful water resources, along with the winter rains, resulted in an abundance of cereal grains, plants, fruits, and animals. For a long time this had been a favorite stopover for groups of hunters and gatherers.[3,4]

Around 8000 BC their unsettled pattern of life changed. At first, the groups

TABLE 1
Development of Food Preservation Methods

Time	Important events	Preservation methods (used, introduced, recognized)
40,000 to 10,000 BC	Fourth glacial age ended; inter-glacial period began; *Homo sapiens neanderthalensis* replaced by *Homo sapiens sapiens*; hunter-gatherers; Paleolithic tools and weapons	**Physical methods** Limited use of heating (cooking), low temperature (freezing and cooling), and reduction of Aw (natural drying)
10,000 to 6000 BC	Ice receded to the poles; use of language for communication; establishment of settled communities; domestication of plants and animals; beginning of weaving and pottery making; Neolithic tools and weapons	**Physical methods** Extensive use of cooking, cooling, and natural drying **Chemical methods** Smoking, salting (also nitrate as contaminants in salt)
6000 to 1000 BC	City civilization in: Middle East, Indus Valley, Egypt, Far East, India; use of metal; transportation and trade; writing; class system	**Physical methods** Baking, modified atmosphere (in large vessels and pits). **Chemical methods** SO_2 (sanitize equipment) **Biological methods** a. Fermentation: grains: beer, bread fruits: wine, vinegar milk: yogurt, cheese, butter b. Spices and herbs (limited use)
1000 BC to 1800 AD	Empire builders; spice trade; dominance of Western civilization; discovery of America; Napoleon at war; spontaneous generation; industrial revolution	**Physical methods** Heating (Appertizing); reduced Aw (artificial drying) **Chemical methods** Pickling, NO_3-NO_2. **Biological methods** a. Ethnic fermented foods b. Fermented vegetables c. Different spices
1800 to 1900 AD	Pasteur discovered microorganisms; expansion in food marketing	**Physical methods** Heating (canning); low heat (mechanical refrigeration); modified atmosphere (waxing) **Chemical methods** Many chemical antimicrobials.

TABLE 1 (continued)
Development of Food Preservation Methods

Time	Important events	Preservation methods (used, introduced, recognized)
1900 to 1970 AD	Mathematical model of microbial death by heat; food regulations; several big wars; home refrigerator/freezer; starter cultures identified; space flight	**Physical method** a. Heating (commercial sterilization, pasteurization, UHT treatment) b. Low heat (extensive use of freezing/refrigeration, instant quick freezing) c. Reduced Aw (spray drying, drum drying, freeze drying, intermediate moisture foods) d. Modified atmosphere (vacuum packaging, gas packing) e. Radiation: UV and ionizing **Chemical methods** a. Many nonsynthetic and synthetic preservatives
1900 to 1970 AD (continued)		b. Antibiotics (limited use). c. Detergent/sanitizers (equipment) **Biological methods** a. Controlled fermentation of dairy and nondairy foods b. Recognition of antimicrobials of starter cultures, including nisin c. Antimicrobial properties of spices.
1970 to 1991 AD	Changes in socioeconomic and demographic patterns; population explosion; convenience foods, health consciousness; new foodborne pathogens; new industrial revolution; microwave cooking	**Physical methods** a. High temperature (minimal heat processing). b. Low temperature (long shelf life under refrigeration) c. Modified atmosphere (increased use of vacuum and gas packaging) d. Radiation (approval of ionizing radiation in certain foods in the U.S.)

TABLE 1 (continued)
Development of Food Preservation Methods

Time	Important events	Preservation methods (used, introduced, recognized)
1979 to 1991 AD (continued)		**Chemical methods** Concern over use of nonfood chemicals **Biological methods** a. Increase in production of fermented foods b. Approval of nisin as a preservative. c. Approval of lactate in food. d. Increase in research for natural antimicrobial food preservatives (biopreservatives).

started staying at a particular campsite for longer and longer periods of time than before, and finally they abandoned their nomadic lifestyle by establishing settled communities. Also, initially, they started harvesting the seasonal wild cereal grains, fruits, and vegetables and stored them to ensure the continued supply of their staple diet until the next season. Important innovations were made during this period, like the weaving of baskets and the making of pottery. Such innovations helped them in storing foods for longer periods of time and in transporting these foods from places of abundance to places of scarcity. The Neolithic people, or the people of the New Stone Age, during this time, with their polished stone tools, began making baskets and hampers from rushes and straws, and also sturdy vessels of baked clay. This increased their capability of preserving foods for longer periods and transporting them where and when needed, as opposed to their Paleolithic ancestors who had limited abilities to transport and store foods and who were still using limited natural storage spaces like gourds, coconuts, and sea shells. Pretty soon the Neolithic people started selecting better varieties of cereal grains and cultivating them by tilling the land. As a result, production of cereal grains like wheat, barley, millet, and corn increased greatly. Cultivation of fruit- and vegetable-bearing plants and trees also increased. During this period when they were learning the domestication of plants, i.e., agriculture, they were consuming more plant foods and supplementing this staple diet with meat that they obtained through hunting. Then between 8000 and 6000 BC, they learned the domestication of sheep, goats, pigs, and cattle, in that order. Soon they learned the secret of breeding domesticated animals and using them in ploughing land for agriculture. To ensure a steady supply of staples, they preserved the grains and fruits by natural drying and stored them in baskets and pots. Many innovations in pottery were made, and its use for cooking cereals (either the grains or the flour for cakes), vegetables and meat became a regular part of life. The excess meat and fish were preserved by smoking and dry salting. Salt was either obtained from the ocean beaches or brought from deposits in the Dead Sea by traders and who visited these settlements. With the settlers, they usually traded salt and luxury items for grains and other preserved foods. During this time, agriculture, animal husbandry, and techniques of food preservation were also developed in the Indian subcontinent, China, and many parts of Europe, Mexico, and Peru.[2–6]

2. *Between 6000 BC and 1000 BC*

Around 6000 BC, with their past 4000 years of knowledge about the innovative development of tools, baskets, and pottery, and knowledge about domesticating plants and animals, our ancestors recognized that inclement weather, pests, the lack of transportation and distribution of foods, and, more importantly, inability to store food for long periods of time made life insecure (Table 1). These factors locked them to the soil and made them live from year to year without any long-term security.[3,5] Some of the settlements, particularly in and around Mesopotamia and in the Indus Valley, started producing more food by irrigating fields with easily available water from the rivers. The capability increased to feed many more people than the small group who originally established these settlements. The ability to produce surplus food, knowledge of better ways to preserve this surplus, the discovery of wheels, and the use of animals to transport food freed them from the agony of starvation and helped them to develop other cultures (professions) along with agriculture.[3] With the passing of time, many settlements enlarged and developed into city civilizations. One of the earliest civilizations, the Sumerian civilization (between 5000 and 4000 BC)[7,8] used canals to irrigate fields, and oxen to plough the land, and harvested large amounts of cereal grains, fruits, and vegetables. The incentive to increase grain and fruit production came from the accidental discovery of the fermentation of barley to beer, and grapes to wine. The Sumerians also recognized the benefits of malting and the use of honey to facilitate alcohol fermentation and, as a result, enjoyed over 30 varieties of beer and wine. Alcohol was not only used for happy pastimes, but also for preserving foods.[3,7,8]

During this period the use of milk from sheep, goats, and cows as human food, was conceived. Herds of these animals were thus bred and kept primarily for their milk. However, storage of unconsumed milk in containers in the warm climate of the Near East soon caused souring to occur. This later paved the way for the production of fermented dairy products such as yogurt, cheese, and butter. Not only did these fermented foods have distinct tastes, but they could also be stored without spoilage for much longer periods of time compared to fresh milk. Later, the natural souring of alcohol to produce vinegar was discovered. Following this discovery came the use of vinegar and alcohol to overcome spoilage of foods. Thus, another dimension in food preservation, the biological method, in the form of fermenting grains, milk, and fruits and the use of alcohol and vinegar for the preservation of foods, along with the already known physical and chemical methods, came into existence to help in the progress of civilization.[3,8]

This was the period when the Stone Age ended and civilization entered into the Metal Age, beginning with the use of copper, followed by tin and bronze. This was also the period when writing in clay tablets began. Initially, this was used to keep account of the trading that was done on food and nonfood items. This was also the period when class systems sprung into being. The society was divided into priests, kings, rich, poor, and slaves. Taxes in the form of produce were levied on the farmers and fishermen by the kings, which were then effectively preserved in large royal warehouses for later distribution among officials and employees of the state.[3,8]

At different periods during 6000 to 1000 BC, new civilizations came into existence, reached their pinnacles, and then either vanished or were engulfed by other civilizations. These included the civilizations of Egypt in the Nile Valley, of Babylonia and Assyria in Mesopotamia, of Palestine and Phoenicia, of Mohenjodaro and Harappa in the Indus Valley, of Minoans in Crete, Xia and other dynasties in China, of Aryans in India, and others.[9–18] During this time span several new innovations in food preservation were made. These included the use of sulfur dioxide by burning sulfur to "sanitize" the equipment used in the fermentation and storage of wine in Egypt.[9] The Egyptians also recognized the benefits of fermentation of flour (dough) in the production of leavened bread. Fermenting rice in the Indian subcontinent and in China came into vogue, as did the practice of fermenting milk from buffaloes and cows in India.[13,14] Finally, towards 1000 BC, spices and herbs were added to food items, not as a preservative ingredient, but more as seasoning additives.[9,13,19] During this time span of 5000 years, the natural calamities in the form of drought and flood, and manmade calamities in the form of war between the states, made the kings, priests, rich, and poor recognize the value of preserving food. Preservation of food by different methods (including physical, chemical, and biological) was used in times of plenty so as to stabilize and enhance the possibility of survival of the civilizations during times of scarcity.[14,19]

3. Between 1000 BC and 1800 AD

During the next 2800 years, starting from 1000 BC (Table 1), some of the city civilizations started expanding their territories on a permanent basis, and in so doing, they established empires. It started with the Greeks and Romans,[19–21] and ended with the British. Transportation and trade improved, both on land and on sea. Plant and animal sources of food increased with time, keeping pace with population growth and other necessities. Techniques of food preservation using the physical, chemical, and biological means inherited from ancestors were made effective to meet these changing needs. Several new techniques were added to the existing practices, to meet more specialized needs. The most important of these was a form of high-heat treatment called

"Appertizing."[22] In the late 1790s Napoleon in France was at war with Germany. A major hindrance facing him was his inability to provide good food to his troops in the battlefield, due to the spoilage of most food items during shipping and storage. The French government offered a reward to anybody who could develop a technique to preserve food. Nicholas Appert, a chef and candy maker, knew from prior experience that heating food in closed containers reduced spoilage. He studied the influence of heating time, using boiling water on food stored in sealed glass containers, and succeeded in preserving food by this method. He called this method "Appertization." Not surprisingly, he won the reward for his successful attempt. Food spoilage was viewed at that time as due to "spontaneous generation," and Appertization was thought to be able to either remove oxygen from the bottle or make it react with food and thus prevent spontaneous generation. Another innovative food preservation method was the artificial drying of vegetables in a hot-air room, developed by Masson and Challet in 1795 in France. Preservation of foods by pickling, which involved the use of both salt and vinegar, was also developed probably in the beginning of this time. The fermentation of many ethnic foods gained popularity in the Middle East, the Indian subcontinent, the Far East, in the Balkans, and in other parts of the world, while in most parts of Europe and America it was primarily practiced to produce alcohol from grains and grapes, and to produce cheese from milk. Finally, the use of spices and herbs in food, probably for culinary delight, spread from Asia to Europe and America.[22,23]

4. Between 1800 AD and 1900 AD

In the next 100 years following 1800 AD (Table 1), several innovations in food preservation were made in Europe and the U.S., following the "Industrial Revolution" that started in the middle of the 18th century in Europe. Following the technique of Appertization, Durand[22] in England developed the "tin can" used for canning foods. Subsequently, canned foods became available in England by 1814 and in Boston by 1820.

The science of food preservation by canning or by any other technique was not actually understood until Louis Pasteur, in 1864, presented evidence proving that microorganisms in foods were the cause of food spoilage, that heat treatment of food killed these microbes, and that sealed containers, by preventing recontamination of the heated products with microbes from the atmospheric air, helped to preserve food. Then, in 1875, the autoclave was invented and thus provided a means to heat-treat foods at temperatures above 212° F. In the late 1880s Prescott and Underwood at the Massachusetts Institute of Technology and Russel at the University of Wisconsin studied the microbiology of the canning process and proved that success in the preservation of canned foods was dependent upon the thermal destruction of heat-resistant bacterial spores.[24]

Another major development was the invention of mechanical refrigeration systems. Low temperature as a means of food preservation was practiced by the hunter-gatherers before 10,000 BC, followed by their predecessors all along, where either the temperature was low or ice was available. In 1842 a patent was granted in England for freezing foods by immersion in ice-salt brine. Soon, both the direct immersion method and an indirect contact method which involved putting a pan on ice-salt brine and then placing the food on the pan, were being used in Europe and in the U.S. to freeze meat and fish.[25] In 1874 Carl Von Linde[26] in Germany developed a mechanical refrigerator and chilled lager beer for aging. By the late 1880s, meat and fish were being transported under refrigeration in ships and in railroad cars, both transcontinentally and intercontinentally.[26,27]

During the 19th century many chemicals were introduced and used as food preservatives[5] including nitrate and nitrite (nitrate/nitrite were used as a curing salt in meat before the 19th century). In addition, waxing was used as a means of modifying the atmosphere in the preservation of some cheeses.

5. *The 20thCentury*

The methods of preservation of food by freezing, drying, smoking, cooking, salting, and fermentation, developed over several millennia, were empirical and were more of an art than a science. Innovative progress was slow. As the biological basis of spoilage of food was understood from Pasteur's studies, the effectiveness of thermal processing of food for the destruction of microorganisms was recognized, and the economic benefits of commercialization of food was anticipated. This resulted in rapid advances in developing appropriate techniques to deal with the causative microbial agents. Thus, from the beginning of the 20th century (Table 1), many innovations were introduced, especially in the U.S. and in some European countries. These have revolutionized the processing, preservation, and storage of foods. Wars have always been one of the major incentives in the development of innovative methods for such processing and preservation of foods. The two world wars, together with the Korean war and the Vietnam war, have played major roles in the development of many modern food preservation techniques in the U.S. Our desire to conquer space has added another dimension to those techniques.

A major advancement was made when Ball in 1928,[28] based on studies by Prescott and Underwood, published the mathematical expression of the microbial destruction rate during thermal processing. This resulted in the preservation of many different types of foods by commercial sterilization. Other major developments in the area of heat processing of food were pasteurization to ensure destruction of vegetative pathogens, and ultra-high temperature (UHT) treatment to destroy vegetative cells and spores of pathogenic bacteria. Pasteurized products are nonsterile and have limited shelf life even under refrigeration, while UHT-treated products are commercially sterile and are stable at room temperature. Finally, microwave heating, which is more of a food preparation method than a preservation method and which does not give a predictable destruction of both pathogenic and spoilage microorganisms, was introduced.

By 1940 the availability of low-cost home refrigerators and freezers, together with the development of affordable refrigeration and freezing facilities for processing, storage, transportation, and final distribution of foods in retail supermarkets, made refrigerated and frozen foods very important commercial items. It started with the selling of frozen raw vegetables in the U.S. by Clarence Birdseye in 1923.[25] By 1944 not only were different raw foods being commercially preserved by freezing, but also frozen prepared foods of considerable varieties, popularly called TV dinners, became available in retail stores. Frozen foods still occupy a major portion of the retail food market in developed countries. One major innovation adding to this was the preservation of foods by instant quick freezing (IQF) using liquid nitrogen or nitrogen vapor as a coolant. This has enabled freeze preservation of many food items that were not feasible before the 1960s. Refrigeration of both raw and semipreserved foods prior to 1970 was regarded to be a method that provided only a limited shelf life as compared to frozen foods.

In the area of preservation using the drying method, the innovations of the 20th century included the artificial drying of fruits, vegetables, and liquids such as milk by drum or spray drying, and freeze-drying.[23] Vacuum packaging of foods was also introduced as a means of preservation by modified atmosphere.[29] One major and important innovation was the use of cold sterilization of foods by ionizing radiation to achieve both commercial sterilization and pasteurization.[30] However, it was not accepted widely by consumers in many countries, including the U.S. During this period (1900 to 1970) many effective chemicals were also used as microbial food preservatives, and the list included chemicals of food origin, nonfood origin, as well as synthetics. In the U.S., since 1950, chemical preservatives that were/are listed in the generally regarded as safe (GRAS) category could be incorporated in foods.[31] The uses of chemical detergents and sanitizers to clean and sanitize processing and other equipment also helped to reduce the microbial load in foods, and to increase their safety and shelf life.

Among the innovative biological methods, specific metabolites of food-grade starter culture bacteria, such as a peptide called nisin,[32,33] were found to have antimicrobial properties against many spoilage and pathogenic bacteria and were used as preservatives. Some research also revealed that many spices have active compounds with antimicrobial properties.[34,34a]

II. CURRENT NEED FOR FOOD BIOPRESERVATION

A. Changes in the Economic and Demographic Patterns and its Impact on Food Preservation

We have come a long way in the preservation of foods since the days of our hunter-gatherer *Homo sapiens sapiens* ancestors around 10,000 BC. In their nomadic lives they used to migrate from one place to another, following the seasonal availability of food. But today, especially in the developed world, it is the food producers and food processors who are constantly on the lookout for opportunities in the market place to meet the varied tastes of consumers.

Further, in the developed countries the abundant supply of food, in combination with the changes in the socioeconomic and demographic patterns, the changes in the consumers' concepts of nutrition, preferences for different types of foods, food selection patterns, along with the technological innovations and the competition among the food processors, have created several unique problems in the area of food preservation. Since 1970, and more in the 1980s, the number of two-income families, singles, working mothers, single-parent households, latch-key children, and the elderly have increased significantly.[35] In the 1980s the number of people over the age of 65 exceeded the number of teenagers in the U.S. In the 1990s the demographic pattern is projected to be as follows: individuals in the age group of 24 to 34 years, viewed by the industry as the most valuable group, will drop in number by about 14%, while the group between 35 and 54 years of age will increase in number by around 28%. Similar large increases are expected to occur in the age groups between 55 and 64, and also those over 64 years. The spending power of the 25-to-34 years age group will decline by almost $80 billion, while that of the 35 and higher age groups (including those above 65) will increase dramatically.[36,36a] These expected changes in the demographic patterns have created a huge demand for convenience foods and for "ultraconvenient" foods that can be eaten with very little preparation. Currently, easy availability of microwave ovens among households has increased the demand for convenience foods. The prediction is that in the near future, cars will have microwave ovens built in, which will be utilized in the preparation of breakfast and lunch, and many of the food items will be in the form of finger foods.[37] Also, if the rapid technological developments are any indication, the era where one can get beef steaks cooked and served by a vending machine is not very far away.[38] For successful marketing of a food product, convenience and several other factors need to be considered. The typical American family is decreasing in size. This will change the demand for prepared foods to meet the needs of single individuals instead of families. In addition, there will probably be more males who will buy and prepare food than before.[39] The older people, who are the seasoned buyers, will go for quality. The environmental concern of the consumers will also have a big impact on the packaging and serving of foods.

B. The Importance of Quality and Nutritional Value of Foods

In addition to convenience, a considerable percentage of the population is interested in the nutritional aspects of foods. This segment of the population is aware of the adverse effects of high levels of fat, calories, cholesterol, and sodium. This group relates to food as an agent of health and well being. They are also concerned about the loss of nutritional value of the harshly "processed foods" and the possible health risk of foods preserved with certain "chemicals." Instead, they are interested in foods that are "natural," fresh, and healthy. A recent survey has indicated that the buyers of healthful foods differed in their reasons for purchase, which ranged from "preventing any future problems" to

or hurdles, in addition to refrigeration, into the product formulation so as to minimize the microbial number, especially the pathogens. These include the lowering of pH with acids, reducing water activity (Aw), and the addition of a suitable preservative. Modified atmosphere packaging (MAP), in the form of vacuum packaging and gas-flush packaging, can be used to enhance shelf life by reducing the growth of spoilage bacteria. However, this will not prevent the growth of pathogens that are anaerobes and facultative anaerobes.[47]

D. The Dilemma with Chemical Preservatives and a Different Approach

The consumers' preference for refrigerated foods that do not contain any preservative(s) is based on their perception of these foods to be nutritious, healthy, and close to "natural'" as opposed to the "harshly" processed and "chemically" preserved foods. This is because the safety of some preservatives (and some other additives) has been questionable, and because a gap exists in the flow of information between the consumers, the food processors, and the regulatory agencies. The introduction of varieties of shelf-stable refrigerated foods without any preservatives has created concerns among both the regulatory agencies and the food processors.[61] On one side is the consumers' concern for foods with preservatives that are perceived to be unsafe, and on the other is the concern of the regulatory agencies for the safety of these foods in the absence of a suitable preservative. The recommendation by the NFPA to include a suitable preservative to enhance the safety of these foods is an indication of those concerns.[47] The food processors and the regulatory agencies are thus in a dilemma and are looking through a crystal ball for an answer.

Antimicrobial chemical preservatives are those that inhibit or retard the growth of spoilage and pathogenic microorganisms and that are used to enhance the safety and shelf life of foods. They produce their antimicrobial effect by interfering with the structural and functional components of microorganisms. Their effectiveness is dependent upon the type of chemical, the concentration of the chemical used, the microorganisms present, the type of microorganisms and their physiological state (vegetative cells or spores), the composition and pH of the food, and the temperatures and duration of storage. There are regulatory requirements for the use of chemical preservatives: namely they are used only when other preservation methods are not effective. In addition, they should not lower the quality of food, should be effective in small quantities, should have a wide antimicrobial spectrum, and, finally should be harmless to the consumers.[62] In the U.S., by law, the antimicrobial food preservatives fall under one of four groups: (a) those that are not defined by law (such as natural organic acids and their salts, spices, and oils, and wood smoke), (b) those that are GRAS (such as propionic acid, sorbic acid, benzoic acid, their salts and derivatives, sulfites, and nitrites), (c) those that have been proved to be safe and approved by the regulatory agencies (FDA in the U.S.), and, finally, (d) those that are not used currently and where proofs have yet to be provided for their safety to animals and humans, before they are formally approved.[31,63,64] Some of these chemicals are naturally present in human foods, without any known undesirable effect (such as lactic acid, acetic acid, and citric acid). Some are not normally found in foods (such as sorbates and sulfites), while others are synthetics or derivatives (such as parabans). Also, some of these chemicals have been questioned for their safety in human consumption (such as NO_2, sulfites, Na-diacetate, beta-propiolactone, and therapeutic antibiotics).[62,63] It has been mentioned before that the consumers are concerned about the safety and wholesomeness of foods that are preserved using many of these "chemicals". These concerns have resulted in the rise of consumer activist groups with their demand for greater control of the use of these chemicals. Are these concerns valid? To answer this, Miller[5] thinks that we need to understand what the real problem is. In the early years the chemical preservatives (and other additives) were introduced only a few at a time, went through the test of time, and were accepted only when they were found to be nontoxic. In recent years the time needed to evaluate the hazardous nature of a chemical preservative has become much shorter, and its use has spread widely within a short period. In the near future the time requirement for the introduction of a chemical may be even shorter, and the chemical may be used by a large segment of the population quickly. Further, in recent years the chemicals used in food items have increased exponentially to several thousands. The effect of actual levels of consumption of these chemicals in multiple products over a substantial length of time should be an important consideration in judging their safety. The perception of safety in the use of a chemical in any food item differs greatly. "The professional offers safety from hazards that are measurable in terms of known pathological conditions; the consumers want safety so that their lives and their descendants' lives can be healthy and happy."[5] The public concern about the safety of chemical preservatives and additives has increased greatly with the confusing and contradictory research findings, and so has the aggressiveness of the news media in providing sensational information to the public before a controversy has been resolved. It will take quite some time before the effects of the long-term use of many chemical preservatives on individuals and on neonates can be resolved with the help of effective and efficient testing procedures (which may be at the molecular level).[5] However, there are other compounds that have antimicrobial properties. Many are now used in foods and have been consumed by humans for a long time, without any known adverse effect. These include natural components found in plant foods (such as spices, citric acid), in some animal foods (such as lysozyme in eggs), and as metabolites of food-grade microorganisms (such as metabolites of starter culture bacteria and some yeasts). The consumers' preference for foods that are not harshly processed and are not chemically preserved is because of their perception that such foods are natural and healthy. Thus, the basis of the selection of antimicrobial preservatives to be used in foods, as recommended by the NFPA,[47] should not only be their effectiveness against both Gram-positive and Gram-negative pathogenic and spoilage microorganisms and other desired characteristics of preservatives, but also their proven safety records and their acceptance by the health-conscious consumers and the regulatory agencies. Among the compounds that have generated considerable interest in recent years are several antimicrobial metabolites of bacteria used to produce, or associated with, fermented foods.[32]

Fermented foods have been consumed by mankind probably for the past 8000 years. They include over 3500 different foods, many of which are ethnic, made from milk, fruits, sap, vegetables, roots, cereals, beans, nuts, lentils, meat, fish, and many other food components.[65] In fermentation the raw materials are converted by the desirable microorganisms (that include bacteria, yeasts, and molds) to products that have acceptable qualities of food. The desirable microorganisms can be present in raw materials coming from the environment, along with many other undesirable microorganisms. To achieve fermentation, the conditions are set in such a way that the desirable microorganisms grow preferentially, produce metabolic products, and give the unique characteristics to the product. This process is called natural fermentation. Sauerkraut is an example of a naturally fermented food item produced by the fermentation of cabbage by *Leuconostoc mesenteroides, Lactobacillus plantarum*, and probably *Pediococcus pentosaceus* and *Lactobacillus brevis,* all of which are naturally present in cabbage. During fermentation they grow and metabolize the carbohydrates in the cabbage to produce lactic acid, acetic acid, CO_2, and some diacetyl and ethanol. However, natural fermentation cannot be depended upon to produce uniform products all the time, and in some batches the desired microorganisms will not grow, resulting in product failure as well as the possible growth of undesirable microorganisms that include pathogens. To overcome this problem, many fermented foods are now produced by controlled fermentation. For this, initially, the microorganisms that are involved in natural fermentation of a food are isolated in pure cultures, identified and tested for their desirable characteristics, and then maintained for use (Table 2). These are called starter cultures. In controlled fermentation they are added to the raw materials in large numbers and then incubated under conditions to stimulate their growth and production of desirable products. An example of a food produced by controlled fermentation is yogurt in which *Lactobacillus delbrueckii* subsp. *bulgaricus* and *Streptococcus salivarious* subsp. *thermophilus* are added to heated, but not sterile, milk at about 10^6 cells per militer and incubated at 42° C. Both bacteria grow favorably at this temperature and metabolize lactose of milk to produce lactic acid, which is responsible for the curdling of milk and the sour taste. They also produce acetaldehyde, which is responsible for the characteristic flavor of yogurt. The lactic acid produced by these desirable bacteria in sauerkraut and yogurt prevents the growth of undesirable microorganisms present in the nonsterile, raw materials and makes the products shelf stable.

Since the 1970s, consumption of fermented foods in the U.S. and other developed countries has increased greatly. The list includes not only products such as yogurt, buttermilk, and fermented sausages, but also many ethnic products such as kefir, kumiss, and tofu. One of the reasons for the increase

TABLE 2
Some Starter Culture Bacteria Associated with Food Fermentation

Bacteria	Fermented foods
Lactococcus lactis subsp. *lactis* Lactococcus *lactis* subsp. *cremoris*	All types of cheeses, buttermilk, sour cream
Lactococcus lactis subsp. *lactis* serovar *diacetylactis*	Sour cream, buttermilk, ripened cream, butter, cheese
Streptococcus salivarious subsp *thermophilus*	Yogurt, Swiss cheese, Italian cheese
Leuconostoc mesenteroides subsp. *mesenteroides*	Fermented vegetables and fermented cereals
Leuconostoc mesenteroides subsp. *cremoris*	Buttermilk, sour cream, cottage cheese, ripened cream, butter
Pediococcus acidilactici *Pediococcus pentosaceus*	Fermented sausage, fermented vegetables, fermented cereals
Latobacillus acidophilus	Fermented acidophilus milk
Lactobacillus lactis *Lactobacillus helveticus*	Swiss cheese
Lactobacillus delbrueckii subsp. *bulgaricus*	Yogurt, Swiss cheese, Bulgarian buttermilk, Italian cheese, kumiss
Lactobacillus casei	Fermented milk
Lactobacillus plantarum	Fermented sausage, sauerkraut, fermented pickle
Lactobacillus brevis	Kefir, sauerkraut, fermented pickle
Lactobacillus sake	Fermented sausage
Lactobacillus sanfrancisco	San Francisco sourdough bread
Bifidobacterium bifidum	Bifidus milk
Propionibacterium freudenreichii	Swiss cheese

in the consumption of fermented foods is that consumers consider them to be healthy and natural. The consumption of live cells of desirable microorganisms in the billions, and their metabolic products in fermented foods, does not cause any panic or loss of interest among the safety-concerned consumers. Also, the proof of the safety of these fermented foods for human health has stood the test of time. The use of these microorganisms and/or their metabolites in foods

TABLE 3
Some Antimicrobial Metabolites of Bacteria Produced in Fermented Foods

Types	Examples
Organic acids	Lactic, acetic, propionic
Alcohols, ketones, aldehydes	Ethanol, diacetyl, acetaldehyde
Bacteriocins (purified peptides)a	Nisin, Pediocin AcH
Bacteriocin-like compounds (Proteinaceous nature)	Sakacin A, Pediocin A
Other small molecules	H_2O_2, reuterin

[a] Purified by one or more methods used in protein purification and characteristics studied with the purified materials.

as preservatives, if found suitable, may be less objectionable to the consumers, the regulatory agencies, and the food processors. Among the metabolites, several have been known to have antimicrobial properties against Gram-positive and Gram-negative bacteria, and yeasts and molds associated with food spoilage and foodborne diseases. However, this effect is dependent upon specific metabolites. This aspect will be discussed in detail in the succeeding chapters of this book. Some of these metabolites, such as lactic acid, acetic acid, propionic acid and their salts, H_2O_2, and nisin (Table 3), have been recommended by the regulatory agencies either as antimicrobial agents or as flavoring agents.[32–34] There are good possibilities that others, if found effective, will also be recommended.

E. Biopreservatives and Biopreservation

The term "biopreservative" in this book has been used to include the antimicrobial compounds that are of plant, animal, and microbial origin, and have been used in or as human food for a long time, without any adverse effect on human health. Their use to enhance the safety and extend shelf life of food can thus be regarded as "biopreservation." Fermented foods are good examples of biopreserved foods in which the starter cultures are allowed to grow so that they can produce antimicrobial metabolites. This book will include discussions only on the biopreservatives of microbial origin.

III. CONCLUDING REMARKS

The concerns of consumers about the safety and wholesomeness of foods containing many nonfood additives, including preservatives are real; and so are the consumer activist groups' increasing demands for greater control or the use of these "chemicals,"[5] and the increasing news media coverage of these concerns. In recent years, as opposed to the past, there has been an exponential rise in the numbers and diversity of use of such chemicals. The time scale for introducing many of these preservatives and other additives has been shortened, and, maybe in some cases, are too short to judge the cumulative effect on human health, resulting from the use of these different types of nonfood chemicals that are consumed through different types of food during one's life span, on the unborn, and maybe even on the next generation. Reports on possible health hazards from the consumption of some preservatives and other additives currently being used, such as nitrite and saccharine, and additives previously used but currently not being permitted for use, such as cyclomate and some dyes, have shaken the faith of consumers, thus prompting them to question the safety and wholesomeness of the foods they eat. However, they have received very few answers of assurance. On the contrary, they have been bewildered by the controversial statements of lobbyists, activists, scientists, self-imposed specialists, legislators, and the food industry.

"The public is willing to accept levels of risk in other aspects of their life that they are unwilling to even consider for food. This is a consequence of the special role that food plays in human society. For many people it is the only component of their environment over which they believe they have control. Under these circumstances the public insists that food products be absolutely safe."[66] This belief, at least in the developed countries, has paved the way for a segment of the consumers to question the safety of the thousands of nonfood preservatives and other additives that are incorporated into foods. It has also encouraged them to voice their feelings against the use of these chemicals in foods and also to look for foods that are "natural," "healthy," and not treated with "harsh contaminants." The producers have started marketing the "new generation refrigerated" foods, the "sous vide" foods, and similar types of foods, without much consideration for the safety of these products to the consumers, especially in the event of temperature abuse. The regulatory agencies are in a dilemma. The NFPA has suggested the use of suitable food preservatives to ensure the safety of these products. However, these preservatives should be safe, effective, and acceptable to the consumers, the regulatory agencies, and the food industry. The dilemma is where to find such preservatives.

Antimicrobial metabolites or biopreservatives of the food-grade starter culture bacteria are in a unique position in this regard, especially in their potential for immediate application. Their safety has been proven through the test of time; they have been used, and are still being used, by consumers, in some form, more probably by the health-concerned consumers. If found effective as preservatives, they will probably be of the least controversy to the regulatory agencies and to the food industry. The biggest question is to determine their effectiveness in food systems. This has to be done by well-designed studies. Controversial results from unplanned, incomplete, and oversimplified studies will not provide the answers that are needed. The metabolites need to be identified, their physical and chemical characteristics determined, their mode of antimicrobial effects elucidated, their stability in the food systems established, and finally, their safety on laboratory animals proven.

While suitable biopreservatives from starter culture microorganisms can be incorporated within a short period of time, the search should be on for safer and better preservatives from both biological and nonbiological sources, including many that are now in use, for solving the problem on a long-term basis.[67] These studies should be directed at the molecular level, to answer questions about the safety of using these preservatives on a long-term basis.

REFERENCES

1. **Durant, W.,** The economic elements of civilization, in *The Story of Civilization: Our Oriental Heritage,* Part I, Simon and Schuster, New York, 1954, chap. 2.
2. **Durant, W.,** The pre-historic beginning of civilization, in *The Story of Civilization: Our Oriental Heritage,* Part I, Simon and Schuster, New York, 1954, chap. 6.

2a. **Stewart, G. F. and Amerine, M. A.,** Evolution of food processing and preservation, in *Introduction to Food Science and Technology,* 2nd ed., Academic Press, New York, 1984, chap. 1.

3. **Farman, C. and Nobel, C.,** The human dawn, in *Time Frame,* Middleton, C., Ed., Time-Life Books, Alexandria, VA, 1990, chap. 1–4.
4. **Stuart, G. S.,** Ice Age hunters: artists in hidden caves, in *Mysteries of the Ancient World,* Grosvenor, G. M., Ed., National Geographic Society, Washington D.C., 1985, 8.
5. **Miller, S. A.,** Additives in our food industry, in *Food and Nutrition in Health and Disease,* Moss, N. H. and Mayer, J., Eds., New York Academy of Sciences, New York, 1977, 397.
6. **White, L.,** Food and history, in *Food, Man and Science,* Walcher, D. N., Kretchmer, N., and Barnett, H. L., Eds., Plenum Press, New York, 1976, chap. 12.
7. **Durant, W.,** The economic elements of civilization, in *The Story of Civilization: Our Oriental Heritage,* Part I, Simon and Schuster, New York, 1954, chap. 7.
8. **Fishbein, S. L.,** The Sumerians of Mesopotamia, in *Splendours of the Past: Lost Cities of the Ancient World,* Crump, D. E., Ed., National Geographic Society, Washington, D.C., 1988, 34.
9. **Durant, W.,** The economic elements of civilization, in *The Story of Civilization: Our Oriental Heritage,* Part I, Simon and Schuster, New York, 1954, chap. 8.
10. **Durant, W.,** The economic elements of civilization, in *The Story of Civilization: Our Oriental Heritage,* Part I, Simon and Schuster, New York, 1954, chap. 9.
11. **Durant, W.,** The economic elements of civilization, in *The Story of Civilization: Our Oriental Heritage,* Part I, Simon and Schuster, New York, 1954, chap. 10.
12. **Durant, W.,** The economic elements of civilization, in *The Story of Civilization: Our Oriental Heritage,* Part I, Simon and Schuster, New York, 1954, chap. 12.
13. **Durant, W.,** The economic elements of civilization, in *The Story of Civilization: Our Oriental Heritage,* Part I, Simon and Schuster, New York, 1954, chap. 14.
14. **Durant, W.,** The economic elements of civilization, in *The Story of Civilization: Our Oriental Heritage,* Part I, Simon and Schuster, New York, 1954, chap. 23.
15. **Kinney, K.,** The age of God-kings, in *Time Frame 3000-1500 BC,* Hyslop, S. G., Jone, R., and Thomson, D. S., Eds., Time-Life Books, Alexandria, VA, 1990, chap. 1–4.
16. **De La Haba, L.,** Roots of the city: Jericho and Catal Huyuk, in *Mysteries of the Ancient World,* Grosvenor, G. M., Ed., National Geographic Society, Washington D.C., 1985, 32.
17. **Eckstrom, C. K.,** Minoans: a joyous people vanish in myth, in *Mysteries of the Ancient World,* Grosvenor, G. M., Ed., National Geographic Society, Washington D.C., 1985, 126.
18. **Ramsay, C. R.,** Mycenaeans: warrior-merchants of Greece, in *Mysteries of the Ancient World,* Grosvenor, G. M., Ed., National Geographic Society, Washington D.C., 1985, 146.
19. **Hyslop, S. G. and Jones, R.,** Barbarian tides, in *Time Frame 1500 to 600 BC,* Hicks, J. and Thomson, D. A., Eds., Time-Life Books, Alexandria, VA, 1990, chap. 1–4.

20. **Hyslop, S. G., Jones, R., Phoanka, B. C., and Thomson, D. S.,** A soaring spirit, in *Time Frame 600 to 400 BC,* Hicks, J., Ed., Time-Life Books, Alexandria, VA, 1990, chap. 1–4.
21. **Hyslop, S. G.and Phoanka, B. C.,** Empires ascendant, in *Time Frame 400 BC to 200 AD,* Hicks, J., Ed., Time-Life Books, Alexandria, VA, 1990, chap. 1–4.
22. **Desrosier, N. W.,** *The Technology of Food Preservation,* 3rd ed., AVI Publishing, Westport, CT, 1970, chap. 6.
23. **Desrosier, N. W.,** *The Technology of Food Preservation,* 3rd ed., AVI Publishing, Westport, CT, 1970, chap. 5.
24. **Labuza, T. P. and Erdman, J. W.,** *Food Sciences and Nutritional Health: An Introduction,* West Publishing, St. Paul, MN, 1984, 279.
25. **Desrosier, N. W.,** *The Technology of Food Preservation,* 3rd ed., AVI Publishing, Westport, CT, 1970, chap. 4.
26. **Anon.,** *Food Preservation,* Encyclopaedia Brittanica, Vol. 9, Encyclopaedia Brittanica, Chicago, 1976, 543.
27. **Labuza, T. P. and Erdman, J. W.,** *Food Sciences and Nutritional Health: An Introduction,* West Publishing, St. Paul, MN, 1984, 295.
28. **Ball, C. O.,** *Mathematical Solution of Problems on Thermal Processing of Canned Foods,* University of California, 1928, 1.
29. **Koski, D. V.,** Is current modified/controlled atmosphere packaging technology applicable to the U.S. food market? *Food Technol.,* 42(9), 54, 1988.
30. **Desrosier, N. W.,** *The Technology of Food Preservation,* 3rd ed., AVI Publishing, Westport, CT, 1970, chap. 10.
31. **Desrosier, N. W.,** *The Technology of Food Preservation,* 3rd ed., AVI Publishing, Westport, CT, 1970, chap. 9.
32. **Daeschel, A. M.,** Antimicrobial substances from lactic acid bacteria for use as food preservatives, *Food Technol.,* 43(1), 164, 1989.
33. **Hurst, A.,** Nisin and other inhibitory substances from lactic acid bacteria, in *Antimicrobials in Foods,* Branen, A. L. and Davidson, P. M., Eds., Marcel Dekker, New York, 1983, chap. 11.
34. **Davidson, P. M., Post, L. S., Branen, A. L., and McCurdy, A. R.,** Naturally occurring and miscellaneous food antimicrobials, in *Antimicrobials in Foods,* Branen, A. L. and Davidson, P. M., Eds., Marcel Dekker, New York, 1983, chap. 13.
34a. **Farrell, K. T.,** *Spices, Condiments and Seasoning,* 2nd ed., Van Nostrand Reinhold, New York, 1990, 3.
35. **Swientek, R. J.,** Refrigerated entrees offer convenience and freshness appeal, *Food Process. (Chicago),* 49(1), 70, 1988.
36. **Duxbury, D. D.,** 90s consumers say no to fats — health and convenience top trends, *Food Process. (Chicago),* 51(7), 92, 1990.
36a. **Bowers, P.,** Age wave: good bye teenagers, hello middle agers, *Meat Process.,* 30(4), 36, 1991.
37. **Aylwood, L.,** What will be cooking in the kitchen coming 2010? *Meat Process.,* 30(3), 30, 1991.
38. **Bowers, P.,** Any time beef machine, *Meat Process.,* 30(3), 34, 1991.
39. **Smith, G.,** Looks good! Tastes good! But will it sell? *Natl. Provisioner,* 204(9), 8, 1991.
40. **Aylwood, L.,** Life in the fast lane, *Meat Process.,* 30(3), 28, 1991.
41. **Anon.,** What motivates buying 'healthy'? *Prep. Foods,* 159(11), 62, 1990.
42. **Dornblaser, L.,** Shelf-stable strikes again, *Prep. Foods,* 160(3), 37, 1991.
43. **Swientek, R. J.,** New generation chilled foods face cold realities, *Food Process. (Chicago),* 51(14), 94, 1990.
44. **Baird, B.,** Sous vide: what's all the excitement about? *Food Technol.,* 44(11), 92, 1990.
44a. **Swientek, R. J.,** Sous vide refrigerated upscale food in a vacuum pouch, *Food Process. (Chicago),* 50(6), 34, 1989.

45. **Eilers, J. R.,** New foods provide new safety challenges, *Food Process. (Chicago),* 51(6), 104, 1990.
46. **Corlett, D. A.,** Refrigerated foods and use of hazard analysis and critical control point principles, *Food Technol.,* 43(2), 91, 1989.
47. **Anon.,** Safety consideration of the new generation refrigerated foods: refrigerated foods and microbiological criteria committee of the National Food Processors Association, *Dairy Food Environ. Sanit.,* 8, 5, 1988.
48. **Snyder, O. P. and Poland, D. M.,** America's safe food — Part II, *Dairy Food Environ. Sanit.,* 11, 14, 1991.
49. **Gilliland, S. E., Michener, H. D., and Kraft, A. A.,** Psychrotrophic microorganisms, in *Compendium of Methods for the Microbiological Examination of Foods,* Speck, M. L., Ed., American Public Health Association, Washington, D.C., 1984, chap. 9.
50. **Jay, J. M.,** *Modern Food Microbiology,* 2nd ed., Van Nostrand, New York, 1970, chap. 20.
51. **Gill, C. O. and Newton, K. C.,** The development of aerobic spoilage flora on meat stored at chilled temperatures, *J. Appl. Bacteriol,* 43, 189, 1977.
52. **Kalchayanand, N., Ray, B., Field, R. A., and Johnson, M. C.,** Spoilage of vacuum-packaged refrigerated beef by *Clostridium, J. Food Prot.,* 52, 424, 1989.
53. **Grau, F. H.,** Inhibition of an anaerobic growth of *Brochothrix thermosphacta* by lactic acid, *Appl. Environ. Microbiol.,* 40, 433, 1980.
54. **Newton, K. G. and Gill, C. O.,** The development of the anaerobic spoilage flora of meat at chill temperatures, *J. Appl. Bacteriol.,* 44, 91, 1978.
55. **Grau, F. H.,** The role of pH, lactate, and anaerobiosis in controlling the growth of some fermentative Gram-negative bacteria on beef, *Appl. Environ. Microbiol.,* 42, 1043, 1981.
56. **Palumbo, S. A.,** Is refrigeration enough to restrain food borne pathogens? *J. Food Prot.,* 12, 1003, 1986.
57. **Kornacki, J. L. and Gabis, D. M.,** Microorganisms and refrigeration temperatures, *Dairy Food Environ. Sanit.,* 10, 192, 1990.
58. **Del Giudice, V. J.,** We must meet the 21st century standards, *Natl. Provisioner,* 204(6), 12, 1991.
59. **Anon.,** Factors to be considered in establishing good manufacturing practices for the production of refrigerated foods: refrigeration of foods and microbiological criteria committee of the National Food Processors Association, *Dairy Food Environ. Sanit.,* 8, 288, 1988.
60. **Bryan, F. L.,** Application of HACCP to ready-to-eat chilled foods, *Food Technol.,* 44(7), 70, 1990.
61. **Shibasaki, I.,** Food preservation with non-traditional antimicrobial agents, *J. Food Saf.,* 4, 35, 1982.
62. **Banwart, G. J.,** *Basic Food Microbiology,* AVI Publishing, Westport, CT, 1979, 596.
63. **Frazier, W. C. and Westhoff, D. C.,** *Food Microbiology,* 4th ed., McGraw-Hill, New York, 1988, chap. 9.
64. **Labuza, T. P. and Erdman, J. W.,** *Food Sciences and Nutritional Health: An Introduction,* West Publishing, St. Paul, MN, 1984, chap. 25.
65. **Campbell-Platt, G.,** *Fermented Foods of the World: A Dictionary and Guide,* Butterworths, Boston, 1987.
66. **Miller, S. A.,** *Quest for Safe Food: Knowledge and Wisdom,* S. B. Hendricks memorial lecture, sponsored by the Agricultural Research Service, U.S. Department of Agriculture and the American Chemical Society, Miami Beach, FL, 1989.
67. **Board, R. G. and Gould, G. W.,** Future prospect, in *Food Preservatives,* Russell, N. J. and Gould, G. W., Eds., AVI Publishing, New York, 1991, chap. 14.

Chapter 2

FOODS AND MICROORGANISMS OF CONCERN

Bibek Ray

TABLE OF CONTENTS

The science of food preservatives is undergoing a quiet revolution. Consumers want attractive, appealing, and safe foods, but they want it as fresh and unprocessed as possible — that is, they want their food untainted by technology. This remains one of the primary stimuli behind the enthusiastic growth of refrigerated or fresh foods. Many of these products are prepared in locations where quality controls are lax and the technical quality of the labor is low, a situation some microbiologists characterize as a ticking time bomb. In 1991, 46% of the 352 companies surveyed indicated that they plan to increase their development of all-natural/no-additive foods, while 43% indicated they would increase efforts to reformulate additives out of their existing food product lines. Many of the food processing establishments currently on the all-natural, fresh-foods kick may be pushing the limits of technology too far. Supermarkets are notorious for their high turnover of personnel — are they equipped to implement the proper safety controls on the manufacture of chilled prepared foods?[1]

I. PRESERVATION METHODS AND PRESERVATIVES

A food can lose its acceptance quality or spoil from the growth of microorganisms or due to nonmicrobial causes such as lipid oxidation. Similarly, a food can be unsafe for consumption because of the growth or presence of pathogenic microorganisms or due to the presence of toxic chemicals.[1] In this chapter some characteristics of the spoilage and pathogenic microorganisms important in refrigerated and low heat-treated foods, and the effectiveness of biopreservatives in their control, will be discussed.

Microorganisms which include bacteria, yeasts, molds, and viruses, gain access to food through various sources. Their level and types vary greatly depending upon their sources, the effectiveness of sanitation used during all phases of handling, and the methods used in the processing and storage of a food. Different physical, chemical, and biological preservation methods are applied with an intention to make a food safer and to extend its shelf life by destroying, and retarding or preventing the growth of pathogenic and spoilage microorganisms that might be present in the food. These include physical methods, such as temperature treatment (both high and low), reduced Aw, and controlled or modified atmosphere and radiation; chemical methods, such as

TABLE 1 (continued)
Antimicrobial Efficiency of Some Preservation Methods and Preservatives

Some preservation methods or preservatives	Major antimicrobial effect(s)
Smokes	Effectiveness is dependent on type and pH, preservation is due to the combined effects of drying from heat and the chemicals in smoke, which include acids, alchols, aldehydes, ketones, phenols, tars, and many others, some of these are carcinogen
Acetic and lactic acids and salts[a]	Can be produced by lactic bacteria, Gram-negative spoilage and pathogenic bacteria are more susceptible, yeasts, mold, and aciduric bacteria are less susceptible, viability loss and growth inhibition are concentration and pH dependent, surviving cells can be injured, undissociated molecules, due to better permeability through membrane are more effective
Propionic acid and salts[a]	Produced by some starter culture bacteria, fungistatic and bacteriostatic
Hydrogen peroxide[a]	Produced by some starter culture bacteria in small amounts, bacteriocidal to many types of microorganisms.,can cause color defects in cooked meat products due to oxidation of myoglobin
Acetaldehyde[a]	Produced by some lactic acid bacteria, has bacteriocidal and fungicidal properties
Ethyl alcohol[a]	Bacteriocidal to most microorganisms, not used as a preservative, but its presence in alcoholic beverages prevents or reduces growth of microorganisms
Bacteriocins of starter culture bacteria[a]	Bacteriocidal to some Gram-positive pathogenic and spoilage bacteria, effectiveness depends upon the bacteriocin and sensitive cells, resistant survivors can grow
Diacetyl[a]	Produced by some starter culture bacteria, has some bacteriocidal and bacteriostatic effect, intense flavor may limit its use in many foods.
Lysozyme	Bacteriocidal to some Ggram-positive spoilage and pathogenic bacteria, high cost may be a limiting factor in its use
Spices	Antibacterial effect is due to essential oils, varies with the spice.

[a] Metabolic products of starter culture bacteria.

II. FOODS IN WHICH BIOPRESERVATION WILL BE USEFUL

A. Frozen and High Heat Treated Foods

Foods, in general, provide a good nutritional environment for the growth of spoilage and pathogenic microorganisms. Many preservation techniques are used to restrict their growth and, in some cases, to destroy a part of the population. The extent of viability loss and/or growth restriction of the viable cells depends upon the microbial types, the food systems, and the preservation techniques used.

Microorganisms present in foods or food ingredients that are either frozen to –20°C and below or have low water activity (Aw 0.6 or below due to drying or binding free water with solutes) will not grow and spoil the foods. However, if these foods harbor pathogens, especially some of the enteric pathogens, they will pose health hazards. These foods and food ingredients, before consumption or use in food formulation, can be subjected to conditions in which the viable microorganisms can multiply and the spores can germinate and outgrow. A suitable biopreservative can be added to these foods or food ingredients as a built-in safety precaution. Similarly, foods that are given high-heat treatment, such as UHT-treated commercially sterile foods, can have thermophilic spores of some spoilage bacteria, which, under normal conditions of storage around 25°C, will not germinate and pose a problem.[12] However, if the temperature of the product increases to above 40°C even for a short duration, these spores can germinate. Subsequently, their growth can occur even if the temperature is reduced to the mesophilic range.[13–16] Such temperature abuse of commercially sterile products can occur especially in the summer months and in places where the normal temperature is high. Presence of a suitable biopreservative in commercially sterile foods will prevent germination and outgrowth of the surviving thermophilic spores.[17] There can be another potential advantage of using an effective biopreservative in commercially sterile foods. The high and prolonged temperature treatment results in some nutritional loss and loss of certain acceptance qualities in these foods. Yet such a treatment is necessary to assure their safety. Several heat-resistant biopreservatives are known to inhibit germination and outgrowth of spores of *Clostridium botulinum*. Incorporation of these as preservatives can aid in reducing the time and temperature currently used in commercially sterile foods and can help to improve their nutritional acceptance and microbial qualities. However, this aspect needs to be thoroughly studied.[17,18]

B. Refrigerated Foods

The greatest application of biopreservatives will be in the enhancement of the safety and shelf life of many traditionally refrigerated foods and more importantly, of the new generation refrigerated foods. The traditional refrigerated foods include raw meats and fish, pasteurized milk, cured and cooked ready-to-eat meats, high-moisture cheeses and spreads, fermented cheeses, yogurt, pickles, perishable delicatessens, and other similar products.[19] The new generation refrigerated or chilled foods include the minimally processed entrees, dinners, meats, salads, pastas and sauces, side dishes, and others.[19–22] Many of the traditional refrigerated processed foods have a backup preservation system (such as nitrite in cured and cooked meats) without which they will spoil quickly and thus warn consumers of potential hazards. In addition, consumers, in general, are acquainted with methods to handle these foods safely. In contrast, the new generation refrigerated foods are minimally processed value-added products, with very little or no preservatives and with an expected extended shelf life (up to 60 d). Vacuum packaging, along with refrigeration, can further increase shelf life by inhibiting the growth of aerobic spoilage bacteria. However, this may not inhibit, but can rather facilitate, growth of some pathogenic bacteria.[21,23] An inadequate distribution system and the possibility of abuse due to improper temperature control can increase the risk of hazards in these foods.[19] A minimum heat treatment that many of these products undergo does not kill the thermoduric spoilage and pathogenic microbial cells and spores (some of which are psychrotrophs), that can grow at normal refrigeration temperatures (Table 2). Proper vacuum packaging with very low oxygen-permeable materials and refrigeration will discourage aerobes, but will stimulate the growth of psychrotrophic anaerobic and facultative anaerobic spoilage and pathogenic bacteria. Any temperature abuse, depending upon the temperature and time of abuse, will also facilitate the growth of some strains of mesophilic anaerobic and facultative anaerobic pathogens and spoilage bacteria that have minimum growth temperatures of 10 to 12°C or below. A national audit report on retail dairy case product temperatures indicated that around 17% of the store coolers and display cases were above 10°C, while around 66% were about 4.5°C (Table 3).[23a] In addition, many of these products are handled extensively after heat treatment; thus, the possibility of post-heat-treatment contamination of these foods with other pathogenic and spoilage bacteria exists. These may include even those microorganisms that are sensitive to heat treatment and that can grow at low temperatures under anaerobic conditions. During extended storage under proper refrigeration and in the absence of a suitable preservative, the psychrotrophic anaerobic and facultative anaerobic pathogenic and spoilage bacteria can grow and adversely affect the safety and quality of these products. Growth of some of these pathogens, even those that can produce toxins, may not produce detectable changes in the acceptance quality of these foods. Such foods, many of which are in the ready-to-eat or heat-and-eat categories, will be unsafe for consumption. On the contrary, incorporation of a biopreservative that is effective against these pathogens and spoilage microorganisms will provide greater safety and longer shelf life to these foods.[10,11]

C. Sous Vide Foods

"Sous vide" foods are foods that are cooked under vacuum at low temperatures. The process involves vacuum sealing fresh vegetables, meat, and

TABLE 2
Microbiology of Commercially Processed Roast Beef

Processing steps	Microbiological concerns
Fresh beef (8 to 10 lb)	Different types of pathogens and spoilage microorganisms (vegetative cells, spores)
Either vacuum-packaged / Or not packaged	
Cooked to internal temperature 145 F (60°C). Temperatures vary, be as low as 130°F (54.5°C) for 121 min	Vegetative cells of thermoduric bacteria and spores of pathogenic and nonpathogenic bacteria survive.
Chilled to 55°F(13°C) in 6 h	
Opened and vacuum packaged again / Vacuum packaged	Recontamination with spoilage and pathogenic bacteria can occur
Refrigerated to 40°F (4.5°C)	
Stored for 2 to 4 d	
Shipped to retail stores or institutions; advised to maintain at 40 F; shelf life expected to be 21 d.	Temperature abuse can occur
	Pathogenic and spoilage bacteria (psychrotrophic anaerobic and facultative anaerobic) can grow, controlling factors: refrigeration, vacuum packaging, meat pH

TABLE 3
Temperature Range in Coolers in Display Cases in Stores[a]

Store type	Number audited	Temperature range (°C) in percent					
		≤1.7	1.8–4.4	4.5–7.2	7.3–10	10.1–12.8	≥12.9
Major stores	352	9	31	30	23	3	4
Independent stores	113	3	19	36	28	9	5
Ma & Pa stores	31	6	26	16	26	10	16
Convenience stores	13	8	38	38	0	8	8
Average		6	28	30	19	9	8

a Table adapted from Reference 23a.

fish in oxygen- and moisture-impermeable plastic pouches, cooking them in bags at low temperatures in circulating water, and, finally, storing them under refrigeration or on ice at a temperature of 0 to 3°C for up to 3 weeks.[24,24a,25] When required, a package is warmed, opened, and served. Currently, entrees, side dishes, soups, sauces, and appetizers are included in the list of sous vide foods that are being catered to restaurants, hotels, banquet halls, and other places. These foods have a health and convenience appeal, and "some specialists are forecasting that sous vide foods, along with the new generation refrigerated foods, will be the fresher alternative to shelf-stable and frozen foods."[24]

However, the regulatory agencies recognize that sous vide foods, like the new generation refrigerated foods, have inherent problems of health risk from the growth of pathogens. Low-heat processing, possible temperature abuse during transportation and storage during the 3 weeks after processing, and anaerobic conditions provide a scope for some surviving pathogens to grow and produce toxins.[24–26] Warming these foods prior to eating them will not ensure their destruction. Currently, in France, the Ministry of Agriculture has issued specific guidelines for food preparation by the sous vide method. In England its preparation is controlled by the existing chilled-food regulations. However, in the U.S. there are no specific regulations for sous vide food preparations and distribution, although the FDA and USDA recognize that this technique is dangerous.[25] A suitable biopreservative can be incorporated into sous vide products prior to vacuum packaging and heating to reduce the health risk of these products while at the same time satisfying the consumers' desire for their freshness and healthfulness.

D. Vacuum-Packaged and Gas-Flush-Packaged Fresh Meat

Vacuum-packaged and gas-flush-packaged fresh meat will be another area where biopreservatives can be used to extend shelf life. In recent years, fabrication of carcasses into wholesale and retail cuts has become a popular means of marketing fresh beef in the national and international markets.[27] Marketing retail cuts of lamb and pork is also gaining in popularity.[28] The wholesale and retail cuts of meat are packaged in a modified atmosphere (which includes vacuum packaging and gas-flushed packaging) individually and stored at refrigeration temperatures (0 to 3°C) until they are sold. This method is not only economical because it reduces weight loss due to evaporation of moisture and reduces transportation cost of the carcasses that include unusable bones and fats, but it also provides a means for ensuring better quality control of the products.[29] The processors expect a shelf life of up to 12 weeks, especially for the products intended for international markets. However, in the U.S. some processors have experienced unusual spoilage of vacuum-packaged refrigerated beef within 2 to 4 weeks due to the growth of a psychrotrophic *Clostridium* species *C. laramie*.[30,31,31a] Spoilage of vacuum-packaged fresh meat can also be from the growth of psychrotrophic lactobacilli, leuconostocs, *Brochothrix thermosphacta*, *Altermonas putrefacience*, and some species/strains from *Enterobacteriaceae*.[32–36] The shelf life of vacuum-packaged and gas-flush-packaged fresh meat can also be extended by using a suitable biopreservative.[37,38]

It appears then that incorporation of a biopreservative preparation that has an antimicrobial property against different types of spoilage and pathogenic bacteria will have application in a wide variety of foods, including fresh, minimally processed, refrigerated, frozen, and commercially sterile foods.

III. PATHOGENIC BACTERIA OF CONCERN

A. Microbial Foodborne Illnesses

Foodborne illnesses result from the consumption of foods contaminated with pathogenic microorganisms, toxic chemicals, or parasites. Among foodborne health hazards in the U.S., 69% of the outbreaks involving 94% of the total cases are of microbial origin, which includes pathogenic bacteria, pathogenic viruses, and toxigenic molds. Foodborne diseases of bacterial origin account for the highest number of outbreaks, involving 91% of the total number of cases.[39–41]

Health hazards from the consumption of contaminated foods have baffled humans even before civilization. However, direct association of the foodborne illnesses with the pathogenic microorganisms was not recognized until the 1860s when investigation by Pasteur lead to the understanding of the role of microorganisms in food spoilage.

A microbial foodborne illness is an event in a sequence of preceding events. There must first exist a source for a pathogen(s) that, in the process of handling, enters the food. Due to favorable conditions, the pathogen grows in the food, and finally, the contaminated food, when consumed in sufficient quantities by humans produces symptoms of an illness.[42,43] Animals, humans, and the environment to which a food is exposed, can all be the source of pathogens. The microbiological quality of both raw and processed foods, and the frequency of foodborne diseases, vary greatly in different countries and depend upon several factors. These include the level of sanitation during all phases of food handling, regulatory requirements, economic condition, availability of food, level of education, and the aesthetic sense of handlers. In general, the number of foodborne illnesses is relatively less in the developed countries as compared to the developing countries. In the U.S., where the processed food supply is probably the safest, between 6 and 24 million people become ill every year from the consumption of food contaminated with pathogens. This could be a conservative estimate, as another report has suggested that as high as 60 million people per year could be involved. The foodborne illnesses result in about 9000 deaths and cost $4 to $14 billion in medical costs, insurance costs, lost wages, and liabilities.[39,44–49] However, only about 3% of the total outbreaks are the result of contaminated foods processed in large commercial processing plants; over 95% are from foods that have been prepared at food service establishments and homes.[39,42,50] The factors that contribute to these outbreaks, in the order of importance, include temperature abuse, prolonged storage,

TABLE 4
Frequency of Contamination of Raw Foods with Pathogens[a]

Pathogen	Food	% Positive
Aeromonas hydrophila	Fresh poultry, fish, red meat, produce (lettuce, celery, kale, etc.)	95 to 100
Bacillus cereus	Raw rice	100
	Raw hamburger and beef	46 to 63
	Dry milk powder	60
Campylobacter jejuni	Fresh chicken, fresh turkey	45 to 64
Clostridium botulinum B	Corn syrup	50
Clostridium perfringens	Pork, chicken (raw, cooked)	39 to 45
Escherichia coli 0157:H7	Fresh beef, pork, poultry	1.5 to 3.7
Listeria monocytogenes	Ground beef, ground pork, ground veal, raw chicken	50 to 100
Listeria spp.	Fresh potato,	26
	Fresh radish	30
Salmonella spp.	Fresh chicken, turkey	40 to 100
	Fresh pork	3 to 20
	Raw shell fish (clams,oysters)	8 to 33
Yersinia enterocolitica	Raw pork	49
	Raw milk	48
	Raw vegetables	46
Staphylococcus aureus	Raw chicken	73
	Raw pork	13 to 33
	Raw beef	16
	Raw seafood	38
Vibrio spp. (*V. cholera*, *V. parahaemolyticus*, *V. vulnifacus*)	Seafood	33 to 46
Enteroviruses (polio, echo, coxsakie, reo, and hepatitis B)	Seafood (clams, mussels, and oysters)	47

[a] Adapted from Reference 44.

infected food handlers, inadequate reheating, contaminated raw materials, improper cleaning, and the use of leftovers.[39,50,51] Considering that consumers in the U.S. consume over $300 billion worth of food each year and a large proportion of this huge volume moves through about 25,000 processors, 35,000 wholesalers, 250,000 food stores, and 275,000 eating establishments involving 5.5 million workers, it is not surprising that pathogens, in some instances, do escape identification, and when the sequence of events is set in motion, they result in illnesses.[50,52] If we add the number of households and people who can be involved in the sequence of events, the potential for the occurrence of foodborne diseases increases greatly.[50] Finally, if we consider the frequency of incidence of pathogens in raw foods (Table 4), it is not surprising that the actual incidence of foodborne diseases can be quite high.[44]

B. Classification of Illnesses

Microbial foodborne illnesses are classified into three groups: intoxication, infection, and toxico-infection.[52–56]

1. Intoxication

Intoxications occur from the ingestion of food containing toxin(s) produced by toxin-producing bacteria or molds. A toxin could be heat labile or heat stable. *Staphylococcus aureus, Clostridium botulinum,* and molds such as *Aspergillus flavus* are examples of toxin-producing microorganisms. Toxins of *S.aureus* are heat stable, while toxins of *C. botulinum* are heat labile.

2. Infection

Infections are caused when live cells of pathogenic bacteria or viruses are ingested through eating contaminated foods, and the pathogens either invade the intestinal mucosa and multiply or pass to other organs. *Listeria monocytogenes Salmonella* spp., *Shigella* spp., *Campylobacter jejuni, Escherichia coli* 0157:H7, *Yersinia enterocolitica, Vibrio parahaemolyticus, Aeromonas hydrophila,* and hepatitis B virus are all examples of infectious microorganisms. *C. jejuni* causes only gastric symptoms, while *L. monocytogenes* can invade other organs and tissues and produce different types of symptoms.

3. Toxico-Infection

In toxico-infection a large number of live cells of pathogens are consumed with food, and they produce or release enterotoxins in the gastrointestinal tract, which then produce symptoms. Examples are *Clostridium perfringens, Bacillus cereus, Vibrio cholera,* and enterotoxigenic *Escherichia coli.*

Many other pathogens, some of which are opportunists, can also be involved in foodborne diseases. However, only some, such as *S. aureus* (26%), *Salmonella* (32%) and probably *Campylobacter jejuni* (same as *Salmonella*) are involved more frequently than others.[51,54–56] Some of these are fatal (*C. botulinum, L. monocytogenes, Salmonella* spp.), while several others produce symptoms that last for a short duration, with uneventful recovery (*C. perfringens, C. jejuni, E. coli*).

C. Pathogens and Biopreservative Barriers

In general, regulatory agencies and industries set up standards or specifications for the tolerance level of these pathogens in foods. Ready-to-eat foods should have a "zero tolerance" level for *C. botulinum, Salmonella* spp and *L. monocytogenes,* but may contain low levels of *S. aureus* ($< 100/g$) or *C. jejuni.* These foods are considered acceptable, provided they are preserved in a way that discourages further growth of *S. aureus* and *C. jejuni.* However, a "zero-tolerance" for a specific pathogen means that the pathogen cannot be present in a given amount of food, as specified by the regulatory agencies, but can be present in low numbers in a much larger volume. These products also need to be preserved properly to discourage growth of (if not to destroy) pathogens that can be present in very low numbers.

Refrigerated foods, in contrast to the conventionally preserved foods (frozen, dried, canned), probably need the greatest consideration in controlling the growth of pathogens. A possible solution can be the incorporation of multiple barriers against these pathogens. One such barrier can be a biopreservative. The pathogens that can grow at refrigeration temperature or those that are psychrotrophs are of greatest concern in refrigerated foods. Until the 1950s it was generally considered that pathogens could not grow in foods stored at 5°C (41°F) in home refrigerators.[57] Since the 1960s, it became apparent that some pathogens (*C. botulinum* E) could grow and produce toxins in foods stored even at 3.3°C (38°F).[57–61] More recently, the ability of some pathogens (*Y. enterocolitica*) to grow at temperatures below 1°C has been demonstrated.[51,55,57] Thus, refrigeration of a food even at 1°C cannot assure its complete safety against the growth of some psychrotrophic pathogens to hazardous levels, especially in foods with a long shelf life. In addition, some mesophilic pathogens (optimum growth at 32 to 37°C) have variants or strains that can grow at or below 12°C (53°F).[53–55,62] From the time of processing until the time of consumption, a refrigerated food can undergo temperature abuse, raising the temperature of the product to 10 or 12°C, enabling some of the mesophilic pathogens to grow and/or produce toxins. Some of the refrigerated foods are vacuum packaged (or gas flushed) to extend their shelf life. Depending upon the oxygen permeability of the packaging materials, this method can prevent, or greatly reduce, the growth of aerobic microorganisms, but will allow the anaerobic and facultative anaerobic spoilage and pathogenic bacteria to grow. There is a suspicion that vacuum packaging might selectively stimulate the growth of the anaerobic and facultative anaerobic pathogens by reducing the competition of the associated aerobic spoilage bacteria.[11,19] The pathogens that are of major concern in vacuum-packaged refrigerated food products are listed in Table 5. Other important characteristics of these pathogens, in relation to food preservation, are included below.[44,50,53,55,57,59,63]

D. Some Pathogens and Their Characteristics

1. Listeria monocytogenes

Main sources: found widely in nature and in the intestinal tract of animals, birds, and humans; has been isolated from many raw and processed foods; not all strains are pathogenic.

Important characteristics: Gram-positive; facultative anaerobic; lowest growth temperature 1.7°C (35°F).

Effect of preservation: killed at pasteurization temperature (mild-heat treatment); present in heat-treated foods due to post-heat-treatment contamination, insufficient heating, or the development of heat resistance due

TABLE 5
Foodborne Pathogens Capable of Growing at Low Temperatures

Pathogen	Gram *stain*	Lowest growth temperature	Other important characteristics
Listeria monocytogenes	+	35(1.7)	Facultative anaerobe
Clostridium botulinum (nonproteolytic B, E)	+	38(3.3)	Anaerobe; spore formers
Yersinia enterocolitica	–	32(0)	Facultative anaerobe
Aeromonas hydrophila	–	38(3)	Facultative anaerobe
Enterotoxigenic *Escherichia coli*	–	36.5(2.5)[a]	Facultative anaerobe
Bacillus cereus	+	41(5)	Facultative anaerobe
Staphylococcus aureus	+	43.8(6.5)	Facultative anaerobe
Salmonella spp.	–	41(5)[a]	Facultative anaerobe
Vibrio parahaemolyticus	–	41(5)	Facultative anaerobe

[a] Only some strains

to sublethal heating; survives adverse conditions; sensitive to low pH; killed by some bacteriocins of lactic acid bacteria.[64–66]

Virulence: virulent strains can be fatal; causes infection; infective dose about 100 to 1000 viable cells; regulatory requirement is "zero-tolerance" in certain volumes of ready-to-eat food.

2. Clostridium botulinum

Main sources: type B is found normally in soil and in raw meat as a contaminant; type E is of marine origin and found in fish and marine products.

Important characteristics: Gram-positive; spore formers; anaerobic; spores germinate, overgrow and multiply at 3.3°C (38°F); produces a toxin in 3 weeks at 8°C, but not at 5°C; growth and toxin production may not be associated with off flavor.

Effect of preservation: cells are killed at pasteurization temperature, but the spores survive; boiling for 6 h is required to ensure the destruction of spores; sensitive to organic acids (low pH), low Aw, some bacteriocins of lactic acid bacteria, nitrites, and salts.

Virulence: the toxin is extremely potent and fatal; it can be destroyed by heating at 85°C for 15 min; regulatory requirement is "zero-tolerance" in certain volumes of food.

Note: spores of some protelytic strains (found as contaminants in raw vegetables, fruits, honey, and meat) can germinate and overgrow, and the cells can multiply and form toxins at 10°C (50°F).

3. Yersinia enterocolitica

Main sources: found in the gastrointestinal tract of animals, especially in swine; in foods such as raw milk, meat, poultry, shellfish, tofu, and vegetables, it is found as a contaminant.

Important characteristics: Gram-negative; facultative anaerobic; grows at 0°C (32°F); generation time is relatively short.

Effect of preservation: killed by pasteurization; present in pasteurized foods as post-heat-treatment contaminants; not killed by bacteriocins of starter cultures; susceptible to organic acids (low pH).

Virulence: only the pathogenic strains cause health hazards; need to consume in very high numbers (10^7/g of food); generally not fatal.

4. Aeromonas hydrophila

Main sources: many foods, such as meat, poultry, fish, shellfish, dairy products, fruits, and vegetables, due to fecal contamination from animals and humans (fish and shellfish are contaminated by polluted water).

Important characteristics: Gram-negative; facultative anaerobic; grows at 3°C (38°F).

Effect of preservation: sensitive to pasteurization temperature; presence in properly pasteurized processed foods is due to post-heat-treatment contamination; not sensitive to bacteriocins of lactic acid bacteria (although some reports indicate that specific strains could be sensitive)[64]; sensitive to organic acids (low pH).

Virulence: may be an opportunistic pathogen.

5. Pathogenic Escherichia coli

Main sources: food contaminated with fecal materials from animals and humans; animals can be carriers of virulent strains.

Important characteristics: Gram-negative; facultative anaerobic; some strains can grow at 2.5°C (36°5 F).

Effect of preservation: destroyed at pasteurization temperature; present in properly pasteurized foods due to post-heat-treatment contamination; not sensitive to bacteriocins of lactic acid bacteria; sensitive to organic acids (low pH).

Virulence: virulent strains are of four types: one type produces infant diarrhea, the second type causes travellers' diarrhea, the third type causes severe dysentery (enterotoxigenic *E. coli* 027:H20), and the fourth type causes bloody diarrhea and hemorrhagic colitis (*E. coli* 0157:H7); need to ingest large number of cells (10^6 to 10^7) to cause illness in healthy individuals, but a much lower number can cause illness in young, old, and sick people.

6. Bacillus cereus

Main sources: present in grains (particularly rice), flour, starch, spices, and animal products contaminated with soil.

Important characteristics: Gram-positive; facultative anaerobic; spore former; grows rapidly; lowest growth temperature 5°C (41°F).

Effect of preservation: pasteurization does not kill spores; killed by some bacteriocins of lactic acid bacteria.[64]

Virulence: virulent types could be either emetic or vomiting; need 5×10^5 cells per gram of food to produce illness.

7. Staphylococcus aureus

Main sources: present in skin, nose, throat, and hair of humans and animals; also found in sores and lesions; contamination common in meats, salads, cheese, cream-filled pie, and canned mushrooms.

Important characteristics: Gram-positive; facultative anaerobic; lowest growth temperature 6°5 C (43°8 F); lowest toxin producing temperature 10°C (50°F); grows in low Aw (0.84), low pH (4.0), and high salt concentration (18%); grows rapidly.

Effect of preservation: killed at pasteurization temperature; present in heat-treated foods as post-heat-treatment contaminants; killed by some bacteriocins of lactic acid bacteria;[17,64] however, toxins remain potent even after heating and after the death of the cell(s); sensitive to organic acids (low pH < 4.5)

Virulence: one of the major causes of foodborne illnesses; need $\geq 10^6$ cells per gram in food to produce illness; toxin is heat stable even after boiling for 25 min.

8. Salmonella spp.

Main sources: over 2000 serotypes, all of which are considered pathogens; found in the intestinal content of animals, birds, pets, and humans; they can also be carriers; major source in food is fecal contamination; the insides of eggs can probably be contaminated with infected ovaries in carrier poultry; cross contamination.

Important characteristics: Gram-negative; facultative anaerobic.

Effect of preservation: killed at pasteurization temperature; survive in dried and frozen foods; present in heated products due to post-heat-treatment contamination; some strains grow at 5°C (41°F); not sensitive to bacteriocins of lactic acid bacteria; sensitive to low pH.

Virulence: one of the major causes of foodborne illness; serotypes differ in virulence; can be fatal, especially in infants, old, and sick people; infection dose varies with strains; usually, for some strains, ingestion at 10^4 cells is necessary to produce salmonellosis; in some conditions four to five cells per 100 grams of food is enough to cause illness; multiplies very rapidly; regulatory requirement is "zero tolerance" for specific quantities of processed food.

9. Vibrio parahaemolyticus

Main sources: found in warm coastal waters and in brackish and polluted waters; common sources are fish and shellfish; cross-contamination of other foods, especially due to an increase in fish consumption, and handling of fish and meat in the same area inside the supermarket (also true for *C. botulinum* E).

Important characteristics: Gram-negative; facultative anaerobic; lowest growth temperature 5°C (41°F).

Effect of preservation: sensitive to pasteurization; presence in heat-processed products is due to post-heat-treatment contamination; grows better in media with 3% salt; multiplies very rapidly; not sensitive to bacteriocins of lactic acid bacteria, but sensitive to organic acids (low pH) and freezing.

Virulence: not all strains are virulent;[67] grows very rapidly; needs ingestion of a large number of cells to cause illness; raw fish and shellfish consumption is the major cause of foodborne illnesses from this species.(*Vibrio cholera* and *Vibrio vulnificus* are two other pathogenic species of marine vibrios. *V. cholera* is responsible for cholera, while *V. vulnificus* is associated with septicemia. Both are highly fatal.)

10. Shigella spp.

Main sources: human intestinal tract (carrier); food contaminated with fecal material.

Important characteristics: Gram-negative; facultative anaerobic; lowest growth temperature 7°C (45°F).

Effect of preservation: killed by proper pasteurization; presence in pasteurized food is through post-heat-treatment contamination; not killed by bacteriocins of lactic acid bacteria; sensitive to organic acids (low pH).

Virulence: highly virulent; 10 to 100 cells can cause disease; can be fatal.

Other important bacteria that can grow in food at higher temperatures (>12°C) and that are associated with foodborne illnesses include *Campylobacter jejuni* (25°C), *Clostridium perfrigens* (15°C), and several others, some of which can be opportunistic pathogens. In addition, pathogenic viruses (such as hepatitis B) and toxigenic molds (such as *Aspergillus flavus,* producing the mycotoxin aflatoxin) can also be involved in foodborne illnesses.[40–43,53,54]

IV. SPOILAGE MICROORGANISMS OF CONCERN

A. Introduction

Loss of acceptance quality or spoilage of food could be of either microbial or nonmicrobial origin. Microbial spoilage can occur both from the metabolic activities of the live cells of bacteria, molds, and yeasts and, in the absence of live cells, from the action of the microbial enzymes, especially heat-stable enzymes in heat-processed foods. In general, the incidence of food spoilage that is of microbial origin is much higher than that of nonmicrobial origin, and among the microorganisms, bacteria (followed by molds) are more frequently involved in spoilage than yeasts. In the U.S., even with the advanced technology available for processing, distribution, storage, and marketing, the total loss of perishable foods, especially refrigerated foods at the retail level, accounts for 2 to 3% of the total production; and in addition, another 10 to 12% is partially lost as a result of the reduction in prices to facilitate quick selling.[68,69] The total dollar value that U.S. consumers spend in a year on food is about $350 billion at the shipment level. If one third of this is produced as perishable products, the total annual dollar loss owing to the spoilage of food can be $5 billion or more. In many other developed countries, the extent of economic loss from food spoilage is likely to be similar, if not more. In the developing countries, food loss due to spoilage will be much higher. To provide food to the growing population on the earth, consideration should be given, not only to control population growth, and increase food production, but also to reduce food loss from spoilage and other causes.

B. Factors Affecting Growth of Microorganisms

Perishable raw and processed foods contain different types of microorganisms; the types and their relative numbers will vary greatly with the nature of the food and the handling method. Foods, in general, can provide nutrients necessary for the growth (metabolism and multiplication) of most microorganisms.[7–9] The potential for growth of the individual species/strain among the mixed microflora is influenced by the intrinsic factors, extrinsic factors, and the processing factors, all of which are interdependent (especially for the processed foods).[10,70] The intrinsic factors include the chemical, physical, and biochemical characteristics, such as nutrient composition, presence of natural antimicrobials, pH, Aw, and oxidation-reduction potential (Eh). The extrinsic factors include environmental factors such as storage temperature during handling, and the storage atmosphere. Processing factors include the survival and growth of microorganisms during processing, such as heating or irradiation, and the colonization of microorganisms in the equipment surface and their subsequent access to food during processing.

Microorganisms differ considerably in their ability to survive and grow when they are exposed to these factors. The response of microorganisms to several of those factors that have important use in controlling food spoilage will be discussed briefly. Among the intrinsic factors, Aw, which is a measure of free water available in a food system, can be used to control microbial growth. Aw of a food can be reduced either by binding with specific inorganic (NaCl) or organic (carbohydrates, proteins) chemicals or by removing part of the free water from a food (drying). Thus, without any other control, many Gram-negative spoilage bacteria, such as *Pseudomonas* spp., will grow in a food with Aw 0.95 and above, but will not grow at Aw 0.94 or below. In contrast, some spoilage Gram-positive bacteria, like lactobacilli, can grow at Aw 0.92, while leuconostocs will not grow below 0.95.

The other important intrinsic factor is pH. The pH of a food is dependent upon the amount of acids present either naturally, formed during fermentation, or that is added in the formulation, and the buffering capability of the food.

Microorganisms also differ greatly in their ability to grow at lower pH. While many spoilage bacteria, such as *Pseudomonas* spp., will not grow at pH below 5.0, some lactic acid bacteria will grow at pH below 4.0; leuconostocs will not grow below pH 4.0; yeasts and molds can grow at pH 2.0. The presence of specific organic acids, either natural or added in the formulation of a food, can also adversely affect microbial growth. This aspect will be discussed in greater detail in later chapters of this book.

The most important extrinsic as well as processing factor is temperature. Microorganisms are arbitrarily divided into three groups on the basis of two characteristics:[62,71–81] (1) the temperature at which the growth rate is maximum, i.e., the optimum growth temperature; and (2) the lowest and highest growth temperatures, i.e., the temperature range of growth. The psychrophiles are those with an optimum growth temperature of 15°C, a maximum growth temperature of 20°C, and a minimum growth temperature of 0°C or below. Mesophiles have an optimum temperature around 37°C and a maximum and a minimum growth temperature of 40° and 20°C, respectively. Thermophiles are those that grow optimally at about 50°C, with a growth temperature range of between 40 and 60°C. However, one should recognize that there is considerable overlap between the optimum temperature and the temperature ranges of the species/strains of microorganisms in these three groups. In addition, two separate terms are used to describe microbial response to temperature, which are very important in food microbiology. One is "thermoduric,"[77] which is used to differentiate the microbial cells that survive pasteurization temperature (or low-heat treatment). The other term is "psychrotrophs"[57,58,62,74–76] and is used to include microorganisms that can grow at 5°C (the temperature inside a domestic refrigerator). For both groups the optimum, as well as the maximum and minimum, growth temperatures are not taken into account. Over the years, different definitions of psychrotrophs have been proposed, depending upon their ability to grow at 7°C, 5°C or even 3°C.[74] However, this group includes all the psychrophiles and some mesophiles as well as some thermodurics. By definition, spores, especially bacterial spores, are all thermoduric; however, they are much more heat resistant than the thermoduric vegetative cells, and some can survive much higher temperature treatments, including commercial sterilization and ultra-high-temperature treatments. Among the microorganisms, probably the thermophilic and psychrotrophic groups are more important in food spoilage; thermophilic for foods that are processed or held at temperatures above 40°C and below 60°C for considerable periods of time, and psychrotrophic for foods that are kept at refrigeration for long periods of time and that can undergo temperature abuse.

The other extrinsic factor is the storage atmosphere of food. A food can be stored in the presence of air or under modified atmosphere (MA). In the presence of air, aerobic microorganisms will grow preferentially, followed by the facultative anaerobes.[70–73,78] MA packaging can be either vacuum packaging only or first drawing a vacuum, followed by gas flushing and sealing of the package. In MA-packed foods the anaerobes and the facultative anaerobes will grow preferentially. However, we should recognize that even under aerobic conditions anaerobes and facultative anaerobes can grow inside the food. Similarly, in MA-packaged foods some aerobes, especially microaerophiles, can grow, although it will depend upon the effectiveness of the vacuum packaging, the nature of the food, and the permeability of the packaging materials to oxygen.[24a] In gas flushing, gas mixtures containing different proportions of CO_2, N_2, and O_2 are used. However, recent studies have indicated that 100% CO_2 is probably better than the other gas mixtures in controlling bacterial growth associated with spoilage of raw or processed meats.[28]

C. Food Spoilage Due to Microbial Growth

Microbial food spoilage is directly associated with the growth of microorganisms or indirectly from the enzymes, including the heat-stable enzymes released by some microorganisms before being killed. Microorganisms grow by metabolizing the carbohydrates, proteins, and lipids in the food, and produce different metabolic end products.

The effectiveness of metabolizing different food components and the nature of the end products produced are dependent upon the chemical nature of the food components, the microbial types, and the metabolic pathways through which a component has been utilized (Table 6). Microbial spoilage can change the flavor (such as sour, bitter, pungent, putrid, rancid, sweet), texture (slimy, mushy), and color (by forming pigments or changing the natural color, such as the brown color defect in meat) of foods and make them unacceptable. Many of these changes are associated with the nature of end products produced through the metabolism by the microorganisms of carbohydrates, proteins, and fats, in food.

Metabolic pathways through which the substrates are converted into end products differ greatly with the chemical nature of a substrate, its concentration, the type of microorganisms, and the availability of oxygen. One example of this is the metabolism of glucose, resulting in the production of CO_2, citric acid, and keto acids, under aerobic conditions, and in the production of alcohol, CO_2, lactic acid, and acetic acid, under anaerobic condition. In general, microorganisms will preferentially utilize metabolizable carbohydrates as an energy source, as opposed to using proteins and lipids. Metabolic end products produced by one microorganism can inhibit or stimulate growth of other microorganisms. The metabolism of carbohydrates by lactic acid bacteria thus produces lactic acid that inhibits the growth of many Gram-negative spoilage bacteria capable of metabolizing proteins, and helps in preventing the spoilage of protein-rich foods (protein-sparing effect). The acid environment can also prevent or reduce the growth of many other bacteria, but not that of yeasts and molds, which can grow and cause spoilage of acid foods.

Microorganisms utilize the food components to supply energy and to synthesize the cellular, structural, and functional components. For this the food molecules are transported inside. However, the three-dimensional configuration

TABLE 6
Metabolic End Products from Microbial Degration of Food Components

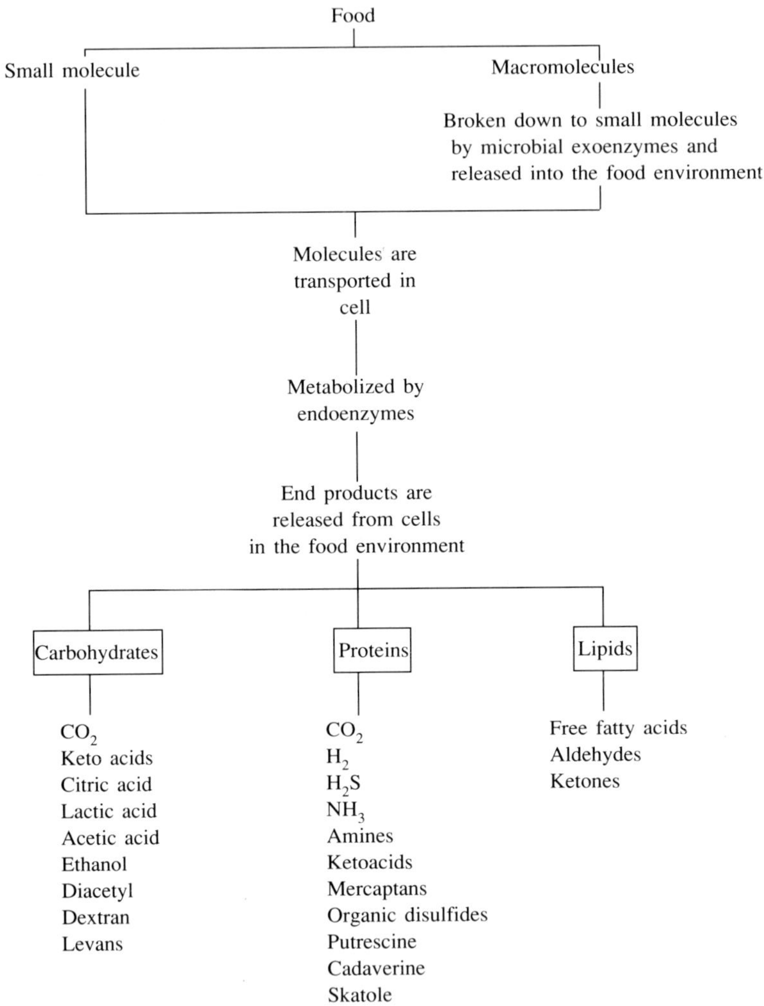

and size of the molecules, in addition to their insolubility (hydrophobicity) and the mechanisms of their transportation, dictate the entrance of the molecules inside the microbial cells for metabolism by the endoenzymes (enzymes inside the cells). Large food molecules such as starch, pectin, casein, and myoglobin or hydrophobic molecules, like triglycerides, cannot enter the microbial cells. However, many microorganisms, including some that are important in food spoilage, are capable of producing extracellular enzymes (exoenzymes). These exoenzymes can hydrolyze the large food molecules (macromolecules) into smaller components in the environment. These smaller components can then be transported inside the cell and utilized by the microorganisms. Even a microorganism that does not have the capability to produce specific exoenzymes can utilize these breakdown components produced by the exoenzymes of another microorganism and produce metabolic end products.

By the action of their endoenzymes and exoenzymes, the live microbial cells bring about some changes associated with the spoilage of foods. In addition, the exoenzymes, after their release from the cells, and the endoenzymes, after the death and lysis of the cells, can continue to react with the specific food molecules to bring about changes in food quality even in the absence of live cells. This is especially important for heat-stable microbial enzymes in many heat-processed foods that are expected to have long shelf life.

D. Microbial Spoilage of Perishable Foods

A perishable food in the initial stages generally contains many different types of microorganisms in different proportions. Some are present in much higher numbers (predominant types) than the others. However, when spoilage occurs, the same food can have just one or two predominant types and which might not even be among the initial predominant types. The intrinsic, extrinsic, and processing factors dictate which types among the mixed population will grow preferentially. One best example is a recent study involving the incidence of spoilage of vacuum-packaged refrigerated fresh beef by a psychrotrophic *Clostridium* spp., *Clostridium laramie*.[30,31] This species was probably present in the meat processing environment all the time. However, a combination of relatively good sanitation (aerobic plate count of the fresh meat was $10^{3\text{-}4}$/g), efficient vacuum packaging with packaging material of very low oxygen permeability, a storage temperature of 1 to 2 °C, and long storage times inhibited or reduced the growth of predominant microorganisms (*Pseudomonas*, lactic acid bacteria), but it facilitated the growth of *C. laramie*, which was originally present in very low numbers (Figure 1 and Table 7). The selection of extrinsic and processing factors facilitated the growth of a spoilage microorganism that under less stringent conditions would not have had a chance to be the dominant type and cause spoilage.

Most refrigerated foods, as compared to dried, frozen, ultra-high-temperature-treated, and commercially sterile foods,are susceptible to microbial spoilage. The new generation of refrigerated foods and sous vide type foods that are processed under low heat are more susceptible to spoilage because they do not have antimicrobial preservatives as an intrinsic factor to control microbial growth during storage. The microbial population in these foods, even when they are heat processed in bags, includes those that survive the low-heat treatment and those that gain entrance as post-heat-treatment contaminants during handling. Among these microorganisms only the psychrotrophs will grow, during extended storage under refrigeration. If the products are temperature abused, depending upon the time period involved and the temperature(s), some strains of mesophiles can also grow, in addition to the psychrotrophs, and probably at a faster rate. If a product is vacuum packaged (or gas flushed), as is customary with most of these products, then both psychrotrophic anaerobes

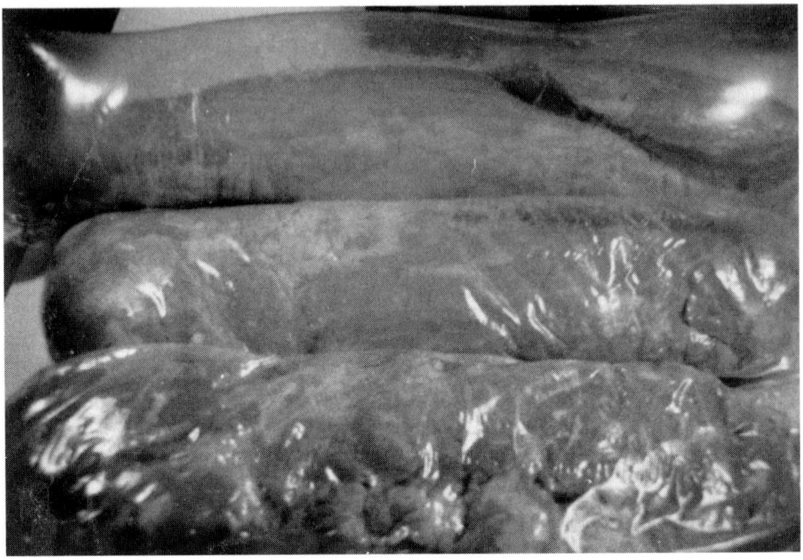

FIGURE 1. Spoilage of vacuum-packaged fresh beef by *Clostridium laramie* at refrigeration storage. The spoilage is characterized by large accumulation of foul-smelling gas and purge, proteolysis, and red coloration of meat.[30,38]

and facultative anaerobes will grow. In addition to the vegetative cells, spores of psychrotrophic *Bacillus* and *Clostridium* can also germinate, outgrow, and undergo cell multiplication.[62,79–81] Growth of yeasts and molds will mostly be inhibited in vacuum-packaged (or gas-flushed with no O_2) products, especially with packaging materials of very low oxygen permeability.[24a] Frozen foods, if subjected to accidental thawing and refreezing, depending upon the time and temperature, can facilitate the growth of psychrotrophiles. The ultra-high-temperature-treated products and commercially sterile canned products can have thermophilic spores of spoilage bacteria. Under normal storage at room temperature, these spores do not germinate. However, if the products are abused to temperatures above 40°C, the spores will germinate. Subsequently, even if the temperature drops to below 40°, cell multiplication can occur from the germinated spores, resulting in spoilage of the products.[12–16]

The aim of using biopreservatives in these products will be to prevent or reduce the growth of these spoilage microorganisms. Some of the common psychrotrophic and spore-forming thermophilic spoilage microorganisms are listed in Table 8. This table also includes other growth characteristics that can be advantageously used as barriers to inhibit or reduce the growth of spoilage microorganisms in foods.

TABLE 7
Characteristics of *Clostridum laramie* Isolated from Spoiled Beef[38]

Growth conditions	
Aerobiosis	•Strict anaerobes in agar media, less in broth
Media	•Pure culture is able to grow in many broths including tryptic soy broth •On streaking, cells form colonies on many types of agar media, including tryptic soy agar, but in fewer numbers •On pour plating of a cell suspension, 0.001% or less formed colonies, even in media recommended for *Clostridium*.
Temperature	•Cells grew at –3 to 20 °C, optimum 15°C, no growth at 25°C, killed at 50°C in 24h (with 10^7 cells failed to grow subsequently at 10°C in 20 d)
pH	•Cells grew at pH 4.5 to 7.5, optimum 6.5.
Sporulation	•2°C in 2 weeks, 15°C in 1 week, in 6 to 8 weeks about 50% cells sporulate, germination/ outgrowth at 2°C in 2 weeks.
Biochemical characteristics	
Acid formation	•Fructose, galactose, glucose, sucrose, raffinose
Other traits	•Hydrolyzed starch; reduced nitrate; digested meat; betahemolysin, lipase-positive, and lecithinase-negative
Inhibition by pediocin AcH	
Cells (10^5) treated with pediocin	•Failed to grow in broth
Spores in broth + pediocin	•Failed to outgrow

V. CONTROLLING MICROORGANISMS WITH BIOPRESERVATIVES

A. Microbial Level

To have a finished product of good microbial quality the initial level of microbial population in raw materials should be low. In general, a low initial population, of microorganisms will require a relatively longer time to multiply and reach the spoilage level ($>10^{6\text{-}7}$/g) than a higher initial population under proper conditions of storage and preservation. A low initial population of pathogens for which a higher cell number is necessary to cause health hazards

TABLE 8
Predominant Psychrotrophic and Thermophilic Food Spoilage Bacteria and Their Characteristics

Species	Gram stain	Aerobiosis	Minimum growth Temp. (°C)	Minimum growth pH	Minimum growth Aw
		Psychrotrophic bacteria			
Pseudomanas spp.	–	Aerobic	1	5.0	0.95
Acenetobacter spp.	–	Aerobic	4	5.0	0.95
Aeromonas spp.	–	Fac. An.c	1	5.5	0.95
Flavobacterium spp.	–	Aerobic	5	5.5	0.95
Enterobacter spp.	–	Fac. An.	5	5.0	0.95
Escherichia coli	–	Fac. An.	5	4.5	0.94
Altermonas putrefaciens	–	Fac. An.	2	5.0	0.95
Lactobacillus spp.[a]	+	Fac. An.	2	3.6	0.93
Leuconostoc spp.	+	Fac. An.	4	4.5	0.95
Brochothrx thermosphacta	+	Fac. An.		4.6	NA
Propionibacterium spp.	+	Anaerobic	2	NA[d]	NA
Enterococcus spp.[a]	+	Fac. An.	5	4.6	0.94
Clostridium spp.[b]	+	Anaerobic	4	4.6	0.90
Clostridium laramie[b]	+	Anaerobic	–3	4.5	NA
Bacillus spp.[b]	+	Fac. An	2	4.2	0.90
		Thermophilic bacteria (spore)			
Bacillus stearothermiphilus	+	Fac. An.	30	5.3	0.90
Bacillus coagulans	+	Fac. An.	15	4.0	0.90
Desulfotomaculum nigrifacience	–	Anaerobic	30	5.6	0.93
Clostridium thermosacchrolyticum	+	Anaerobic	37	4.1	0.93

[a] Some are thermoduric. Some are classified as *Carnobacterium*.
[b] Spore formers.
[c] Facultative anaerobic.
[d] NA, Not available.

will also require a longer time to attain risk level. For pathogens for which there is a definite standard, such as zero tolerance in a specified volume of food, the initial population level has to be very low.

A low initial microbial load in a product can be achieved by selecting good quality raw products and using effective sanitation methods during all phases of food handling.

B. Hazard Analysis Critical Control Point (HACCP)[11,19,24a,82–85]

For a processed food the sources of microbial contamination, their possible growth and points of recontamination should be identified beginning with the raw materials at all phases of processing and subsequent handling. Corrective measures through actual tests should be included to determine the effectiveness of controls.

C. Selection of Effective Biopreservatives

A biopreservative preparation should meet the criteria set by the regulatory agencies for food preservatives. It should not be unsafe and should not affect the acceptance quality of a food. It preferably should have a wide host range, i.e., it should be effective against many Gram-positive and Gram-negative bacteria, as well as bacterial spores, yeasts, molds, and possibly viruses. But more specifically, it should be effective against the microorganisms that can grow in a food under the condition it is stored in. It should be able to kill a high percentage of cells and spores and inhibit or retard the growth of survivors. The antimicrobial activity of the preservative should be heat stable, stable at wide pH ranges, have relatively long storage stability, remain active in the food environment, and be effective in liquid, semisolid, and solid foods. In addition, for effective use, the chemical nature, the mode of action, and the influences of its concentrations on antimicrobial property should be known. Finally, methods for its application in different food systems should be developed.[37,38]

D. Use of Multiple Barriers[10,11,19,20]

For proper control of microbial growth, as many of the intrinsic, processing, and extrinsic factors as possible should be used along with an effective biopreservative. Some of these intrinsic factors include a reduced pH, a reduced Aw, heat treatment at maximum temperatures for long periods of time that can be applied to a food without affecting its acceptance and nutritional qualities, the use of the lowest storage temperatures possible, the control of storage temperatures, and the application of the modified-atmosphere packaging. It is well known that microbial growth can be greatly controlled by taking advantage of the interaction of the different factors. These factors have additive effects; by combining the suitable growth-limiting factors, each at the subinhibitory level, the antimicrobial effectiveness of a biopreservative in ensuring safety and in enhancing the shelf life of a food can be greatly increased.

VI. CONCLUDING REMARKS

The spoilage and pathogenic microorganisms that can grow in food, especially at refrigeration temperature, include yeasts, molds, and gram-positive and gram-negative bacteria. Although their growth rate is slow at refrigeration temperature, temperature abuse, which can occur from the time of production to the time of consumption, can accelerate their growth rate. In fact, some mesophilic microorganisms can grow in foods that are even marginally temperature abused (10 to 12°C). Effective sanitation and HACCP during all stages of production and processing are of the utmost importance to reduce the initial microbial load in these foods. To control or reduce the growth of the survivors to assure extended shelf life and safety of these products, multiple barriers have to be included in the food system. A biopreservative should be considered as one of the barriers. As better control systems are introduced, new types of spoilage and pathogenic microorganisms, which previously could not compete well with others, will gain predominance. A different set of barriers will need to be developed for their control. Thus, the quest for effective preservatives will be a continuous process.

REFERENCES

1. **Anon.,** Showcase: preservatives, *Prep. Foods,* 160, 101, 1991.

1a. **Ray, B.,** Ed., *Injured Index and Pathogenic Bacteria: Occurrence and Detection in Foods, Water, and Feeds,* CRC Press, Boca Raton, FL, 1989.

2. **Hurst, A. and Nasim, A.,** Eds., *Repairable Lesions in Microorganisms,* Academic Press, New York, 1984.

3. **Andrew, M. H. E. and Russell, A. D.,** Eds., *The Revival of Injured Microbes,* Academic Press, New York, 1984.

4. **Ray, B. and Speck, M. L.,** Freeze-injury in bacteria, *Crit. Rev. Clinic. Lab. Sci.,* 4, 161, 1973.

5. **Ray, B.,** Methods to detect stressed microorganisms, *J. Food Prot.,* 42, 346, 1976.

6. **Ray, B. and Adams, D. M.,** Repair and detection of injured microorganisms, in *Compendium of Methods for the Microbiological Examination of Food,* 2nd ed., Speck, M. L., Ed., American Public Health Association, Washington, D.C., 1984, 112.

7. **Frazier, W. C. and Westhoff, D. C.,** *Food Microbiology,* 4th ed., McGraw-Hill, New York, 1988, chap. 5–10.

8. **Banwart, G. J.,** *Basic Food Microbiology,* AVI Publishing, Westport, CT, 1979, chap. 10–12.

9. **Jay, J. M.,** *Modern Food Microbiology,* 2nd ed., D Van Nostrand, New York, 1970, chap. 9–13.

10. **Scott, V. N.,** Interaction of factors to control microbial spoilage of refrigerated food, *J. Food Prot.,* 52, 431, 1989.

11. **Anon.,** Safety considerations of the new generation refrigerated foods: refrigerated foods and microbiological criteria committee of the National Food Processors Association, *Dairy Food Sanit.,* 8, 5, 1988.

12. **Denny, C. B.,** Thermophilic organism involved in food spoilage: introduction, *J. Food Prot.,* 44, 144, 1981.

13. **Ashton, D. H.,** Thermophilic organisms involved in food spoilage: thermophilic anaerobes not producing H_2S, *J. Food Prot.,* 44, 146, 1981.

14. **Speck, R. V.,** Thermophilic organisms in food spoilage: sulfide spoilage anaerobes, *J. Food Prot.,* 44, 149, 1981.

15. **Thompson, P. J.,** Thermophilic organisms involved in food spoilage: aciduric flat-sour spore forming aerobes, *J. Food Prot.,* 44, 154, 1981.

16. **Ito, K. A.,** Thermophilic organism in food spoilage: flat sour aerobes, *J. Food Prot.,* 44, 157, 1981.
17. **Hurst, A.,** Nisin, *Adv. Appl. Microbiol.,* 27, 85, 1981.
18. **Branen, A. L. and Davidson, P. M.,** *Antimicrobials in Foods,* Marcel Dekker, New York, 1983.
19. **Corlett, D. A.,** Refrigerated foods and use of hazard analysis and critical control point principles, *Food Technol.,* 43(2), 91, 1989.
20. **Swientek, R. J.,** Refrigerated entrees offer convenience and freshness appeal, *Food Process.(Chicago),* 49(1), 70, 1988.
21. **Eilers, J. R.,** New foods provide new safety challenges, *Food Process.(Chicago),* 51(6), 104, 1990.
22. **Swientek, R. J.,** New generation chilled foods face cold realities, *Food Process.(Chicago),* 51(4), 94, 1990.
23. **Del Giudice, V. J.,** We must meet 21st century sanitation standards, *Natl. Provisioner,* 204(6), 12, 1991.
23a. **Harris, R. D.,** Kraft builds safety into next generation refrigerated foods, *Food Process.(Chicago) (Chicago),* 50(13), 111, 1989.
24. **Baird, B.,** Sous vide: what's all the excitement about? *Food Technol.,* 44(11), 92, 1990.
24a. **Beauchemin, M.,** Sous vide technology, in Proc. 43rd Reciprocal Meat Conf., American Meat Science Association National Livestock and Meat Board, Chicago, IL, 1991, 103.
25. **Swientek, R. J.,** Sous vide — refrigerated upscale food in a vacuum pouch, *Food Process.(Chicago),* 49(6), 34, 1989.
26. **Anon.,** Microbial dangers of MAP — Much more research needed, *Food Process.(Chicago),* 49(6), 88, 1989.
27. **Mutkoski, S. A. and Schurer, M. L.,** *Meat and Fish Management,* Breton, North Scituate, MA, 1981, chap. 1.
28. **Warburton, D. J. and Gill, C. O.,** New Zealand effort leads to extend meat shelf life, *Natl. Provisioner,* 202(8), 33, 1989.
29. **Harr, J. W.,** Cryogenic meat processing cost and quality, in *Proc. Meat Industry Research Conf.,* 1977, 7.
30. **Kalchayanand, N., Ray, B., Field, R. A., and Johnson, M. C.,** Spoilage of vacuum-packaged refrigerated beef by *Clostridium, J. Food Prot.,* 52, 424, 1989.
31. **Ray, B., Kalchayanand, N., and Field, R. A.,** Isolation of a *Clostridium* spp. from spoiled vacuum-packaged refrigerated beef and its susceptibility to bacteriocin from *Pediococcus acidilactici,* in *Proc. 35th Int. Cong. Meat Sci. and Technology,* Danish Meat Research Institute, Denmark, 1989, 285.
31a. **Kalchayanand, N., Ray, B. and Field, R. A.,** Characteristics of *Clostridium laramie,* a new psychrotrophic species associated with spoilage of vacuum-packaged fresh beef, Proc. Western Section, *Am. Soc. Anim. Sci.,* 42, 26, 1991.
32. **Russell, A. D. and Fuller, R.,** *Cold Tolerant Microbes in Spoilage and the Environment,* Academic Press, New York, 1979, 83.
33. **Egan, A. F., Shay, B. J., and Rogers, P. J.,** Factors affecting the production of hydrogen sulfide by *Lactobacillus sake* L13 growing on vacumm-packaged beef, *J. Appl. Bacteriol.,* 67, 255, 1989.
34. **Shaw, B. C. and Harding, C. D.,** *Leuconostoc gelidum* sp. nov. and *Leuconostoc carnosum* sp. nov. from chill stored beef, *Int. J. Syst. Bacteriol.,* 39, 217, 1989.
35. **Gill, C. O.,** *Advances in Meat Microbiology,* Vol. 2, AVI Publishing, Westport, CT, 1986, 49.
36. **Egan, A. F. and Shang, B. J.,** Significance of lactobacilli and film permeability in the spoilage of vacumm-packaged beef, *J. Food Sci.,* 47, 1119, 1982.
37. **Rozbeh, M.,** *Extension of Shelf Life of Vacuum-packaged Refrigerated Beef,* M.S. thesis, University of Wyoming, Laramie, WY, 1990.
38. **Kalchayanand, N.,** Extension of Shelf life of Vacuum-Packaged Refrigerated Beef by Bacteriocins of Lactic Acid Bacteria, Ph. D. thesis, University of Wyoming, Laramie, 1990.

39. **Gravani, R. B.,** Food science facts, *Dairy Food Sanit.,* 10, 20, 1987.
40. **Sours, M. E. and Smith, D. G.,** Outbreak of food borne disease in the United States, 1972-78, *J. Infect. Dis.,* 142, 122, 1980.
41. **Center for Disease Control,** *Foodborne Disease Outbreaks Annual Summary,* 1983, p. 1980.
42. **Bryan, F. L.,** Foodborne diseases, in *Survey of Contemporary Toxicology,* Vol. 1, Tu, A. T., Ed., John Wiley Sons, New York, 1980, 189.
43. **Bryan, F. L.,** *Guide to Investigate Foodborne Disease Outbreaks and Analyzing Surveillance Data,* Center for Disease Control, Atlanta, GA, 1973, 90.
44. **Snyder, O. P. and Poland, D. M.,** America's safe food, *Dairy Food Environ. Sanit.,* 10, 719, 1990.

44a. **Snyder, O. P. and Poland, D. M.,** America's safe food, *Dairy Food Environ. Sanit.,* 11, 14, 1991.

45. **Roberts, T.,** Bacterial foodborne illness cost in the USA, *Food Lab. News,* 19, 53, 1990.
46. **Roberts, T. and Van Ravenswaay, E.,** The economics of safeguarding the US food supply, *USDA Agric. Info. Bull.,* No. 566, 1, 1989.
47. **Todd, E. C. D.,** Preliminary estimates of cost of food borne diseases in the United States, *J. Food Prot.,* 52, 595, 1989.
48. **Bennett, J. V., Scott, D. H., Rogers, M. F., and Solomon, S. L.,** Infections and parasitic diseases, in *The Burden of Unnecessary Illness,* Amler, R. W. and Dull, H. B., Eds., Oxford University Press, New York, 1987.
49. **Eilers, J. R.,** US food poisoning cases greatly under-reported, *Food Process.(Chicago),* 51(6), 110, 1990.
50. **Liska, B. J.,** New bacteria in the news: a special symposium, *Food Technol.,* 40(6), 17, 1986.
51. **Bryan, F. L.,** Control of foodborne diseases, in *The Safety of Foods,* 2nd ed., Graham, H. D., Ed., AVI Publishing, Westport, CT, 1980, 55.
52. **Bryan, F. L.,** Risk of practices, procedures and processes that lead to outbreaks of foodborne diseases, *J. Food Prot.,* 51, 663, 1988.
53. **Frazier, W. C. and Westhoff, D. C.,** *Food Microbiology,* McGraw-Hill, New York, 1988, chap. 24–26.
54. **Banwart, G. J.,** *Basic Food Microbiology,* AVI Publishing, Westport, CT, 1979, chap. 6.
55. **Gravani, R. B.,** Bacterial foodborne diseases, *Dairy Food Sanit.,* 10, 77, 1987.
56. **Bryan, F. L.,** *Diseases Transmitted by Foods — A Classification and Summary,* 2nd ed., Center for Disease Control, Atlanta, GA, 1982, 101.
57. **Palumbo, S. A.,** Is refrigeration enough to restrain foodborne pathogens? *J. Food Prot.,* 49, 1003, 1986.
58. **Eddy, B. P.,** The use and meaning of the term psychrotrophs, *Appl. Microbiol.,* 23, 189, 1960.
59. **Cann, D. C., Wilson, B. B., Hobbs, G., and Shewan, J.,** The growth and toxin production of *Clostridium botulinum* in certain vacuum-packaged fish, *J. Appl. Bacteriol.,* 28, 431, 1965.
60. **Lucke, F. K., Hechelman, H., and Leisteur, L.,** The relevance to meat products of psychrotrophs Strains of *Clostridium botulinum,* in *Psychrotrophic Microorganisms in Spoilage and Pathogenicity,* Roberts, T. A., Hobbs, G., Christin, J. H. B., and Skovguard, N., Eds., Academic Press, London, 1982, 491.
61. **Notermans, S., Dufrenne, J., and Lund, B. M.,** Botulinum risk of refrigerated, processed foods of extended durability, *J. Food Prot.,* 53, 1020, 1990.
62. **Kornacki, J. L. and Gabis, D. A.,** Microorganisms and refrigeration temperature, *Dairy Food Environ. Sanit.,* 10, 192, 1990.
63. **Liston, J.,** Microbial hazard of seafood consumption, *Food Technol.,* 44(12), 56, 1990.
64. **Bhunia, A. K., Johnson, M. C., and Ray, B.,** Purification, characterization and antimicrobial spectrum of a bacteriocin produced by *Pediococcus acidilactici, J. Appl. Bacteriol.,* 65, 261, 1988.

65. **Benkerrom, N. and Sandine, W. E.,** Inhibitory action of nisin against *Listeria monocytogenes*, *J. Dairy Sci.*, 71, 3237, 1988.
66. **Schillinger, U. and Lucke, F. K.,** Antibacterial activity of *Lactobacillus sake* isolated from meat, *Appl. Environ. Microbiol.*, 55, 1901, 1989.
67. **Hackney, C. R., Kleeman, E. G., Ray, B., and Speck, M. L.,** Adherence as a method of differentiating virulent and avirulent strains of *Vibrio parahaemolyticus*, *Appl. Environ. Microbiol.*, 40, 652, 1980.
68. **Ray, B.,** Unpublished information, University of Wyoming, Laramie.
69. **Mullery, M.,** 1989 Food industry economic outlook, *Food Process.(Chicago)*, 50(2), 26, 1989.
70. **Anon.,** International Commission on Microbiological Specification for Foods, *Microbiological Ecology of Food*, Vol. 1, Academic Press, New York, 1980, chap. 7.
71. **Banwart, G. J.,** *Basic Food Microbiology*, AVI Publishing, Westport, CT, 1979, chap. 4.
72. **Frazier, W. C. and Westhoff, D. C.,** *Food Microbiology*, 4th ed., McGraw-Hill, New York, 1988, chap. 1.
73. **Jay, J. M.,** *Modern Food Microbiology*, 2nd ed., D Van Nostrand, New York, 1978, chap. 3.
74. **Cousin, M. A.,** Presence and activity of psychrotrophic microorganisms in milk and dairy products: a review, *J. Food Prot.*, 45, 172, 1982.
75. **Kraft, A. A. and Ray, C. R.,** Psychrotrophic bacteria in foods: an update, *Food Technol.*, 33(1), 66, 1979.
76. **Gilliland, S. E., Michener, H. D., and Kraft, A. A.,** Psychrotrophic microorganisms, in *Compendium of Methods for the Microbiological Evaluation of Foods*, 2nd ed., Speck, M. L., Ed., American Public Health Association, Washington, D.C., 1982, 173.
77. **Nelson, F. E.,** Thermoduric microorganisms, in *Compendium of Methods for the Microbiological Evaluation of Food*, 2nd ed., Speck, M. L., Ed., American Public Health Association, Washington, D.C., 1982, 179.
78. **Banwart, G. J.,** *Basic Food Microbiology*, AVI Publishing, Westport, CT, 1979, chap. 8.
79. **Shehata, T. E. and Collins, E. B.,** Isolation and identification of psychrophilic species of *Bacillus* from milk, *Appl. Microbiol.*, 21, 466, 1971.
80. **Shehata, T. E., Duran, A., and Collins, E. B.,** Influences of temperature on the growth of psychrophilic strains of *Bacillus*, *J. Dairy Sci.*, 54, 1579, 1972.
81. **Bhadsvle, C. H., Shehata, T. E., and Collins, E.B.,** Isolation and identification of psychrophilic species of *Clostridium* from milk, *Appl. Microbiol.*, 24, 699, 1972.
82. **Anon.,** Factors to be considered in establishing good manufacturing practices for the production of refrigerated foods. Refrigerated foods and microbiological criteria committee of the National Science Foundation, *Dairy Food Sanit.*, 8, 288, 1988.
83. **Smith, G.,** HACCP: safety tool or just another acronym? *Natl. Provisioner*, 204(9), 10, 1991.
84. **Snyder, P.,** Food safety 2000: applying HACCP for food safety assurance in the 21st century, *Dairy Food Environ. Sanit.*, 10, 197, 1990.
85. **Bryan, F. L.,** Application of HACCP to ready-to-eat chilled foods, *Food Technol.*, 44(7), 70, 1990.

Chapter 3

PROCEDURES TO DETECT ANTIMICROBIAL ACTIVITIES OF MICROORGANISMS

Mark A. Daeschel

TABLE OF CONTENTS

I. INTRODUCTION

Within the last 8 to10 years there has been a marked renewal of interest in microbial metabolites that possess antagonistic activities toward spoilage and pathogenic microorganisms associated with foods. This interest is evidenced by the publication of many research articles in the scientific literature, as well as the appearance of reviews and monographs on the topic.

From a historic viewpoint, the study of food biopreservatives of microbial origin had its beginnings with fermented foods and the observation that many primary microbial metabolites, such as ethanol, lactic acid and acetic acid imparted a preservative effect to foods that had undergone fermentation. It was not until the Second World War, with the discovery and application of penicillin, that investigations focusing on antagonistic activities of secondary microbial metabolites began in earnest. The enormous clinical therapeutic potential of microbial metabolites fueled very large and well-funded investigations into the search for, and characteriztion of, these metabolites.

Today, there exist more than seven thousand known antibiotic compounds of microbial origin.[1] From this group of metabolites no more than a half dozen have been considered for use in food preservation. Why? There is a historical reluctance — and it is likely justifiable — to use antibiotics in foods. The fear is that antibiotic-resistant microorganisms of clinical importance may appear through repeated exposure to antibiotics in food stuffs. Consequently, the only "antibiotic" type of microbial metabolites evaluated for food use were ones determined unsuitable for clinical use.

An example of such a metabolite is nisin, a bacteriocidal polypeptide produced by *Lactococcus lactis*. Some may argue that nisin is not an antibiotic, since it has no therapeutic use and is more correctly classified as a bacteriocin. Investigations into bacteriocins as food preservatives have appeared sporadically in the literature since the initial characterization of nisin in the 1940s and early 1950s. The current renewal of interest in bacteriocins is likely the result of a combination of several events, some of which are listed in Table 1.

The scope of current investigations into potential food biopreservatives is very broad, and addresses such specifics as DNA sequences, or is as general as screening for previously unrecognized metabolites. In any event, investigators need to be cognizant of the techniques needed to detect and quantify antimicrobial metabolites. We can borrow much from the field of clinical antibiotics; however, there exist many nuances in technique that are germane only to the study of food biopreservatives. It is the intent of this chapter to relate to investigators some of these nuances as well as general methodologies. The reader will be advised of sources of more in-depth coverage of particular topics as they are presented.

II. SOURCES OF MICROORGANISMS

Sufficient consideration must be given to the types of microorganisms one wishes to evaluate for potential food biopreservatives. So-called "food grade"

TABLE 1
Factors Contributing to the Increasing Number of Investigations on Bacteriocins

1. Safe and efficacious use of nisin during the past 30 years
2. Recent FDA approval of nisin as a "GRAS"substance in certain applications
3. Consumer resistance to traditional "chemical" preservatives
4. Concerns over the safety of existing food preservatives such as sulfites and nitrites
5. Realization that bacteriocinogenicity is not a rare occurrence within the lactic acid bacteria
6. Feasibility of using bacteriocin production and immunity as selectable genetic markers in starter culture bacteria
7. Availability of molecular biology tools to transfer, clone, and sequence the genetic determinants and to engineer genetic variants of bacteriocins
8. Willingness of federal funding agencies, food commodity groups, and food processing corporations to fund both basic and applied researches

microorganisms are a preferred source of strains because of their presumed nontoxicity for humans. "Food-grade" microorganisms may be viewed as those that have been intentionally used in foods and/or those that naturally occur with foods and whose action results in desirable characteristics being imparted to such foods. Examples are the addition of *L. Lactis* to milk for cheese making and the natural occurrence and growth of *Leuconostoc, Pediococcus,* and *Lactobacillus* during the fermentation of cabbage. A list of microorganisms that are naturally part of, or have been intentionally used in, food manufacture is given in Table 2.

A. Culture Collections

Sources of potential microorganisms include the culture collections of the world (Table 3). These collections provide a readily available, bona fide source of known microorganisms. These collections have served as sources of microorganisms for screening of antagonistic activities. For example, two relatively newly characterized bacteriocins were identified from culture collection strains. These were Planticin B, a bacteriocin produced by *Lactobacillus plantarum* NCDO 1193,[2] and Pediocin PA-1 produced by *Pediococcus acidilactici* NRRL B-5627.[3,4] With the renewal of interest in antagonistic activities starting in the early 1980s, it is likely that the culture collections have been fairly well "picked over" However, they still remain a valuable resource for those wanting strains for comparison of activities, as well as a source of sensitive indicator strains.

B. Natural Environments

Many environments provide selective conditions in which one can isolate microorganisms that may be sources for potential food biopreservatives. Many

TABLE 2
Microorganisms That are Naturally Part Of, or Have Been Intentionally Used in Food Manufacture

Bacteria	Yeast	Fungi
Bifidobacterium	*Saccharomyces*	*Penicillum*
Leuconostoc	*Candida*	*Aspergillus*
Lactobacillus	*Hansenula*	*Rhizopus*
Pediococcus	*Schizosaccharomyces*	*Mucor*
Lactococcus	*Kluyveromyces*	*Monascus*
Propionibacterium		
Streptococcus		
Zymomonas		
Acetobacter		
Brevibacterium		
Micrococcus		

TABLE 3
Major Microbial Culture Collections

1. Agricultural Research Service Culture Collection (NRRL), Northern Regional Research Center, 1815 N. University St. , Peoria, IL 61604, U.S.
2. American Type Culture Collection (ATCC), 12301 Parklawn Dr., Rockville, MD 20852, U.S.
3. Centraalbureau voor Schimmelcultures (CBS), Yeast Division, Julianalaan 67A, 2628 BC Delft, the Netherlands
4. Culture Collection, Department of Food Science and Technology (UCD), University of California, Davis, CA 95616, U.S.
5. Department of Type Cultures of Microorganisms, Institute of Biochemistry and Physiology of Microorganisms, USSR Academy of Sciences, Pushino, Moscow region, Russia, 142292
6. Deutsche Sammlung von Mikroorganismen (DSM), Grisebachstrasse 8, D-3400 Göttingen, Germany
7. Institute for Fermentation (IFO), 17-85 Juso-honmachi 2-chome, Yodogawa-ku, Osaka 532, Japan
8. Japan Collection of Microorganisms (JCM), RIKEN, Hirosawa, Wako-shi, Saitama 351-01, Japan
9. National Collection of Food Bacteria, AFRC Institute of Food Research, Reading Laboratory, Shinfield, Reading RG2 9AT, U.K.
10. National Collection of Yeast Cultures (NCYC), AFRC Food Research Institute, Colney Lane, Norwich NR4 7UA, U.K.
11. Yeast Genetic Stock Center (YGSC), Department of Biophysics and Medical Physics, University of California, Berkeley, CA 94720, U.S.

fermentative microorganisms such as yeast and lactic acid bacteria (LAB) can easily be isolated from plant sources. These microorganisms will multiply to large populations in decaying vegetation that contain sufficient amounts of free sugar. An environment from which our laboratory, on several occasions, had isolated both yeast and LAB, was the parking lot behind our laboratory. Each fall native crab apples and blackberries fall to the pavement and are macerated by passing automobiles. The resulting pulp has yielded in excess of 10^8 LAB and 10^7 yeast per gram.

Yeast and LAB are indigenous to the surfaces of living plant material. Their numbers vary from 10^1 to 10^3 per gram, depending on plant species and climate. One can select for these microorganisms simply by taking leafy vegetation, chopping it finely, adding 3% NaCl w/w, packing it under anaerobic conditions, and holding it at room temperature for 3 to 5 d. Invariably a strong lactic fermentation will commence with populations of LAB exceeding 10^8 per gram. For yeast, substitute 5% sucrose for the salt and use preferably noncitrus fruits, i.e., apple, instead of leafy vegetation.

C. Food and Beverage Processing Environments

Food-processing environments provide a plethora of substrates for growth and proliferation of a variety of microorganisms. The type of food being processed will dictate which populations of microorganisms will dominate. Fruit-and-vegetable processing facilities will harbor many yeast and LAB species, whereas meat-processing operations can provide conditions conducive to the growth of putrefactive microorganisms. Many fermented food processes employ commercially produced starter cultures. If one is interested in obtaining previously uncharacterized strains, it probably is best not to attempt to isolate strains from processing facilities where starters are used, i.e., dairies, wineries, etc.

III. SCREENING OF MICROORGANISMS FOR ANTAGONISTIC ACTIVITIES

Rarely is one as fortunate as Alexander Fleming was with penicillin in observing antagonism without systematically looking for it. In looking for new potential food biopreservatives, a screening process must be one that will eliminate previously characterized compounds, but at the same time will not overlook metabolites produced in small quantities or only under specific growth conditions. A generalized screening process is given in Figure 1 that represents one possible approach. More specific approaches will evolve as a result of experience and specific needs or objectives.

A large variety of techniques have evolved to assess the ability of microorganisms to produce antagonistic substances. Many of these techniques have their origin in the innumerable studies on compounds with clinical antibiotic potential. The two essential features for any screening method to be successful were put forward by Berdy 1 as (1) the sensitivity and selectivity of the basic screening method applied and (2) the early recognition of known and undesired compounds. The latter feature is especially relevant for industrial scientists where time and economics cannot allow the "rediscovery" of compounds. The techniques described in the following paragraphs will attempt to focus on procedures commonly used by researchers working in the fields of food preservatives and bacteriocins.

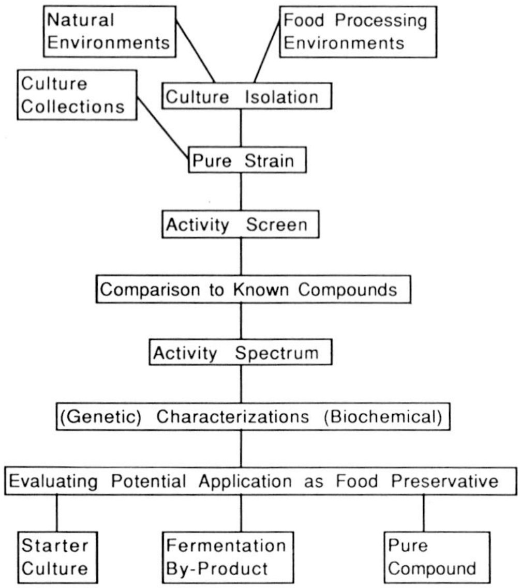

FIGURE 1. Flow diagram of steps in screening and characterizing antimicrobial metabolites.

A. Media Selection

The selection of a medium and conditions of growth, for either the microorganism to be screened for antimicrobial activity or for using sensitive indicator strains, is critical for successful screening. A strain may need certain environmental or nutritive conditions to synthesize and excrete antimicrobial compounds in detectable amounts. A growth variable that has been observed to be important in regard to bacteriocin production is pH. An example of pH-dependent bacteriocin production was observed by Muriana and Klaenhammer [5] in which the production of lactacin F was maximal at pH 7.0, but was essentially undetectable at either pH 6.6 or 7.6. A pH of 4.2 was found to be optimal for expression of K_2 Killer toxin by *Saccharomyces cerevisiae* K_1 with negligible activity at pH 3.3 or 5.4.[6] Time and temperature of incubation will influence the growth, and subsequently the detection of, antimicrobial metabolites. A common phenomenon observed is that bacteriocin production accumulates to a maximum during late logarithmic or early stationary phase, and then detectable activity drops off somewhat rapidly. This has been observed with bacteriocin-producing strains of *Lactobacillus plantarum*,[7] *Streptococcus zymogenes*,[8] *Lactobacillus sake*, [9] and *Lactobacillus acidophilus*.[10] Possible explanations that have been discussed for this observation are (1) appearance of proteolytic enzymes or other types of inactivators late in the growth cycle, (2) destabilization of bacteriocins as a result of acid accumulation, and (3) adsorption of bacteriocins onto either producer cells or components of the medium.

The effect of temperature on antagonist production has not been a specific research focus but rather is reported as part of a series of descriptive characteristics of a particular substance. Intuitively one assumes that antagonist production is greatest at or near the maximal specific growth rate of the producing strain. However, numerous exceptions have been documented. Hastings and Stiles [11] reported the amount of inhibitor produced by *Leuconostoc gelidum* was approximately the same when grown at 5 or 25°C for 70 h at pH 6.0 or 6.5. Moore [6] observed that killer toxin production (activity) by *S. cerevisiae* K_1 was highest at 15°C , least at 30°C, and none at 37°C. It was observed by Foulds and Shemin[12] that bacteriocin production by a strain of *Serratia marcesens* was maximal at 39°C, but not detectable at 37°C. They concluded that a bacteriocin inhibitor was concomitantly produced with the bacteriocin at temperatures less than 39°C, but that the inhibitor loses its own activity at 39°C.

Many genetic determinants for bacteriocin production have been shown to be localized on plasmids.[13] Daeschel and Klaenhammer[14] reported that curing of a native plasmid of *Pediococcus pentosaceus* by growing at a temperature 4°C above optimum (38°C) resulted in populations unable to produce bacteriocin. Likewise, Davey and Pearce[15] reported the occurrence of non-diplococcin-producing variants of *Streptococcus cremoris* isolated after growth of the strain at elevated temperatures.

Although not reported in the literature, it is conceivable that volatile low-molecular inhibitors (e.g., diacetyl, ethanol, propionic acid) may volatilize off at higher incubation temperatures and may not be detected at levels originally produced. The effect of atmospheric conditions on production of antagonistic metabolites by fermentative-type microorganisms has not been systematically studied. Generally speaking, static conditions in ambient atmosphere are employed in screening tests with these microorganisms. However, anaerobic conditions have been used with more oxygen-sensitive strains of LAB in investigating bacteriocinogenicity.[10] It is conceivable that intentional aeration of fermentative microorganisms may generate metabolites with antagonistic activity. An example may be the generation of hydrogen peroxide either by non-enzymatic or enzymatic reduction of oxygen by lactic acid bacteria.[16]

Medium composition has been shown to influence production and detection of antimicrobial metabolites of lactic acid bacteria. As discussed previously, pH can have a marked influence on inhibitor production and subsequent detection. Interestingly, Geis et al.[17] in 1983 reported on a study comparing the production of bacteriocin from different media by Lactococci, that unbuffered medium (Lactic broth) gave much higher bacteriocin titers than Brain Heart Infusion, MM medium, synthetic medium, or milk. Although not discussed by the authors,[17] perhaps these particular bacteriocins were stabilized by a lower pH environment. Many investigators have determined simply by trial and error, what medium gives maximum inhibitor production. Many of these media are commercially available, and are used as is or are modified to suit particular needs.

Media such as MRS, APT, and TSB are widely used with the nondairy

LAB, whereas M-17, Ellikers Broth, and skim milk are used with the dairy LAB and *Bifidobacteria*. *Propionibacteria* benefit from the inclusion of sodium lactate into dairy-type media or as part of a minimal medium. Fermentative yeast are typically evaluated for inhibitory metabolites in yeast-nitrogen base (Difco Co.) with glucose or YEPD (Yeast Extract Peptone Dextrose).

Typical modifications to media include increasing or decreasing the type and amount of sugars and/or buffers. In our laboratory we have observed[18] that the emulsifier "Tween 80," greatly enhances the activity of nisin, perhaps by facilitating the uptake of the molecule by susceptible bacteria.

The detection of bacteriocin activity can sometimes be easily demonstrated in an agar culture system, but not necessarily in a broth system.[2,10,17] Why this occurs has not been experimentally explained. Perhaps the agar matrix prevents interaction of these bacteriocins with inactivating components whose source is the medium or the bacteriocin-producing microorganism.

B. Sensitive Indicator Strains

Overwhelmingly, the most commonly used method to detect antimicrobial activity is the microbiological assay, sometimes known simply as the bioassay. Although chemical and physical methods have been developed to detect and quantitate antagonism, the bioassay remains the method of choice because it directly measures the property of interest (antagonism).

The bioassay also finds application in the detection of growth-enhancing compounds (i.e., vitamins). The bioassay, simply defined, measures the response of selected microorganisms to introduced substances. Results are recorded by visually observing the extent of the growth response. The two basic techniques of bioassay are the agar diffusion assay and the turbidimetric tube assay, both to be described in more detail in later paragraphs.

Paramount to successful bioassay results is the use of an appropriate indicator or test organism. A list of suggested criteria for selection of indicator strains for detecting LAB bacteriocin is given in Table 4. Generally speaking, it is usually prudent to choose an indicator organism that is of the same species or at least the same genus as the potential inhibitor-producing species. A large number of bacteriocins have a narrow spectrum of activity; this spectrum is many times limited to the same species. Thus, choosing a closely related indicator strain will enhance the probability of detecting antagonism. Examples of indicator strains that have been used to detect antagonism are given in Table 5.

C. Detection of Antimicrobial Metabolites

Many variations of the agar diffusion assay have evolved, reflecting need as well as creative thinking. However, they all share certain essential components. The inhibition zone is the result of several dynamic events that concurrently progress to equilibrium. These are the rate of inhibitor diffusion and subsequent dilution, the growth rate of the indicator, and the kinetics of inhibition or killing. Mathematical models describing these events have been proposed and have been discussed by Linton,[47] Hewitt, and Vincent.[48]

TABLE 4
Suggested Criteria for Selection of Indicator Strains for Detecting Bacteriocins Produced by Starter Cultures

- Nonfastidious
- Nonbacteriocinogenic
- produces uniform dense lawns of suitable contrast
- Sensitive to amounts of bacteriocin naturally excreted by the strains
- Does not harbor significant populations of cells that are spontaneously bacteriocin resistant
- Insensitive to organic acids in the concentrations produced by test inhibitor strains
- Easily distinguished and isolated from bacteriocin-producing strains
- Reproducible growth responses to bacteriocins
- Taxonomically closely related to inhibitor producing-strains

Typically, for screening a number of strains for antagonistic activities, one of two methods are employed. The first method, called "direct antagonism," simply entails providing conditions for the simultaneous growth of the inhibitor-producing and indicator strains. The indicator organism can either be directly seeded into an agar growth medium, swabbed onto the surface of an agar plate, or applied as a seeded overlay agar. Inhibitor-producing strains are applied directly as 1 to 3 μl "spots" onto the surface of seeded plates or plates that will be overlaid. Alternatively, wells can be cut into agar with a suitable tool and inhibitor strains added to the wells, or they can be applied as paper discs saturated with liquid culture. The direct antagonism test in its most simple form is the "cross streak" in which indicator and antagonist are streaked perpendicular and across each other on the agar surface. Deferred antagonism is a variation in which inhibitor-producer cells are either inactivated after producing inhibitor or are physically isolated from the indicator. Exposure to chloroform is commonly used to inactivate cells, and "flipping" agar surfaces, spotted or streaked with inhibitor-producer culture, will physically isolate them from the indicator subsequently applied as an overlay to the reversed surfaces. Physical separation of producer and indicator will provide a control to discount the possibility that inhibition was the result of phage infection.

To efficiently screen large numbers of potential inhibitor strains, a replicator device (Figure 2) can be utilized to spot up to 48 cultures on a standard 100 mm × 15 mm culture disk. Potential inhibitor cultures are transferred to a tissue-culture well plate and then transferred by replicate spotting onto plates seeded with various test organisms in an assortment of culture media. This method can also be used to eliminate duplicate inhibitor strains as well as strains producing inhibitors that have been previously characterized. Investigators can take advantage of the fact that inhibitor-producing strains are invariably immune to the levels of the inhibitor that they produce. Thus, duplications and known inhibitors may be presumptively screened out based on sensitivity patterns.

TABLE 5
Microorganisms That Have Been Used as Sensitive Indicator Strains for Detection of Bacteriocins Produced by Lactic Acid Bacteria, Propionibacteria, and for Detection of Yeast Killer Toxin

Indicator strain	Inhibitor-producing strain or inhibitor	Ref.
Carnobacterium L66	*Carnobacterium piscicola* LV61	19
Carnobacterium LV13	*Leuconostoc gelidium* VAL-187	11
Lactobacillus brevis B18	*Lactobacillus brevis* B87	20
Lactobacillus caseii B109	*Lactobacillus caseii B80*	21
Lactobacillus delbreuckii NCFB 1979	*Lactobacillus acidophilus* LAPT 1060	22
Lactobacillus delbreuckii subsp. *lactis 4797*	*Lactobacillus acidophilus* 88 *Lactobacillus acidophilusN2*	5, 23, 24
Lactobacillus fermentum F1	*Lactobacillus fermentum 466*	25
Lactobacillus bulgaricus 1489	*Lactobacillus helveticus* 481 (Helviticin J)	26
Lactobacillus helveticus LS18	*Lactobacillus helveticus* (Lactocin 27)	27
Lactobacillus plantarum WS0	*Pediococcus pentosaceus* FBB61 (Pediocin A)	14
	Pedicoccus acidilactici, (pediocin AcH)	28, 29
Lactobacillus plantarum NCDO 965	*Lactobacillus plantarum C-11* *P. acidilactici, Pediocin AcH*	7 28, 29
Lactobacillus sake LB90	*Lactobacillus sake 706*	9
Lactobacillus fermentum 1750	*Lactobacillus acidolphilus 88*	5
Lactococcus cremoris 480B$_1$	*Lactococcus lactis* subsp. *cremoris 346* Diplococcin	30, 31
Lactococcus cremoris 1P5	*Lactococcus cremoris 202*	32
Lactococcus lactis BUZ-60	*Lactococcus lactis*, subsp. *cremoris 9B4*	33
Lactococcus lactis IL 403	*Lactococcus lactis*, subsp. *cremoris 9B4*	34
Lactococcus lactis, subsp. *cremoris NS1*	*Lactococcus lactis*,subsp. *diacetylactis S50*	35
Lactococcus lactis, subsp. *diacetylactis* 18–16	*Lactococcus lactis*, subsp. *diacetylactis WM4*	36
Listeria monocytogenes 19113	*Pediococcus acidilactici PO2*	37
Leuconostoc mesenteroides 8293	*Pediococcus acidilactici* PO2, B5627 and PC	38
Listeria monocytogenes DSM 20600	*Carnobacterium piscicola* LV61	19
Pediococcus acidilactici PAC1.0	*Lactobacillus sake L45*	39
Pediococcus damnosus 1832	*Lactobacillus plantarum* NCDO 1193	2
Pediococcus pentosaceus FBB61-2	*Nisin*	40

TABLE 5 (continued)
Microorganisms That Have Been Used as Sensitive Indicator Strains for Detection of Bacteriocins Produced by Lactic Acid Bacteria, Propionibacteria, and for Detection of Yeast Killer Toxin

Indicator strain	Inhibitor-producing strain or inhibitor	Ref.
Propionibacterium acidipropionici P.5	*Propionibacterium thoenii P127*	*41*
Saccharomyces cerevisiae 381	*S. cerevisiae 28*	42
Saccharomyces cerevisiae IAM 4274	*S. cerevisiae H-1, Y-2*	43
Saccharomyces cerevisiae ATCC 48499	*S. cerevisiae (K-1 Marquee)* (V-1116) (L 2056)	6
Streptococcus agalactiae ATCC 14363	Nisin	44
Micrococcus flavus NC1B 8166	Nisin	45
Pediococcus pentosaceus FBB-39	*Pediococcus pentosaceus* FBB-63	46

Once inhibitor strains are identified, attempts are initiated to isolate and characterize the metabolite(s) associated with antagonism. Culture supernatants are usually the starting materials in this process, and are generally evaluated for activity (after various manipulations) in an agar well diffusion system. In this respect, either cell-free supernatants or culture broths containing inactivated cells are applied to the surface of, or into wells, cut into agar. Commonly, a purposefully constructed device or cork borer (Figure 3A and B) is used. The resulting plugs are extracted with a hooked micro spatula (Figure 3C). Holding the plate upside down will allow gravity to facilitate the procedure. The reservoir wells may or may not be sealed with 10 μl of tempered 2.5% agar solution. In our laboratory we have foregone this step after purposefully evaluating its necessity. As long as contact between the dish bottom and agar is uniform, sealing wells is not necessary.

Reproducible quantification of inhibitor activity, whether present as a pure solution or as a culture supernatant, can be achieved with the diffusion assay. It has been established (as discussed by Hewitt and Vincent[48]) that the most linear relationship-inhibitor zone size and inhibitor concentration is the plot of the logarithm of inhibitor concentration vs. the square of the inhibition zone width. The zone width is defined as that which is between the zone edge and the adjacent well edge. In practice, the zone width can be directly measured or obtained by measuring the diameter of the entire inhibition zone, subtracting the diameter of the well, and dividing by two. Enhancement of inhibition zone size can be achieved by allowing antagonistic substances time to diffuse through the agar matrix before allowing outgrowth of the indicator. This can be done by applying samples to wells in indicator-seeded agar and then holding the plates at 4°C for a period of time. The length of time of prediffusion will dictate zone size (Figure 4). Measurement of zones is easily accomplished with

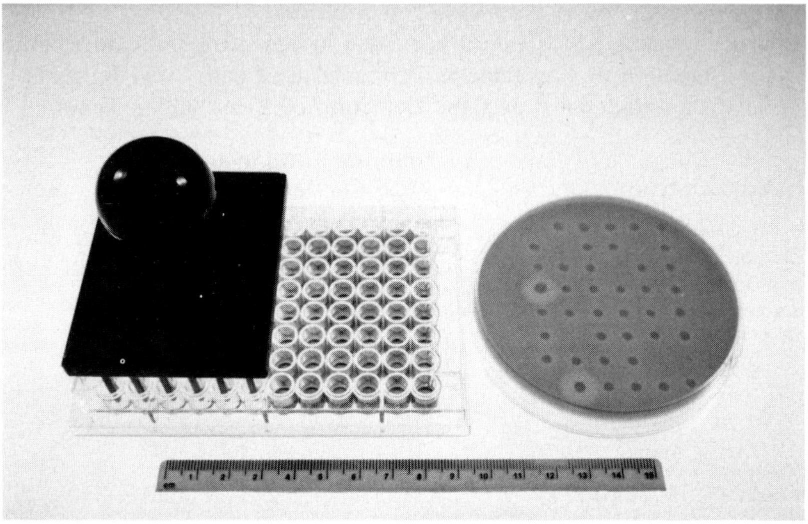

FIGURE 2. Replicator device and 96-well tissue culture plate for rapid screening of microorganisms for antagonistic properties.

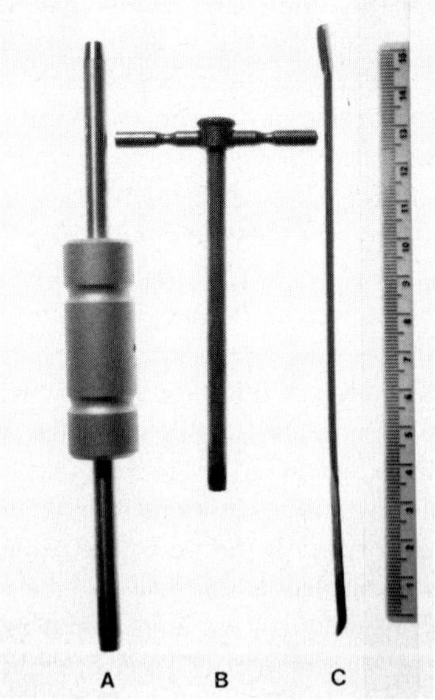

FIGURE 3. Agar well-cutting tools (A and B) and microspatula with hooked end to remove agar plugs (C).

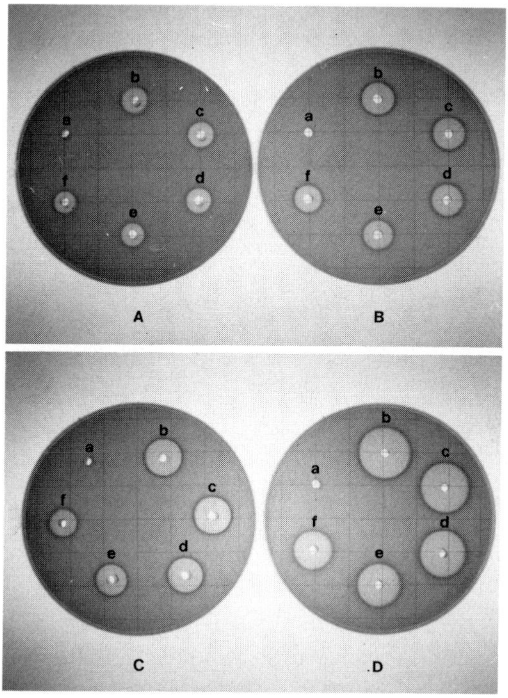

FIGURE 4. Effect of different prediffusions times at 4°C on inhibition zone size produced by varying amounts of nisin in a seeded agar well plate. (A) 24h, (B) 48h, (C) 72h, (D) 96h. Well a = 0, b = 200, c = 100, d = 50, e = 20, f = 10 units nisin/ml; 100 μl volumes added to wells. *P. pentosaceus* FBB-61 is the indicator strain.

dial-type calipers with the plate resting on an illuminated light box to give sufficient contrast (Figure 5). Alternatively, one can photograph the diffusion plates and measure the zones at a more convenient time. Photography also provides a permanent record of the plate for inclusion either in the lab notebook or in publication. It is important to remember to include a size reference in the photograph. Many petri plate manufacturers offer products that have a 1- or 2-cm grid pattern that can serve as a size reference (e.g., Figure 4). To photograph plates, one can use a polaroid-type camera "gun" equipped with prefocused hoods or a standard 35mm camera. The plates are photographed on a light box surface to give maximum contrast and sufficient illumination.

Assignment of activity or arbitrary units (AU) to quantitate bacterium activity is usually accomplished with a liquid-tube assay in which the amount

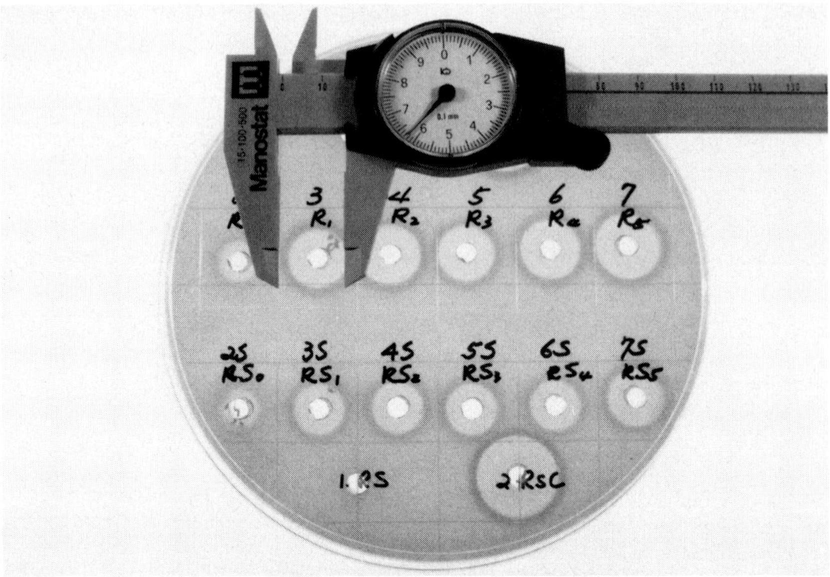

FIGURE 5. Dial caliper for measurement of inhibition zone size on typical experimental well plate.

of microbial growth (measured turbidi-metrically) is dependent on the concentration of inhibitor. The "titer" of a bacteriocin-containing solution is, by convention, the reciprocal of the highest dilution of a solution that gives a certain degree of inhibition (commonly 50%) to an indicator strain under standardized conditions. Several variations of this method are possible, and are discussed by Mayr Hartig et al.[49] Another method to determine AU/ml is to make serial dilutions of the preparation and transfer a specific amount (5 μl or more) from each dilution on the surface of an agar plate seeded with indicator. After incubation, the reciprocal of highest dilution that gave a definite zone of inhibition of at least 1 to 2mm is multiplied by a conversion factor (200 if 5μl is used; 1 ml divided by 5 ml) to obtain AU/ml of the original preparation.[3,28,29]

Other methods that have been described for detecting and quantifying inhibitors from LAB are electrical conductance that was correlated with bacteriocin inhibition of culture,[50] Enzyme Linked Immunoadsorbent Assay for quantification of nisin at concentration not detected by bioassay,[51] and measurement of suppression of β–galactosidase activity in *Escherichia coli* as a screening method for inhibitor-producing lactobacilli.[52]

IV. PRELIMINARY CHARACTERIZATION OF ANTAGONISTIC METABOLITES

A. Strain Identification

Sometime during the course of inhibitor characterization, it becomes desirable to either presumptively identify or confirm the taxonomic status of microorganisms under study. This will further reduce the likelihood of working with previously characterized inhibitors, as well as provide for comparative experimentation with inhibitor producing microorganisms of the same genus or species. The known metabolic characteristics of species will also allow one or more rapidly identify antagonistic activities. Many microorganisms can be rapidly identified on a presumptive basis by using a simple series of morphological andbiochemical tests. Rapid, miniaturized, carbohydrate-utilization tests are available for both yeast and lactic acid bacteria (API, Rapid CH System, Analytab Prod., Plainview, NY) and these tests include keys for species identification.

A presumptive identification will provide a basis for naming inhibitors that appear to be new and previously uncharacterized. No rules exist for nomenclature regarding bacteriocins or low-molecular weight, nonproteinaceous inhibitors. By convention, investigators usually name a new bacteriocin by adding the suffix "in" to the specie name, i.e., brevicin from *Lactobacillus brevis*.. However, some bacteriocins have been named using the genus, i.e., "pediocin" from *Pediococcus* sp. In the event that more than one bacteriocin is identified from a species or genus already existing, usually a letter or number is appended, i.e., pediocin A, pediocin A1, pediocin AcH. The low-molecular weight nonproteinaceous inhibitor from *Lactobacillus reuterii* is called reuterin following the pattern used for bacteriocins.

B. Determination of the Spectrum of Activity

Determination of the spectrum of activity for an inhibitor is accomplished by systematically evaluating groups of microorganisms of different taxonomic status. Most bacteriocin-like inhibitors possess either a "broad" spectrum or a "narrow" spectrum of activity, both of which are generally limited to either Gram-negative or Gram-positive bacteria. Narrow-spectrum bacteriocins are generally limited in activity to the same species, and/or species of the same genera, whereas the broad-spectrum bacteriocin transcend genera boundaries. Examples of broad-spectrum LAB bacteriocin are nisin,[53] pediocin PO_2,[45,46] pediocin A,[14] and pediocin AcH.[28] Narrow-spectrum LAB bacteriocins include helveticin J[26] and casecin 80.[21] A recommended approach to determine the extent of the activity spectrum is first to evaluate closely related species and genera, and then progressively evaluate strains less related. If it appears that a bacteriocin is broad-spectrum, a limited evaluation of yeast, fungi, and bacteria of the opposite Gram reaction may be revealing. Killer toxins from yeast invariably have very narrow activity spectra. Reuterin, a low-molecular weight (<500 mol wt) inhibitor has an extremely broad activity spectrum including antimycotic and antiprotozoal activity.[54] At some stage, purified or partially purified materials should be examined to determine the activity spectra of the compound being studied.

C. Classifying Inhibitors

Inhibitors from "food grade microorganisms" are usually either low-molecular weight, nonproteinaceous compounds or higher molecular weight compounds

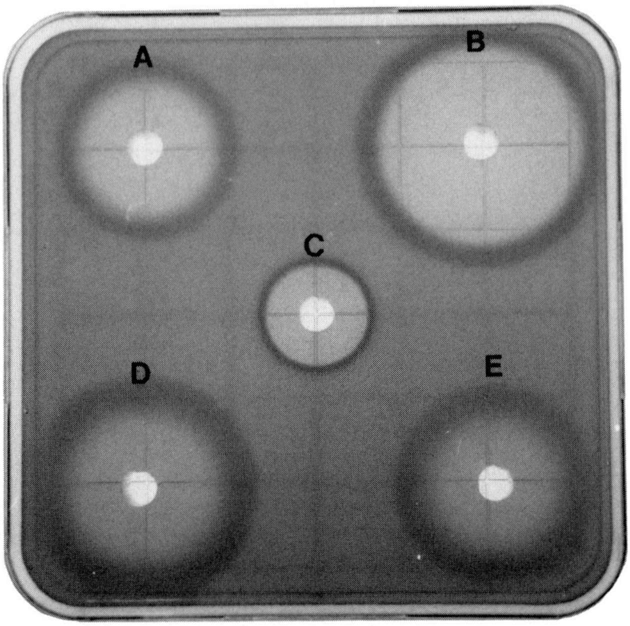

FIGURE 6. Influence of different inhibitors on inhibition zone definition. (A) 1% H_2O_2, (B) 3% H_2O_2, (C) 500 units/ml nisin, (D) 30% acetic acid, (E) 15% acetic acid; 100 μl volumes added to wells. *P. pentosaceus* FBB-61 is the indicator strain.

containing protein. The latter are termed bacteriocins if derived from bacteria and killer toxins if derived from yeast. Bacteriocins also carry the other criterion of being bacteriocidal in action. Lactic acid bacteria, propionibacteria and bifidobacteria can produce as primary metabolites several low-molecular weight compounds that exhibit antimicrobial activity toward certain microorganisms. These compounds include acetic acid, lactic acid, propionic acid, ethanol, diacetyl, and also hydrogen peroxide. To distinguish between these types of compounds and proteinaceous inhibitors, techniques have been developed to take advantage of certain fundamental differences between the two classes. The organic acids produced by fermentative bacteria are bacteriostatic at most food pH values rather than bacteriocidal. Agar diffusion assays can be set up using media that are buffered to the extent that acids produced by fermentative bacteria are noninhibitory. This is especially desirable when evaluating the sensitivity of microorganisms that are relatively more acid sensitive. The amounts of organic acids necessary to inhibit lactic acid bacteria in an agar diffusion system are far greater than amounts that would be naturally produced. This is illustrated in Figure 6 with wells D and E, in which 100 μl of 15% and 30% of acetic acid were applied. It should also be noted that it is generally observed that inhibition-zone definition is less sharp, and a gradient of inhibition is evident, whereas with a bacteriocin inhibitor (Figure 6,well C) such as nisin, the inhibition-zone boundary reflects both diffusion rates,

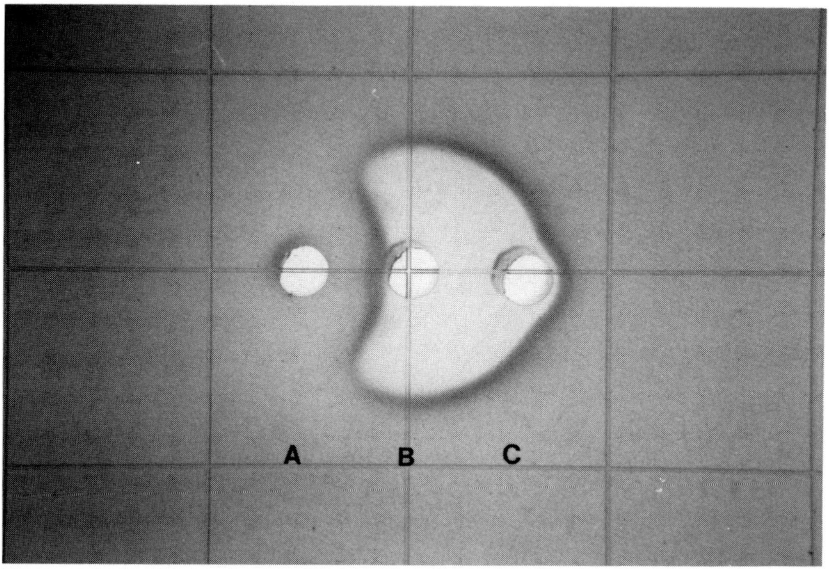

FIGURE 7. Degradation of the activity of nisin by proteolytic enzymes in an agar well diffusion system. (A) 100 ul of 1 mg/ml Pronase, (B) 100 ul of 100 units of nisin, (C) 100 μl of 0.1 M phosphate buffer, pH 6.5. *P. pentosaceus* FBB-61 is the indicator strain.

differences between large- and small- molecular weight compounds and the difference between bacteriocidal and bacteriostatic activity. Hydrogen peroxide production by LAB can occur if oxygen is present.[16] Hydrogen peroxide is inhibitory to a large variety of microorganisms at fairly low concentrations (less than 0.5%). Inhibition zones generated by hydrogen peroxide are better defined than those seen with acetic acid (Figure 6, wells A and B), perhaps indicating a more lethal effect. Hydrogen peroxide as a source of inhibition can be reduced by incubating the agar diffusion plates under anaerobic conditions. Alternatively, catalase preparations can be used to degrade H_20_2, hence providing a control to discount the possibility of the inhibitor under investigation being H_20_2. Proteolytic enzymes are frequently used to provide evidence that an inhibitor is proteinaceous or its activity is dependent on a proteinaceous component. Typically, a broad spectrum protease preparation such as Pronase (Sigma Co., St. Louis, MO) is either directly incorporated into the agar diffusion base, or placed into a well adjacent to a well containing inhibitor (Figure 7). Alternatively, inhibitor-containing preparations are incubated with protease and placed into wells or impregnated discs, and are placed on seeded agar plates (Figure 8). To enhance detection of bacteriocin-type inhibitors, and to preclude organic acids and other low-molecular weight components, culture supernatants can be dialyzed and concentrated against polyethylene glycol in dialysis tubing (Figure 9). Occasionally, spontaneous

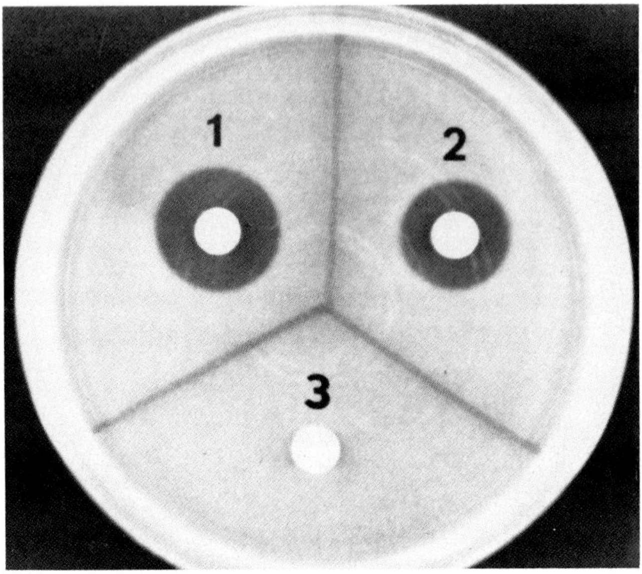

FIGURE 8. Disc assay of antimicrobial activity of curde antimicrobial peptide (pediocin AcH) from *P. acidilactici* H against *L. plantarum* WSO; (1) untreated, (2) sample after heating at 121°C for 15 min, (3) sample after treatment with trypsin (200 μg/ml, 1 h at 37°C). Each disc contained 40 μl of sample. (From Bhunia, A. K., Johnson, M.C., and Ray, B., *J. Ind. Microbiol.*, 2, 319, 1987. With permission.)

inhibitor-resistant mutants of the senstive strain will appear in the inhibition zone (Figure 10). In our laboratory we have observed this phenomenon only with bacteriocin-type inhibitors and not with low-molecular weight, nonproteinaceous compounds. Bacteriocins can be typed according to their sensitivity to a series of proteolytic enzymes which may be useful for eliminating duplications and previously characterized inhibitors. An example is nisin, which is inactivated by chymotrypsin, but not trypsin.[53] Bhunia et al.[55] developed a novel procedure in which trypsin-treated and nontreated bacteriocin preparations were electrophoresed on SDS-page gels (Figure 11) that were subsequently overlaid with indicator-seeded agar. A zone of inhibition appeared and was correlated with a single band on the gel (Figure 12), but was absent with the trypsin-treated sample. This procedure provided a rapid way to identify and assign a molecular weight to a bacteriocin that was part of a nonpurified preparation.

The relative heat tolerance of inhibitors is a parameter frequently included in characterization schemes. The majority of broad- and narrow-spectrum bacteriocins from LAB are resistant to boiling temperatures at100°C for periods of 5 to 60 min.[56] Several, such as pediocin AcH and lacticin F, can maintain

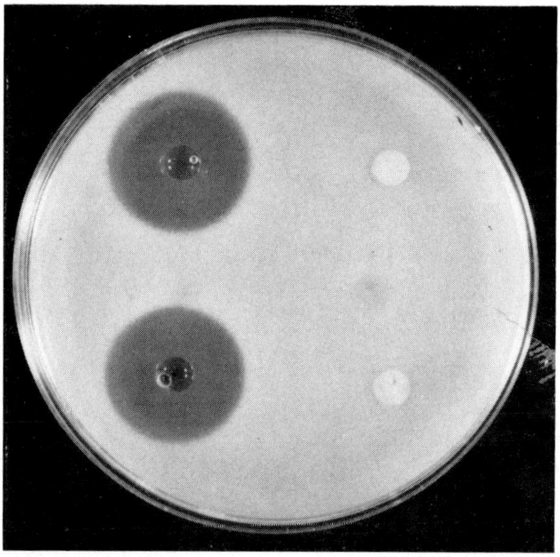

FIGURE 9. *Lactobacillus plantarum* C-11 cell-free cultrue broth concentrated approx. 50× in membrane dialysis tubing with a molecular weight cut-off range of 6000 to 8000. Concentrate (50 μl) was applied to wells in agar and allowed to difuse for 1 h prior to overlaying with *Lactobacillus plantarum* 965. Left = untreated concentrate; right = concentrate treated with 1 mg/ml protease for l h at 37°C. (From Daeschel, M.A., McKenney, M.C., and McDonald, L.C., *Food Microbiol.*, 7, 91, 1990. With permission.)

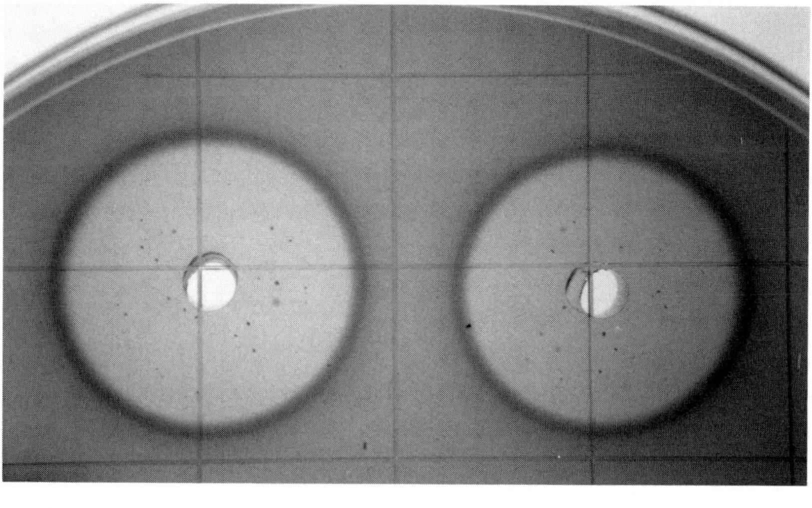

FIGURE 10. Appearance of spontaneous nisin-resistant colonies of *Pediococcus pentosaceus* FBB-61-2 in zones of inhibition. Plate was incubated for 4 d at 37°C. Grid pattern is 2 cm^2.

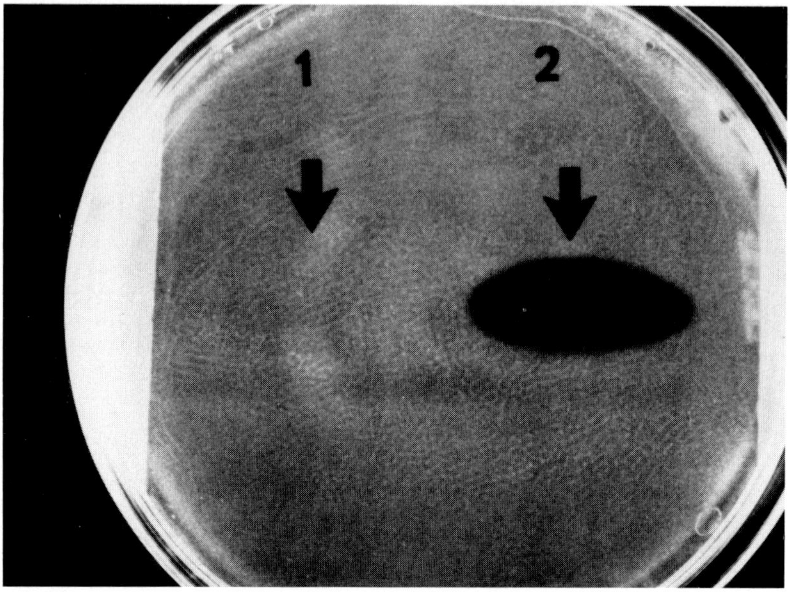

FIGURE 11. The stained half of the SDS-PAGE gel showed bands formed by serveral proteins present in the preparation (Lane 2). The molecular weight of the upper most band is about 50,000, and of the bottom one about 2,700 (see arrow). Lane 1 contains the low-molecular weight standards, (a) 16,950, (b) 14,400, (c) 8160, (d) 6210, (e) 2510, which were used to calculate the molecualr weight of the lower bands. High-molecualr weight standards were used to calculate the molecular weight of the upper bands (not presented). (From Bhunia, A.K., Johnson, M.C., and Ray, B., *J. Ind. Microbiol.*, 2, 319, 1987. With permission.)

activity after being subjected to sterilization conditions (121°C/15 min) The heat stability of LAB bacteriocins is a desirable characteristic when potential applications are to be considered. One of nisin's major uses was to provide a safe mechanism to reduce the total thermal input needed to process canned foods. Heat stability of nisin and other similar bacteriocins make them suitable preservatives in that regard. High-heat stability also provides a simple mechanism for sterilization of culture supernatants destined for further experimentation. Further characterization of the structure and function of bacteriocins invariably requires a purified compound. Methodologies for bacteriocin purification generally fall under peptide and protein purification strategies, and the reader is referred to the myriad of available protocols.

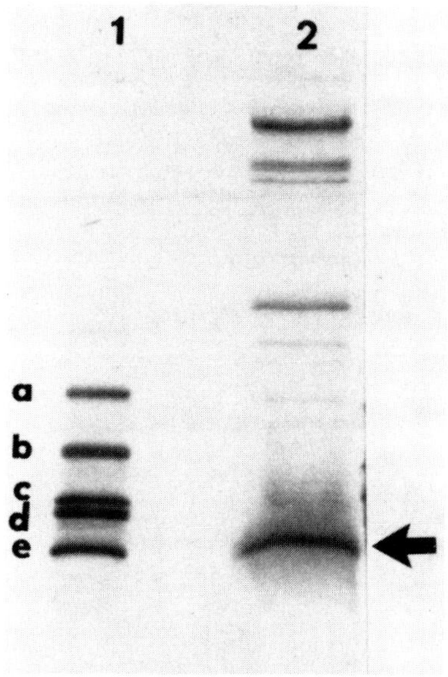

FIGURE 12. The other half of the gel was overlaid with *Lactobacillus plantarum* to determine which bands corresponds to the antimicrobial activity. Lane 2, showing the zone growth inginbition, corresponded with the lowermost band (mol wt 2,700); other bands did not show antimicrobial activity. Lane 1 contained trypsin-treated sample and showed no growth inhibition. (From Bhunia, A.K., Johnson, M.C., and Ray, B., *J. Ind. Microbiol.*, 2, 319, 1987. With permission.)

REFERENCES

1. **Berdy, J.**, Screening, classification and identification of microbial products, in *Discovery and Isolation of Microbial Products,* Verral, M. S., Ed., Ellis Horwood, Chichester, England, 1985, 9.
2. **West, C. A. and Warner, P. J.,** Plantacin B, a bacteriocin produced by *Lactobacillus plantarum* NCD0 1193, *FEMS Microbiol. Lett.*, 49, 163, 1988.
3. **Gonzalez, C. F. and Kunka, B. S.,** Plasmid associated bacteriocin production and sucrose fermentation in P*ediococcus acidilactici.*, Appl. Environ. Microbiol., 53, 2534, 1987.
4. **Gonzalez, C. F. and Kunka, B. S.,** Plasmid transfer in *Pediococcus* spp.: intergeneric and intrageneric transfer of pIP 501, *Appl. Environ. Microbiol.*, 46, 81, 1983.
5. **Muriana, P. M. and Klaenhammer, T. R.,** Conjugal transfer of plasmid-encoded determinants for bacteriocin production and immunity in L*actobacillus acidophilus* 88, *Appl. Environ. Microbiol.*, 53, 553, 1987.
6. **Moore, K. J.,** Yeasts from Arkansas wine grapes: sulfite resistance and killer toxin production as a function of pH and temperature, Ph.D. thesis, University of Arkansas, Fayetteville, 1989.
7. Da**eschel, M. A., McKenney, M. C., and McDonald, L. C.,** Bacteriocidal activity of L*actobacillus plantarum* , C-11, *Food Microbiol.*, 7, 91, 1990.
8. **Davie, J. M. and Brock T. D.**, Effect of techoic acid on resistance to the membrane-lytic agent of S*treptococcus zymogenes, J. Bacteriol.*, 92, 1623, 1966.

9. **Schillinger, U. and Lucke, F. K.**, Antibacterial activity of *Lactobacillus sake* isolated from meat, *Appl. Environ. Microbiol.*, 55, 1901, 1989.
10. Barefoot, S. **F. and Klaenhammer, T. R.**, Detection and activity of lactacin B, and bacteriocin produced by *Lactobacillus acidophilus*, *Appl. Environ. Microbiol.*, 45, 1808, 1983.
11. **Hastings, J. W. and Stiles, M. E.**, Antibiotics of *Leuconostoc gelidum* isolated from meat,*J. Appl. Bacteriol.*, 70, 127, 1991.
12. **Foulds, J. D. and Schemin, D.**, Concomitant synthesis of bacteriocin and bacteriocin inactivation from *Serratia marcesens*, *J. Bacteriol.*, 99, 661, 1969.
13. **Klaenhammer, T. R.**, Bacteriocins of lactic acid bacteria, Biochimie. 70, 337, 1988.
14. **Daeschel, M. A. and Klaenhammer, T. R.**, Association of a 13.6-megadalton plasmid in *Pediococcus pentosaceus* with bacteriocin activity, *Appl. Environ. Microbiol.*, 50, 1538, 1985.
15. **Davey, G. P. and Pearce, L. E.**, The use of *Streptococcus cremoris* strains cured of diplococcin production as cheese starters, *N.Z. J. Dairy Sci. Technol.*, 15, 51, 1980.
16. **Fleming, H. P. McFeeters, R. F., and Daeschel, M. A.**,The lactobacilli, pediococci and leuconostocs: vegetable products, in *Bacterial Starter Cultures for Foods*, Gilliland, S. E., Ed., CRC Press Boca Raton, FL., 1985, 104.
17. **Geis, A., Singh, J., and Teuber, M.**, Potential of lactic Streptococci to produce bacteriocin, *Appl. Environ. Microbiol.*, 45, 205 , 1983.
18. **Jung, D. S., Bodyfelt, F. W., and Daeschel, M. A.**, Influence of fat and emulsifiers on the efficacy of nisin in inhibiting *Listeria monocytogenes* in fluid milk, *J. Dairy Sci.*, 75, 387, 1992.
19. **Schillinger, V. and Holzapfel, W. H.**, Antibacterial activity of Carnobacteria, *Food Microbiol.*, 7, 305, 1990.
20. **Rammelsberg, M. and Radler, F.**, Antibacterial polypeptide of *Lactobacillus* spp., *J. Appl. Bacteriol.*, 69, 177, 1990.
21. **Rammelsberg, M., Muller, E., and Radler, F.**, Casecin 80: purification and characterization of a new bacteriocin from *Lactobacillus casei*, *Arch. Microbiol.*, 154, 249, 1990.
22. **Toba, T., Yoshiko, E., and Itoh, T.**, Acidophilucin A, a new heat labile bacteriocin produced by *Lactobacillus acidophilus* LAPT 1060, *Lett. Appl. Microbiol.*, 12, 106, 1991.
23. **Muriana, P. M. and Klaenhammer, T. R.**, Purification and partial characterization of Lactacin F, a bacteriocin produced by *Lactobacillus acidophilus* 11088, *Appl. Environ. Microbiol.*, 57, 114, 1991.
24. **Barefoot, S. F. and Klaenhammer, T. R.**, Purification and characterization of the *Lactobacillus acidophilus* bacteriocin Lactacin B, *Antimicrob. Agent Chemother.*, 26, 328, 1984.
25. **DeKlerk, H. C. and Smit, J. A.**, Properties of *Lactobacillus fermenti* bacteriocin, *J. Gen. Microbiol.*, 48, 309, 1967.
26. **Joerger, M. C. and Klaenhammer, T. R**., Characterization and purification of helveticin J and evidence for a chromosomally determined bacteriocin produced by *Lactobacillus helveticus* 481, *J. Bacteriol.*, 167, 439, 1986.
27. **Upreti, G. C. and Hinsdel, R. D.**, Production and mode of action of Lactocin 27: bacteriocin from a homofermentative *Lactobacillus*, *Antimicrob. Agent Chemother.*, 7, 139, 1975.
28. Bhunia, A. **K., Johnson, M. C., and Ray, B., Purification, characterization and antimicrobial spectrum of a bacteriocin produced by** *Pediococcus acidilactici*, *J. Appl. Bacteriol.*, 65, 261, 1988.
29. **Bhunia, A. K., Johnson, M. C., Ray, B., and Kalchayanand, N.**, Mode of action of pediocin AcH from *Pediococcus acidilactici* H on sensitive bacterial strains, *J. Appl. Bacteriol.*, 70, 25, 1991.
30. **Davey, G. P. and Richardson, B. C.**, Purification and some properties of diplococcin from *Streptococcus cremoris* 346, *Appl. Environ. Microbiol.*, 41, 84, 1981.

31. **Davey, G.,** Plasmid associated with diplococcin production in *Streptococcus cremoris, Appl. Environ. Microbiol.*, 48, 895, 1984.
32. **Zajdel, J. K., Ceglowski, P., and Dobrzanski, W. T.,** Mechanism of action of lactostrepcin 5, a bacteriocin produced by *Streptococcus cremoris* 202, *Appl. Environ. Microbiol.*, 49, 969, 1985.
33. **van Belkum, M. J., Hayema, B. J., Geis, A., Kok, J., and Venema, G.,** Cloning of two bacteriocin genes from a lactococcal bacteriocin plasmid, *Appl. Environ. Microbiol.*, 55, 1187, 1989.
34. **van Belkum, M. J., Hayema, B. J., Jeeninga, R. E., Kok, J., and Venema, G.,** Organization and nucleotide sequences of two lactococcal bacteriocin operons, *Appl. Environ. Microbiol.*, 57, 492, 1991.
35. **Kojic, M., Svircevec, J., Banina, A., and Topisirovic, L.,** Bacteriocin producing strain of *Lactococcus lactis* ssp. *diacetylactis* S50, *Appl. Environ. Microbiol.*, 57, 1835, 1991.
36. **Scherwitz, K. M., Baldwin, K. A., and McKay, L. L.,** Plasmid linkage of a bacteriocin-like substance in *Streptococcus lactis* subsp. diacetylactis strain WM4: transferability to *Streptococcus lactis, Appl. Environ. Microbiol.*, 45,1506, 1983.
37. Hoover, D. **G., Dishart, K. J., and Hermes, M. A.,** Antagonistic effect of Pediococcus spp. against *Listeria monocytogenes, Food Biotechnol.*, 3, 183, 1989.
38. **Hoover, D. G., Walsh, P. M., Kolaetis, K. M., and Daly, M. M.,** A bacteriocin produced by *Pediococcus* species associated with a 5,5-megadalton plasmid, *J. Food Prot.*, 51, 29, 1988.
39. **Mortvedt, C. I., Nissen-Meyer, J., Slettern, K., and Nes, I. F.,** Purification and amino acid sequence of lactocin S, a bacteriocin produced by *Lactobacillus sake* L45, *Appl. Environ. Microbiol.*, 57, 1829, 1991.
40. **Daeschel, M., Jung, D. S., and Watson, B. T.,** Controlling wine malolactic fermentation with nisin and nisin-resistant strains of *Leuconostoc oenos, Appl. Environ. Microbiol.*, 57, 601, 1991.
41. **Lyon, W. J. and Glatz, B. A.,** Partial purification and characterization of a bacteriocin produced by *Propionibacterium thoenii, Appl. Environ. Microbiol.*, 57, 701, 1991.
42. **Schmitt, M. and Radler, F.,** Molecular structure of the cell wall receptor for killer toxin KT in *Saccharomyces cerevisiae., J. Bacteriol.*, 170, 21921, 1988.
43. **Kitano, K., Sato, M., Shimazaki, T., and Hara, S.,** Occurrence of wild killer yeasts in Japanese wineries and their characteristics, *J. Ferment. Technol.*, 62, 1, 1984.
44. **Hirsch, A.,** The assay of antibiotic nisin, *J. Gen. Microbiol.*, 4, 70, 1950.
45. **Tramer, J. and Fowler, G. G.,** Estimation of nisin in foods, *J. Sci. Food Agric.*, 15, 522, 1964.
46. **Graham, D. C. and McKay, L. L.,** Plasmid DNA in strains of *Pediococcus cerevisiae* and *Pediococcus pentosaceus, Appl. Environ. Microbiol.*, 50, 532, 1985.
47. **Linton, A. H.,** *Antibiotics: Assessment of Antimicrobial Activity and Resistance,* Russel, D. A. and Quensel, L. B., Eds., Academic Press, London, 1983, 19.
48. **Hewitt, W. and Vincent, S.,** *Theory and Application of Biological Assay,* Academic Press, London, 1989, 1.
49. **Mayr-Hartig, A., Hedges, A. J. and Berkeley, R. C. W.,** Methods for studying bacteriocins, in *Methods in Microbiology,* Vol. 7A, Norris, J. R. and Ribbons, D. W., Eds., Academic Press, London, 1972, 316.
50. Girarffa, **G., Neviani, E., and Veneroni, A.,** Use of conductance to detect bacteriocin activity, *J. Food Prot.*, 53, 772, 1990.
51. **Falahee, M. B., Adams, M. R., Dale, J. W., and Morris, B. A.,** An enzyme immunoassay for nisin, *Int. J. Food Sci. Technol.*, 25, 590, 1990.
52. **Axelsson, L. T., Chung, T. C., Dobrogosz, W. J., and Lindgren, S. E.,** Production of a broad spectrum antimicrobial substance by *Lactobacillus reuterii, Microb. Ecol. Health Dis.*, 2, 131, 1989.

53. **Hurst, A.,** Nisin, Adv. *in Appl. Microbiol.*, 27, 85, 1981.
54. **Chung, T. C., Axelsson, L., Lindgren, S. E., and Dobrogosz, W. J.,** In vitro studies on Reuterin synthesis by *Lactobacillus reuterii*, *Microb. Ecol. Health Dis.*, 2, 131, 1989.
55. **Bhunia, A. K., Johnson, M. C., and Ray, B.**, Direct detection of an antimicrobial peptide of *Pediococcus acidilactici* in Sodium dodecyl sulfate-polyacrylamide gel electrophoresis, *J. Ind. Microbiol.*, 2, 319, 1987.
56. **Schillinger, U.,** Bacteriocins of lactic acid bacteria, in *Biotechnology and Food Safety*, Bills, D. and Kung, S., Eds., Butterworths, London, 55, 1990.

Chapter 4

CELLS OF LACTIC ACID BACTERIA AS FOOD BIOPRESERVATIVES

Bibek Ray

TABLE OF CONTENTS

The tremendous ability of the lactic acid bacteria to inhibit a variety of undesirable microorganisms should not be allowed to go untapped by food microbiologists. We have been too slow to extend our knowledge of these extremely versatile bacteria to new products and too conservative in our efforts to use them in food fermentations and the development of new food sources. Another important aspect relates to the lack of refrigeration in underdeveloped (developing) countries. Food preservation by controlled population dominance may be very useful here.[1]

I. INTRODUCTION

Cells of food-grade starter culture microorganisms are added in foods principally for three objectives. The most important one is to produce safe, shelf-stable, and desirable fermented foods. The second one is to indicate product abuse, especially temperature abuse of some refrigerated foods. The third objective is to control the growth of spoilage and pathogenic bacteria in semipreserved refrigerated foods. While in the first two objectives the growth of starter cultures is necessary, the mere presence of cells is effective in the third objective. The starter culture strains are selected on the basis of these specific objectives. In addition, they are expected not to affect the quality of the food adversely.

II. USE OF STARTER CULTURES TO INHIBIT GROWTH OF PATHOGENIC AND SPOILAGE BACTERIA IN FERMENTED FOODS

Several thousand fermented foods are prepared from milk, meat, fish, vegetables, grains, lentils, and fruits, all over the world. Only a limited number of these foods are processed by controlled fermentation.[2–9] In general, food fermentation is a nonsterile process, and even when good sanitary practices are followed, the raw materials can contain both pathogenic and spoilage microorganisms, along with those associated with fermentation. The major objective is to set the correct conditions under which the desirable microorganisms will multiply rapidly and produce antimicrobial metabolites to cause viability loss and/or growth inhibition of the undesirable microorganisms. In controlled fermentation this is achieved by adding very high numbers of viable and active cells of desirable starter culture(s) to the raw material and incubating it under optimum growth conditions. Depending upon a particular product, single or mixed strains and species of lactic acid bacteria (from genera *Lactococcus, Leuconostoc, Pediococcus, Streptococcus, Lactobacillus,* and *Bifidobacterium),* propionic acid bacteria, micrococci, and some yeasts and molds are used in very high population levels (generally at 1% level to give over 10^6 cells per mililiter or gram of raw materials) (Table 1). Depending upon the raw materials and the microbial strains, the starter cultures can produce many types of antimicrobial compounds including organic acids (lactic, acetic, propionic), alcohols, diacetyl, H_2O_2, reuterin, bacteriocins, and other undefined metabolites.[10]

TABLE 1
Some Microorganisms with Current and Previous Names Used in Controlled Fermentation of Foods

Current Name	Previous Name(s) (if changed)	Used to produce fermented foods
Bifidobacterium bifidum	*Lactobacillus bifidus*	Bifidus milk
Lactobacillus acidophilus		Acidophilus milk
Lactobacillus brevis		Soda cracker
Lactobacillus casei subsp. *casei*	*Lactobacillus casei*	Fermented milk
Lactobacillus delbrueckii subsp. *delbrueckii*	*Lactobacillus delbrueckii*	Soda cracker
Lactobacillus delbrueckii subsp. *bulgaricus*	*Lactobacillus bulgaricus*	Yogurt
Lactobacillus delbrueckii subsp. *lactis*	*Lactobacillus lactis*	Yogurt Fermented milk
Lactobacillus fermentum		Soda cracker
Lactobacillus helveticus		Swiss cheese
Lactobacillus plantarum		Fermented sausage and vegetables
Lactobacillus sake		Fermented sausage, rice wine
Lactobacillus sanfrancisco		Sour bread
Lactococcus lactis subsp. *cremoris*	*Streptococcus cremoris*	Cheese
Lactococcus lactis subsp. *lactis*	*Streptococcus lactis*	Cheese, butter, sour cream
Lactococcus lactis subsp. *lactis* serovar *diacetylactis*	*Streptococcus diacetylatis*, *Streptococcus lactis* subsp. *diacetylatis*	Cottage cheese, buttermilk
Leuconostoc lactis		Cheese, buttermilk
Leuconostoc mesenteroides subsp. *cremoris*	*Leuconostoc cremoris*; *Leuconostoc citrovorum*	Cottage cheese, fermented milk

TABLE 1 (continued)
Some Microorganisms with Current and Previous Names Used in Controlled Fermentation of Foods

Current Name	Previous Name(s) (if changed)	Used to produce fermented foods
Leuconostoc mesenteroides subsp. *mesenteroides*	*Leuconostoc mesenteroides*	Cheese, buttermilk
Leuconostoc oenos		Wine
Micrococcus varians		Fermented sausage
Pediococcus acidilactici	*Pediococcus cerevisiae*	Fermented sausage
Pediococcus pentosaceus		Fermented sausage and vegetables
Propionibacterium freudenreichii subsp. *freudenreichii*	*Propionibacterium freudenreichii*	Swiss cheese
Propionibacteria freudenreichii subsp. *shermanii*	*Propionobacterium shermanii*	Swiss cheese
Streptococcus carnosus		Fermented sausage
Streptococcus salivarus subsp. *thermophilus*	*Stretococcus thermophilus*	Yogurt,Swiss cheese
Saccharomyces cerevisiae		Alcoholic beverages and bread
Penicillium roquefertii		Roqueforte and blue cheese
Penicillium camembertii		Camembert cheese

A. During Growth in Broths

Growth inhibition of many foodborne pathogens and food spoilage bacteria by starter culture bacteria has been studied in different laboratory media as well as in many food systems. Some of these observations are presented in Table 2. Daly et al.[1] reported that *Streptococcus diacetylactis* strain 18-16 inhibited the growth of *Pseudomonas* spp., *Alcaligenes* spp., *Escherichia coli, Serratia marcescens, Streptococcus liquefaciens, Salmonellas* spp., *Staphylococcus aureus, Clostridium perfringens,* and *Vibrio parahaemolyticus* by as much as 99.99% during 24-h growth in broth and milk at optimum temperatures of the test organisms. The inhibition was produced mainly by the lactic and acetic acids and some other unknown inhibitory compounds produced by the *S. diacetylactis* strain. Similar growth inhibition by *Leuconostoc*

TABLE 2
Growth Inhibition of Spoilage and Pathogenic Bacteria by Lactic Acid Bacteria in Broth Media

Lactic acid bacteria	Spoilage or pathogenic bacteria	Growth condition	% Inhibition	Ref.
Lactobacillus L-3	*Pseudomonas* pIII	1% Peptone, 30°C	>99.9 in 24 h	13
Streptococcus diacetylactis	*Pseudomonas* spp	Lactic broth, 21°C	>99.9 in 24 h	1
Streptococcus diacetylactis	*Salmonella* spp.	Lactic broth, 30°C	>99.9 in 24 h	1
Streptococcus diacetylactis	*Staphylococcus aureus*	Lactic broth, 30°C	>99.9 in 24 h	1
Streptococcus diacetylactis	*Clostridium perfringens*	Lactic broth, 30°C	>99.9 in 24 h	1
Streptococcus diacetylactis	*Vibrio parahaemolyticus*	Lactic broth, 30°C	>99.9 in 24 h	1
Lactobacillus acidophilus NCFM	*Staphylococcus aureus*	MRS broth, 37°C	96.8 in 6 h	18
Lactobacillus acidophilus NCFM	*Salmonella typhimurium*	MRS broth, 37°C	87.4 in 6 h	18
Lactobacillus acidophilus NCFM	*Clostridium perfringens*	MRS broth, 37°C	98.1 in 6 h	18
Streptococcus lactis	*Listeri monocytogenes*	Internal pH control, 21°C	>99 in 3 h	18a

citrovorum against *Salmonella gallinarum,*[11] by *Streptococcus citrovorus, S. diacetylactis,*[12] and *Lactobacillus* spp.[13] against *Pseudomonas* sp., and by *Lactobacillus*[14] and other lactic acid bacteria[15–17] against *Staphylococcus aureus,* in broth cultures and food systems due to acids, H_2O_2, and unidentified metabolites, have been reported. Gilliland and Speck[18] also reported the antagonistic actions of *Lactobacillus acidophilus* strains against *S. aureus, Salmonella typhimurium, Clostridium perfringens,* and enteropathogenic *E. coli,* in broth media. Depending upon the pathogenic strains and growth conditions, they observed 62 to 98% inhibition during growth at 37°C for 6 h. The antibacterial action of *L. acidophilus* strains against the Gram-positive and Gram-negative pathogens was due to a combination of factors that include acid, H_2O_2, and other inhibitory metabolites.

B. During Fermentation of Meat Products

Goepfert and Chung[19] observed that both *Pediococcus cerevisiae* and a *Lactobacillus* spp. were able to inhibit growth and reduce viability of *Salmonella typhimurium* during fermentation in a simulated thuringer sausage (Table 3); the low pH of the product from the acids produced by the starter bacteria was the principal antibacterial agent, and the addition of NaCl enhanced the antibacterial effect of low pH. However, a combination of low pH and NaCl

TABLE 3
Growth Inhibition of Spoilage and Pathogenic Bacteria by Lactic Acid Bacteria in Meat Products

Lactic acid bacteria	Spoilage or pathogenic bacteria	Growth condition	% Inhibition	Ref.
Pediococcus cerevisiae and *Lactobacillus plantarum*	*Salmonella spp.*[a]	Thuringer sausage and Beaker sausage at 30°C, 18 h	90 to 99	19
Lactobacillus plantarum	*Staphylococcus aureus*	Beaker sausage, 37°C, 50 h	99.9	23
Lactobacillus plantarum and *Pediococcus cerevisiae*	*Staphylococcus aureus*	Beaker sausage, 37°C, 50 h	99.9	23
Pediococcus JD	*Listeria monocytogenes* (Scott A)	Semi dry sausage 37.8°C, 14 h	95.0	36
Pediococcus acidalactici	*Listeria monocytogenes*[a]	Pepperoni, 35.6°C, 12 h	99.5	32
Pediococcus cerevisiae and *Lactobacillus plantarum*	*Clostridium botulinum*[b]	Summer sausage, 28°C, 112 d	25.0	28
Lactobacillus plantarum	*Clostridium botulinum*[b]	Bacon, 27°C, 42 d	95.0	30
Streptococcus diacetylactis	*Staphylococcus aureus*	Ham sandwich spread, 25°C, 24 h	99.5	1
Streptococcus diacetylactis	*Staphylococcus aureus*	Ground beef, 25°C, 24 h	99.3	1

[a] More than one strain was tested.
[b] % Inhibition was calculated from the differences in number of toxic samples out of total samples tested.

could not be used to make the product salmonellae-free. Heat treatment to free the products from this pathogen was recommended. In another study, a mixture of *P. cerevisiae* and *Lactobacillus plantarum* freed Lebanon bologna of *S. typhimurium* and *Salmonella dublin* from the initial population of 10^4/g after 4-d fermentation, with the product pH dropping to 4.6 or lower[20]. Controlled fermentation of pepperoni with *P. cerevisiae* and *L. plantarum* also failed to eliminate *S. typhimurium* from an initial level of 10^4/g at the end of fermentation, with the pH level at 4.7; but after subsequent drying, no viable pathogen was detected. In contrast, natural fermentation, with the final pH of 5.0, failed to free the product of salmonellae even after drying. Cooking to internal temperatures of 60°C was found to be necessary to ensure safety.[21] Other studies reported the survival of salmonellae in dry fermented turkey sausage[22] and European dry sausage,[23] either produced with starter cultures of *P. cerevisiae, Lactobacillus* and *Micrococcus* strains in controlled fermentation or produced by natural fermentation.

The growth and toxin production of *Staphylococcus aureus* during fermentation of different meat products with starter cultures has been well studied as these products have been involved in many staphylococcal food poisoning outbreaks.[23–24] Many conditions used in the manufacture of fermented sausage allow occurrence, competition, and growth of *S. aureus* strains, which include raw meat, a high salt level, reduced water activity, lack of application of smoke on the surface, and higher incubation temperature.[24] In thuringer *P. cerevisiae* inhibited the growth and enterotoxin production of *S. aureus* in the core (anaerobically), but not in the outer 1/8 inch of the product (aerobically) during fermentation. The inhibition of the anaerobic growth of S. *aureus* was due to the combined effects of the starter cultures and nitrite used in fermentation.[25] *S. aureus* inoculated in dry fermented turkey sausages was not inactivated during fermentation by *P. cerevisiae,* followed by heating at 46°C for 5 h and then drying. In European dry sausages, mixed starter cultures of *Lactobacillus* and *Micrococcus* species did prevent formation of staphylococcal enterotoxin A at the final pH of 5.0.[26] Use of a mixed culture of *P. cerevisiae* and *L. plantarum*, along with glucose in the formulation, was found to destroy *S. aureus* during sausage fermentation; the death rate and the level of destruction were dependent upon the glucose concentrations. With 2% glucose an initial *S. aureus* population of 10^8/g was reduced to <0.3/g in 58 h at 35°C, but required 130 h with 0.5% glucose. No destruction was observed in the absence of glucose.[27] The rapid killing of *S. aureus* during fermentation at high glucose concentrations was due to the rapid production of acid by the starter cultures. It was suggested that strict sanitation is necessary for the control of growth and toxin production of *S. aureus* in fermented sausage, by limiting the initial load of this pathogen. The use of starter cultures with rapid acid-producing ability and the addition of the high level of fermentable carbohydrates can assist in controlling the growth and toxin production of *S. aureus*.[23] The use of starter cultures capable of producing bacteriocins and H_2O_2 (along with catalase-positive bacteriocin-resistant *Micrococcus* spp. to counteract color defects by H_2O_2) could also help to induce an antagonistic effect on staphylococcal growth and enterotoxin formation during fermentation of sausages.[24]

Toxin production by *Clostridium botulinum* in meat and meat products is influenced by several factors, namely, oxidation-reduction potential, pH, Aw, salt, nitrite, and temperature.[24,28–30] In controlled fermentation the lactic acid bacteria, in combination with either glucose or sucrose and nitrite, effectively inhibited toxin production by *C. botulinum* (Table 4). In the presence of 2% glucose and 50 ppm nitrite, even the natural flora present in meat was able to reduce the pH to such an extent that toxin production was inhibited by *C. botulinum.* In the presence of fermentable carbohydrates, nitrite concentration can also be reduced greatly without risking the safety of the products. The addition of *P. cerevisiae* during the processing of fermented turkey sausage

TABLE 4
Incluence of Addition of Starter Cultures, Carbohydrates, and Nitrate on Toxin Production by *C. Botulinum* in Meat Products

Starter culture[a]	Carbohydrate[a]	Nitrite (ppm)	Number of toxic samples/in total
–[b]	–	0	20/25
–[b]	–	50	21/25
–[b]	+	0	8/25
–[b]	+	50	0/25
+[b]	–	0	22/25
+[b]	–	50	22/25
+[b]	+	0	2/25
+[b]	+	50	0/25
+[c]	–	0	26/27
+[c]	+	0	1/49
+[c]	+	40	0/30

[a] Added, +, or not added, –
b Adapted from Christiansen et al.[28]: Starter culture was a mixture of *P. cerevisiae* and *L. plantarum*; carbohydrate was 2% glucose; toxicity in summer-sausage was tested after 112-d storage.
c Adapted from Tanaka et al.[30]: Starter culture was *L. plantarum*; carbohydrate was 0.9% sucrose; toxicity in bacon was tested after 6-weeks storage.

did not completely eliminate an initial inoculation level of 10^2 to 10^6/g of *Clostridium perfringens*.[22]

Listeria monocytogenes strains have been isolated from both fresh and processed meats, including fermented sausages.[31] Several studies have indicated that starter culture bacteria during fermentation of different types of sausages can reduce the growth of *L. monocytogenes* strains, due to reduction in pH; however, drying at effective temperatures can be used to eliminate the pathogens from the products.[32–34] The use of bacteriocin-producing strains of *Lactobacillus sake*[35] or *Pediococcus* spp.[36] during fermentation reduced the *L. monocytogenes* levels much more than was obtained by lowering the pH alone with the nonbacteriocin-producing strains.

Starter culture bacteria also failed to destroy Coxsackie virus, echovirus, and poliovirus, during fermentation of sausage.[37,38] In the mold-ripened salami, molds come as natural contaminants. Limited inoculation studies with mycotoxin-producing strains of *Penicillium* and *Aspergillus* spp. failed to produce a high level of mycotoxin in mold-ripened salami.[39,40]

C. During Fermentation of Dairy Products

Growth inhibition of pathogenic and spoilage bacteria in milk during fermentation with dairy starters has been reported by several researchers (Table 5). Gilliland and Speck4[1] reported that lactic streptococci normally used to produce cultured dairy products, while growing in milk with *S. aureus* or *S. typhimurium,* inhibited the growth of the pathogens by 86% to 99% within

TABLE 5
Growth Inhibition of Spoilage and Pathogenic Bacteria by Lactic Acid Bacteria in Milk and Milk Products

Lactic acid bacteria	Spoilage or pathogenic bacteria	Growth condition	% Inhibition	Ref.
Streptococcus diacetylactis	*Pseudomonas* spp.[a]	Milk, 21°C, 24 h	99.9	1
Streptococcus diacetylactis	*Salmonella tennessee*	Milk, 30°C, 24 h	70.0	1
Streptococcus diacetylactis	*Staphylococcus aureus*	Milk 30°C 24 h	99.9	1
Streptococcus diacetylactis	*Clostridium perfringens*	Milk, 30°C, 24 h	99.0	1
Lactobacillus bulgaricus	*Pseudomonas fragi*	Milk, 7°C, 24 h	99.9	66
Streptococcus lactis	*Salmonella typhimurium*	Milk, 32°C, 6 h	93.4	41
Streptococcus lactis	*Staphylococcus aureus*	Milk, 32°C, 6 h	98.6	41
Streptococcus cremoris	*Listeria monocytogenes*[a]	Cottage cheese reparation, varied, 10 h	99.9	55
Streptococcus thermophilus	*Listeria monocytogenes*[a]	Yogurt, 45°C,5 h	100.0	53
Streptococcus cremoris and Streptococcus lactis	*Listeria monocytogenes*	Milk, 21°C, 15 h	98.0	54
Streptococcus spp.	*Enterobacter aerogens*	Buttermilk, 7.2°C, 10 d	95.0	58a
Streptococcus spp.	*Escherichia coli*	Sour cream, 7.2°C, 10 d	99.0	58a

6 h at 32°C. The antagonistic effect was produced, but at a slightly reduced level even when the organisms were grown at a constant pH of 6.5. They suggested that the antimicrobial property of the starter cultures was due to lactic and acetic acids, H_2O_2, and probably other unknown compounds. During the processing of Cheddar and Colby cheeses, a strain of *S. ceremoris* was found to be more inhibitory than a strain of *S. lactis* to salmonellae, and 3% inoculum was more inhibitory than 0.5% to 2% inoculum. Similar variation in inhibition to salmonellae was observed by strains and species of *S. cremoris, S. lactis, S. diacetylactis, Leuconostoc cremoris, Streptococcus thermophilus,* and *Lactobacillus bulgaricus*. Lactic acid was the most important metabolite for the inactivation of salmonellae.[42–44]

Enteropathogenic *E. coli* was also found to be inhibited by the starter culture bacteria during preparation of Camembert cheese, due to low pH (4.5).[45,46] However, the viable cells of the pathogen survived refrigerated storage over 7 weeks.[47]

S. lactis, S. thermophilus, L. acidophilus, and *L. bulgaricus,* during growth in milk, were able to reduce the growth and spore germination of *Bacillus cereus.*[48] The level of inhibition was more when the cells and spores were added to the fermented milk than when added at the beginning of fermentation. Also, Cheddar cheese made from milk inoculated with *B. cereus* had high numbers of spores after 52 weeks of refrigerated storage; the spores were not destroyed by the low pH of the fermented products.[49]

S. aureus cells were found to grow in milk during fermentation with starter culture bacteria used in Cheddar, Colby, Swiss, and Limburger cheeses.[50] However, their numbers decreased during ripening. When the acid production was low or the initial population of *S. aureus* was high, the cheeses had enterotoxin.[51]

C. botulinum growth and toxin production do not appear to be a problem in fermented dairy products, including cheeses. However, growth and toxin production of *C. botulinum* can occur in pasteurized cheese spreads.[52]

Inoculation of *Listeria monocytogenes* in milk during fermentation with thermophilic and mesophilic lactic acid bacteria revealed that growth of this pathogen was inhibited as the pH dropped below 4.7; however, the organisms remained viable.[53,54] *L. monocytogenes* strains, inoculated in milk during the manufacture of Cheddar, cottage, and Camembert cheeses also survived for a long time during refrigerated storage.[55–57]

Limited studies have shown that viruses were not destroyed by the starter cultures during the manufacture and ripening of cheeses.[58]

III. USE OF STARTER CULTURES IN REFRIGERATED FOODS TO INHIBIT GROWTH OF PATHOGENS

The process involves addition of cells of suitable mesophilic lactic starter cultures, in high numbers, to perishable refrigerated foods that may harbor pathogens (Table 6). In the event a food is temperature abused, the mesophilic bacteria will grow and produce acid in sufficient amounts to lower the pH to a level so as to prevent or reduce the growth of pathogenic (or spoilage) bacteria. In addition, the acid may cause a loss in the acceptance quality of the food. Daly et al.[1] reported that addition of viable cells of *S. diacetylactis* at the level of about 10^6/g to several different types of foods inhibited the growth of *S. aureus* (inoculated in foods) by 2 to 3 logs during incubation at 25°C or above. Addition of 1 to 2% glucose to semipreserved meat products provided much better safety against the growth of C. *botulinum* and *S. aureus* when the products were temperature abused. Lee et al.[59] used both viable and irradiated (killed) cells of *P. cerevisiae*. The use of viable cells, especially if used in small numbers (1% or lower inoculum), has some disadvantages. These include the ability of the cells, in the event of temperature abuse, to survive during storage, multiply, and produce enough acid rapidly (faster than the growth of pathogens). Irradiated cells, present in high numbers in the product, can convert the added glucose to lactic acid rapidly in case of temperature

TABLE 6
Influence of Addition of Sucrose and *Lactobacillus Plantarum* in Meat Products to Reduce pH and Inhibit Toxin Production by *Clostridium Botulinum* at Abusive Temperatures[a]

NaNO2 ppm	Sucrose %	L. Plantarum inoculation	pH at 27°C in weeks 0	2	4	# sample positive for toxin of total samples at 4 weeks
0	0.0	+	6.0	6.5	NTb	3/3
0	0.9	–	6.2	4.9	NTb	3/3
0	0.9	+	6.2	4.5	4.6	0/3
120	0.0	–	6.3	6.1	NTb	3/3
120	0.0	+	6.3	5.9	NTb	3/3
120	0.9	–	6.1	4.6	4.5	0/3
120	0.9	+	6.1	4.4	4.5	0/3

[a] Table adapted from Reference 30.
[b] Not tested as all samples were positive for *C. botulinum* toxin.

abuse without any growth. Frozen cell concentrates of *P. acidilactici* at 10^{12}/ml were killed at 0.75 Mrad from a Cobalt-60 source, diluted in glucose solution to have 10^9 cells per milliliter, and used in a series of inoculated-pack studies containing meat. The pH of the meat on incubation at 10 to 37°C temperatures dropped to below 5.0 in a few hours and prevented the growth initiation of *C. botulinum, C. perfringens, and Salmonella* spp. The irradiated preparations of *P. cerevisiae* remained stable in meat (ham and smoked turkey) for months at refrigerated temperature and did not produce any change in flavor and color. However, increasing the storage temperature caused a rapid drop in pH and prevented growth and toxin production by *C. botulinum* or *S. aureus*. Also, the meats had a buttermilk flavor, which could be offensive to people and thus prevent its consumption. In temperature-abused samples without the irradiated cells, both pathogens produced toxins.[59] The use of viable *P. cerevisiae* cells in cured ham also inhibited the growth of *S. aureus*.[60] A mixture of *P. cerevisiae* and *L. plantarum* was inoculated in summer-style sausage, along with *C. botulinum* spores, and stored at 27°C. In the presence of only 50 ppm of nitrite the product did not have toxin even after 112 d. Thus, addition of 2% glucose reduced the need for using higher levels of nitrite to ensure safety against botulisms.[28]

Cooking bacon containing nitrite can result in the formation of carcinogenic nitrosamines from the reaction of the nitrite and secondary amines. Reduction of the nitrite level to minimize nitrosamine formation, however, will increase the risk of *C. botulinum* growth and toxin formation, if the bacon is subjected to temperature abuse. However, the addition of *L. plantarum* cells, along with

0.9% sucrose, to bacon prevented toxin formation by the *C. botulinum* spores at 27°C during 7 weeks of storage, even at only a 40-ppm nitrite level. The use of starter strains that can rapidly produce acid from the sucrose at abusive temperatures was suggested. *P. acidilactici* strains have also been found to be efficient in preventing toxin production by *C. botulinum* in bacon containing fermentable carbohydrates and lower quantities of nitrite.[61]

IV. ADDITION OF LIVE STARTER CULTURE BACTERIAL CELLS IN REFRIGERATED FOODS TO CONTROL BACTERIAL GROWTH

The objective is to use viable cells of mesophilic or thermophilic lactic acid bacteria in high numbers in refrigerated foods to enhance their safety and shelf life (Table 7). The cells of lactic acid bacteria do not grow, yet inhibit growth and reduce viability of the spoilage and pathogenic bacteria that might be present in food. As lactic acid and other acids are not produced by the nongrowing lactic acid bacteria, it is assumed that the live cells probably either act as carriers of or produce other antimicrobial compounds.

A. Dairy Products

Leuconostoc cremoris cells added to raw milk used in the manufacture of cheese and butter were found to inhibit the growth of Gram-negative psychrotrophic bacteria during storage at 3.5 or 7°C. The researchers concluded that the citrate fermenting *L. cremoris* produced bacteriocidal and bacteriostatic effects on *Pseudomonas* and coliforms, by their inherent ability to produce H_2O_2.[62,63] However, in a later study cells of *S. lactis* ssp. *diacetylactis* and *L. cremoris,* both citrate-fermenting bacteria, in raw milk stored in farm bulk tanks, did not produce satisfactory inhibition.[64] This could be due to the differences in strains and testing methods used, as a more recent study reported that the seeding of raw milk with cells of *S. lactis* ssp. *diacetylactis, L. cremoris,* and *Leuconostoc dextranicum* was effective in slowing down the growth of Gram-negative psychrotrophic bacteria.[65] The addition of *S. diacetylactis* cells at a 1% inoculum level also prevented the growth of *Pseudomonas putrefaciens* (current designation: *Alteromonas putrefaciens)* in cottage cheese and in sterile milk at 7.3°C for 21 d. Other studies showed that H_2O_2-producing strains of *Lactobacillus bulgaricus* at a level of 2.5×10^7 cells per milliliter exerted both bacteriocidal and bacteriostatic effects on *Pseudomonas fragi* and other Gram-negative motile rods in raw as well as inoculated sterile milk, at 5°C, and the effectiveness of the antimicrobial action of a strain was directly proportional to the amount of H_2O_2 produced.[66,67]

As many *Pseudomonas* spp. and other psychrotrophic spoilage bacteria during growth produce heat-stable extracellular proteinases and lipases, the loss of curd yield during cheese making, rancidity, and other quality defects of products can occur from their growth in raw milk. Thus, growth inhibition of these psychrotrophs in milk is deemed important. Currently, some culture

TABLE 7
Growth Inhibitation of Spoilage and Pathogenic Bacteria in Refrigerated Foods by Latic Acid Bacterial Cells

Lactic acid bacteria (cell concentration)	Spoilage or pathogenic bacteria	Growth condition	% Inhibition	Ref.
Lactobacillus bulgaricus (1×10^8/ml)	Gram-negative rods	Milk, 5°C, 10 d	95.0	66
Lactobacillus bulgaricus (2.5×10^7/ml)	Gram-negative rods	Milk, 5.5°C, 6 d	82.0	67
Lactobacillus strains C2 (2.4×10^7/ml)	*Pseudomonas putida*	Milk, 7°C, 48 h	90.0	68
Leuconostoc citrovorum (10% inoculum)	*Pseudomonas* spp.	Ground beef, 7°C, 7 d	90.0	72
Streptococcus diacetylactis (2% inoculum)	*Pseudomonas* spp.	Ground beef, 7.5°C, 7 d	99.5	1
Streptococcus lactis and *Leuconostoc citrovorum* (10% inoculum)	*Pseudomonas* spp.	Beef steak, 7°C, 4 d	90.0	73
Pediococcus sp.[a] (10^6/g)	*Listeria monocytogenes*	Sausage mix, 37.8°C, 12 h	95.0	36
Pediococcus cerevisiae and *Lactobacillus plantarum* (2×10^6/g)	*Pseudomonas* spp., Altermonas spp., *Salmonella* spp., *Staphylococcus aureus*	Mechanically deboned poultry meat, 3 to 15°C, 5 to 7 d	99.0	74
Lactobacillus bulgaricus (4.1×10^8/g)	*Pseudomonas spp.*	Crab meat, 5°C, 4 d	75.0	66
Lactobacillus plantarum and *Pediococcus cerevisiae* (10^9/ml)	*Pseudomonas* spp., *Salmonella* spp.	Liquid whole egg, 3°C, 7 d	90.0	76

[a] A bacteriocin producing strain of *Pediococcus* did not grow under experimental conditions.

suppliers provide concentrated lactic cultures specifically for the purpose of inoculation of raw milk to inhibit the growth of psychrotrophic spoilage bacteria during prolonged refrigeration storage. Reports from industry suggest that seeding milk with these lactic acid bacteria has increased yields in Cheddar, Mozzarella, and cottage cheeses and has extended the shelf life of pasteurized milk.[68] A recent study with the commercial cultures containing mesophilic *Lactococcus* strains revealed that inoculation of cells at 2.5×10^6 or less per milliliter did not significantly inhibit the growth of *Pseudomonas putida* in raw milk at 7°C for 48 h. However, at an inoculation level of 25×10^6/ml the growth of *Pseudomonas* was reduced by a factor of 10.[69] This inhibition was not due to the production of acid or bacteriocins, but could be due to the activation of a lactoperoxidase-thiocyanate system by H_2O_2 in raw milk.[69]

Raw milk contains the lactoperoxide-thiocyanate system, which can be activated by H_2O_2 of lactic acid bacteria. In the presence of thiocyanate (SCN^-) and H_2O_2, the lactoperoxidase system generates a hypothiocyanate ($OSCN^-$) anion at neutral pH. The $OSCN^-$ anion exists in equilibrium with hypothiocyanous acid (HOSCN), pK_a of 5.3. Both $OSCN^-$ and HOSCN are strong oxidizing agents and can oxidize sulfahydryl groups in proteins.[69a] They cause damage of the cytoplasmic membrane (inner membrane), especially of Gram-negative bacteria.[67,70] The lactoperoxidase-thiocyanate system is inactivated by heat treatment. Thus, the inhibition of spoilage bacteria by lactic acid bacteria in heat-treated milk should be due to the direct effect of the H_2O_2 produced by them.[67] However, these cultures need to have sufficient amounts of H_2O_2 in milk to produce either a bacteriostatic or bacteriocidal effect against spoilage and pathogenic bacteria.[70,71]

B. Meat and Meat Products

Several studies have been conducted to determine the effectiveness of using viable cells of starter culture bacteria to inhibit the growth of aerobic spoilage bacteria in an effort to extend the shelf life of refrigerated ground beef. Ordinarily, the shelf life of ground beef at refrigeration temperature is relatively short due to the high initial microbial load. About 90% of these are from genera *Pseudomonas, Alteromonas, and Achromobacter,* which can grow at refrigeration temperature and have the ability to degrade proteins and fats rather rapidly.[72] Reddy et al.[72] observed that the addition of 10% cultures of *L. citrovorum* was effective in preventing the growth of aerobic bacteria in ground beef stored at 7°C for 7 d. Frozen concentrate cells of *S. lactis* were also effective in inhibiting the growth of spoilage bacteria, but only when 1% lactose was added to meat. The pH was reduced to about 5.0 to 4.5 in the culture-added meat. However, ground beef acidified with lactic acid to the same level had much higher counts of spoilage bacteria. Thus, the reduced growth of the spoilage bacteria could be due to the combined effects of low pH and other inhibitory actions of the culture bacteria. The addition of cultures did not affect the color and aroma of the ground beef. In a later study Reddy et al.[73] observed that the addition of a 10% mixture of *S. lactis* and *L. citrovorum* to ground beef reduced the count of psychrotrophic Gram-negative bacteria by over 99% during 4 d storage at 7°C. Results of other studies with cultures of *S. diacetylactis* (at 2% inoculum of an 18-h culture) and *L. bulgaricus, P. cerevisiae, or S. lactis*[66] (about 5×10^8/g) also indicated that the growth of psychrotrophic Gram-negative spoilage bacteria in refrigerated ground beef can be significantly reduced by the starter culture bacteria. In these studies, as the lactic acid bacteria were not growing, their antimicrobial property was thought to be due to H_2O_2.[66]

Spraying lactic cultures (*S. lactis and L. citrovorum*) on the surfaces of fresh beefsteak also was found to reduce the growth of Gram-negative psychrotrophic spoilage bacteria, during storage at 7°C for 4 d.[73] The addition of cells of a bacteriocin-producing *Pediococcus* strain to sausage mix was also found to reduce the viability of inoculated *L. monocytogenes* cells, even in the absence of growth of the starter bacteria. This inhibition was thought to be due to the bacteriocins of the specific *Pediococcus* strain used.

In cooked, mechanically deboned poultry meat, the addition of about 2×10^8 cells per gram of either *P. cerevisiae* or *Lactobacillus plantarum*,or a 50:50 mixture of both, inhibited the growth of *P. fragi*, *Pseudomonas fluorescens*, and *A. putrefaciens* by 90% over a period of 5 d at 3°C; the 50:50 mixture resulted in highest inhibition. The growth of *S. typhimurium* and *S. aureus* was completely inhibited by the starter culture mixture, even at 11 and 15°C, respectively in 7 d.[74] In another study, the same researchers showed that addition of the starter culture mixture in both mechanically deboned poultry breast meat resulted in shelf life extension of the products at 3°C. As the lactic acid bacteria did not grow at 3°C, the inhibitory effect was considered to be due to the antimicrobial metabolites other than organic acids.[75]

C. Other Foods

Limited studies have been conducted on the extension of shelf life of refrigerated foods (other than some dairy and meat products) by the addition of starter culture bacteria under nongrowing conditions. Raccach and Baker[76] observed that the addition of 10^9 cells per milliliter of either a *P. cerevisiae* strain or *L. plantarum* strain, or a 50:50 mixture of both strains, in pasteurized liquid whole egg had a repressive action on the growth of *P. fragi, P. fluorescens, and A. putrefaciens,* during 3°C storage for 7 d. Slight growth inhibition was also observed against *S. typhimurium*. In all studies a 50:50 mixture of the two lactic acid bacteria and a pH of 6.8 of the liquid egg produced the highest inhibition.[76] Addition of *L. bulgaricus* at 4.1×10^8/g to freshly picked crab meat also inhibited growth of psychrotrophic Gram-negative bacteria by over 70% during 4 d storage at 5°C.[77]

V. CONCLUDING REMARKS

The active growth of starter culture microorganisms, especially lactic acid bacteria, associated with metabolisms of food components helps to produce large varieties of desirable fermented foods. In addition, by inhibiting growth and causing viability loss of the associated spoilage and pathogenic bacteria, they produce products that are shelf stable and relatively safe. These characteristics are more predictable in controlled fermentation, as opposed to natural fermentation. However, for use in controlled fermentation, the strains should be selected not only for their desirable traits, such as rapid acid production, production of desirable flavor, and other characteristics, but also for their ability to produce specific antimicrobial metabolites and their inability to produce undesirable metabolites, such as biologically active amines. Other important considerations should include the economic production of high cell mass, high survival rate by freezing and drying, and their ability to retain desirable properties under storage conditions.

When used as an indicator of temperature abuse, the specific strains shall not only be able to prevent the growth of undesirable organisms (especially pathogens), but will also have the characteristics to alert the food handler and consumers that the product has been temperature abused. Acid production by lactic acid bacteria has currently been used to discourage the growth of pathogens, by reducing the pH of a food, and as an indicator of the loss of product quality, from the acid flavor and sour taste. But this requires the opening of a package and even tasting it. In a food where even a slight growth of *C. botulinum* could be risky, the selection and development of starter cultures that can produce visible changes in unopened packages when they are temperature abused will be advantageous over the current method.

The use of nongrowing cells of starter culture bacteria to inhibit the growth of psychrotrophic spoilage and pathogenic bacteria in refrigerated semipreserved foods will be another important area, especially for increasing shelf life and assuring the safety of minimally processed new generation refrigerated foods, including the sous vide type foods and the refrigerated raw foods. However, the starter culture cells should be able to produce antimicrobial effects on both Gram-positive and Gram-negative bacteria that can grow at refrigeration temperatures (up to 5°C) and also in products that have been subjected to some temperature abuse (10 to 12°C). This area has been studied the least, and the results are confusing. Most of these results have suggested that the nongrowing starter culture bacteria inhibit the growth of spoilage and pathogenic bacteria, not by producing organic acids, but by producing other antimetabolites. Several studies have suggested that H_2O_2 of some lactic acid bacterial strains, either directly or indirectly (by activating lactoperoxidase-isothiocyanate system), produced the antimicrobial effect. However, the amount of H_2O_2 that is required for the bacteriostatic and bacteriocidal effects (about 6 and 30 μg/ml, respectively, for *S. aureus*) of pathogenic and spoilage bacteria might not be produced by many strains used in these studies (Table 8). Also, during the growth of the starter culture bacteria, H_2O_2 produced by the cells is mostly released in the environment, and very little is retained in the cells. Thus, the amount of H_2O_2 in cells, when they are added to a food, may not be high. Finally, H_2O_2 in lactic acid bacteria is produced by the oxidation of a specific substrate. How these cells, when added to food at low temperatures, can produce H_2O_2 needs to be elucidated.

Most researchers have suggested that the antimicrobial effect of the starter culture bacteria is associated with other undetected metabolites, especially

TABLE 8
Hydrogen Peroxide Formation by Some Lactic Acid Bacteria

Lactic acid bacteria (cell concentration)	Storage or growth environment	H_2O_2 concentration (μg/ml)	Ref.
Lactobacillus lactis[a] (1.2 × 10^8/ml)	Cells stored at 5°C in dextrose broth for 15 d	12.5	14
Lactobacillus spp.[b]	Grown in 1% peptone at 30°C for 5 d	12.0	13
Lactobacillus bulgaricus F1 (7.9 ×10^7/ml)	Shaken in 10% milk for 22 h at 5.5°C	6.9	57
Lactobacillus bulgaricus F2 (1.1 × 10^8/ml)	Shaken in 10% milk for 24 h at 5.5°C	2.2	57
Lactobacillus plantarum (1 ×10^9/ml)	APT broth, 18 h 30°C	0.85	71
Pediococcus cerevisiae	APT broth, 18 h 35°C	0.85	71

[a] H_2O_2 was bacteriostatic at 6 μg/ml and bacteriocidal at 25 to 35 μg/ml to S. *aureus* strains.

[b] H_2O_2 was bacteriostatic at 2 to 8 μg/ml and bacteriocidal at 25 to 40 μg/ml to *Pseudomonas* spp.

bacteriocins. These inhibitors have not been identified; nor were the starter strains known to produce such inhibitors used in the studies. To develop procedures that can be used to obtain predictable results, these aspects (identity of the antimicrobials and their characteristics) need to be studied and understood.

The use of cells of food-grade starter culture microorganisms in foods offers a promising means for the biological control of undesirable microorganisms, without resorting to more severe physical treatments. This method can also be used to avoid the need for adding chemical preservatives, which have become increasingly unpopular as the problem of food additives receives more and more scrutiny from the regulatory agencies.[77] However, for their effective use, the mechanism(s) of their antimicrobial action should be understood. This involves identification of the antimicrobial compound(s) in a strain; determination of its physical, chemical, and biological characteristics; elucidation of environmental and genetic control systems in its production; and, finally, measurement of its antimicrobial effectiveness in the food system and its safety to the consumers. This information can then be used to standardize methods for the application of antimicrobial agents in food to produce predictable results.

REFERENCES

1. **Daly, C., Sandine, W. E., and Elliker, P. R.,** Interactions of food starter cultures and foodborne pathogens: *Streptococcus diacetylactis* versus food pathogens, *J. Milk Food Technol.,* 35, 349, 1972.
2. **Campbell-Platt, G.,** *Fermented Foods of the World,* Butterworths, Boston, 1987.
3. **Kosikowski, F. V.,** *Cheese and Fermented Milk Foods,* 2nd ed., F. V. Kosikowski & Association, Brooktondale, NY, 1977.
4. **Pederson, C. S.,** *Microbiology of Food Fermentation, 2nd ed., AVI Publishing,* Westport, CT, 1979.
5. **Steinkraus, K. H.,** Ed., *Handbook of Indigenous Fermented Foods,* Marcel Dekker, New York, 1983.
6. **Wood, B. J. B.,** Ed., *Microbiology of Fermented Foods,* Vols. 1 and 2, Elsevier, New York, 1985.
7. **Rose, A. H., Ed.,** *Fermented Foods,* Academic Press, New York, 1982.
8. **Reddy, N. R., Pierson, M. D., and Salunkhe, D. K., Eds.,** *Legume-Based Fermented Foods,* CRC Press, Boca Raton, FL, 1986.
9. **Davies, F. L. and Law, B. A.,** Eds., *Advances in the Microbiology and Biochemistry of Cheese and Fermented Milk,* Elsevier, New York, 1984.
10. **Daeschel, M. A.,** Antimicrobial substances from lactic acid bacteria for use as food preservatives, *Food Technol.,* 43(1), 164, 1989.
11. **Sorrells, K. M. and Speck, M. L.,** Inhibition of *Salmonella gallinarum* by culture filtrates of *Leuconostoc citrovorum, J. Food Sci.,* 53, 239, 1970.
12. **Pinheira, A. J. R., Liska, B. J., and Parmelee, C. E.,** Properties of substances inhibitory to *Pseudomonas fragi* produced by *Streptococcus citrovorus* and *Streptococcus diacetylactis, J. Dairy Sci.,* 51, 183, 1968.
13. **Price, R. J. and Lee, J. S.,** Inhibition of *Pseudomonas* species by hydrogen peroxide producing lactobacilli, *J. Milk Food Technol.,* 13, 33, 1970.
14. **Dahiya, R. S. and Speck, M. L.,** Hydrogen peroxide formation by lactobacilli and its effect on *Staphylococcus aureus, J. Dairy Sci.,* 51, 1568, 1968.
15. **Iandolo, J.J., Clark, C. W., Bluhm, L., and Ordal, Z. J.,** Repression of *Staphylococcus aureus* in associative culture, *Appl. Microbiol.,* 13, 646, 1965.
16. **Kao, C. T. and Frazier, W. C.,** Effect of lactic acid bacteria on growth of *Staphylococcus aureus, Appl. Microbiol.,* 14, 251, 1966.
17. **Reiter, B., Fewings, B. G., Fryer, T. F., and Sharp, M. E.,** Factors affecting the multiplication and survival of coagulase positive staphylococci in Cheddar cheese, *J. Dairy Res.,* 31, 261, 1964.
18. **Gilliland, S. E. and Speck, M. L.,** Antagonistic action of *Lactobacillus acidophilus* toward intestinal and foodborne pathogens in associative cultures, *J. Food Prot.,* 40, 820, 1977.

18a. **Wenzel, J. M. and Marth, E. M.,** Behavior of *Listeria monocytogenes* in the presence of *Streptococcus lactis* in a medium with internal pH control, *J. Food Prot.,* 53, 918, 1990.

19. **Goepfert, J. M. and Chung, K. C.,** Behavior of *Salmonella* during the manufacture and storage of a fermented sausage product, *J. Milk Food Technol.,* 33, 185, 1970.
20. **Smith, J. L., Palumbo, S. A., Kissinger, J. C., and Huhtanen, C. N.,** Survival of *Salmonella dublin* and *Salmonella typhimurium* in Lebanon bologna, *J. Milk Food Technol.,* 38, 150, 1975.
21. **Smith, J. L., Huhtanen, C. N., Kissinger, J. C., and Palumbo, S. A.,** Survival of salmonellae during pepperoni manufacture, *Appl. Microbiol.,* 30, 759, 1975.
22. **Baran, W. L. and Stevenson, K. E.,** Survival of selected pathogens during processing of a dry fermented turkey sausage, *J. Food Sci.,* 40, 618, 1973.

23. **Smith, J. L. and Palumbo, S. A.,** Microorganisms as food additives, *J. Food Prot.,* 44, 936, 1981.
24. **Bacus, J. N. and Brown, W. L.,** Use of microbial cultures in meat products, *Food Technol.,* 35(1), 74, 1981.
25. **Barber, L. E. and Deibel, R. H.,** The effect of pH and oxygen tension on staphylococcal growth and enterotoxin production in fermented sausage, *Appl. Microbiol.,* 24, 891, 1972.
26. **Niskanen, A. and Nurmi, E.,** Effect of starter culture on staphylococcal enterotoxin and thermonuclease production in dry sausage, *Appl. Microbiol.,* 11, 34, 1976.
27. **Smith, J. L. and Palumbo, S. A.,** Injury of *Staphylococcus aureus* during sausage fermentation, *Appl. Microbiol.,* 45, 135, 1978.
28. **Christiansen, L. N., Tompkin, R. B., Shaparis, A. B., Johnston, R. W., and Kautter, D. A.,** Effect of sodium nitrite and nitrate on *Clostridium botulinum* growth and toxin production in summer style sausage, *J. Food Sci.,* 40, 488, 1975.
29. **Bowen, V. G. and Deibel, R. H.,** Effects of nitrite and ascorbate in bacterial toxin formation in wieners and bacon, in *Proc. Meat Industry Research Confonference,* American Meat Institute Foundation, Washington, D.C., 1974, 63.
30. **Tanaka, N., Traisman, E., Lee, M. H., Cossens, R. G., and Foster, E. M.,** Inhibition of botulinum toxin formation in bacon by acid development, *J. Food Prot.,* 43, 450, 1980.
31. **Farber, J. M., Tittiger, F., and Gour, L.,** Surveillancc of raw fermented (dry cured) sausages for the presence of *Listeria* spp., *Can. Inst. Food Sci. Technol. J.,* 21, 430, 1988.
32. **Glass, K. A. and Doyle, M. P.,** Fate and thermal inactivation of *Listeria monocytogenes* in beaker sausage and pepperoni, *J. Food Prot.,* 52, 226, 1989.
33. **Johnson, J. L., Doyle, M. P., Cassens, R. G., and Schoeni, J. L.,** Fate of *Listeria monocytogenes* in tissues of experimentally infected cattle and hard salami, *Appl. Environ. Microbiol.,* 54, 497, 1988.
34. **Juntilla, J., Hirn, J., Hill, P., and Nurmi, E.,** Effect of different levels of nitrite and nitrate on the survival of *Listeria monocytogenes* during the manufacture of fermented sausage, *J. Food Prot.,* 52, 158, 1989.
35. **Schillinger, U. and Lucke, F. K.,** Antibacterial activity of *Lactobacillus sake* isolated from meat, *Appl. Environ. Microbiol.,* 55, 1901, 1989.
36. **Berry, E. D., Liewen, M. B., Mandigo, R. W., and Hutkins, R. W.,** Inhibition of *Listeria monocytogenes* by bacteriocin producing *Pediococcus* during the manufacture of fermented semi-dry sausage, *J. Food Prot.,* 53, 194, 1990.
37. **Kantor, M. A. and Patter, N. N.,** Persistence of echovirus and poliovirus in fermented sausage: effect of sodium nitrite and processing variables, *J. Food Sci.,* 40, 968, 1975.
38. **Herrmann, J. E. and Cliver, D. O.,** Enterovirus persistence in sausage and ground beef, *J. Milk Food Technol.,* 36, 426, 1973.
39. **Bullerman, L. B., Hartman, P. A., and Ayers, J. C.,** Aflatoxin production in meats. II. Aged dry salamis and aged country cured hams, *Appl. Microbiol.,* 18, 718, 1969.
40. **Ciegler, A., Mintzlaff, H. J., Weisleder, D., and Leistner, L.,** Potential production and detoxification of penicillic acid in mold-fermented sausage (salami), *Appl. Microbiol.,* 24, 114, 1972.
41. **Gilliland, S. E. and Speck, M. L.,** Interaction of food starter cultures and foodborne pathogens: lactic streptococci versus *Staphylococcus* and salmonellae, *J. Milk Food Technol.,* 35, 307, 1972.
42. **Hargrove, R. E., McDonough, F. E., and Mattingly, W. A.,** Factors affecting survival of *Salmonella* in Cheddar and Colby cheese, *J. Milk Food Technol.,* 32, 480, 1969.
43. **Park, H. S. and Marth, E. H.,** Behavior of *Salmonella typhimurium* in skim milk during fermentation by lactic acid bacteria, *J. Milk Food Technol.,* 35, 482, 1972.
44. **Rubin, H. E. and Vaughan, F.,** Elucidation of the inhibitory factors of yogurt against *Salmonella typhimurium, J. Dairy Sci.,* 62, 1873, 1979.

45. **Park, H. S., Marth, E. M., and Olson, N. F.,** Fate of enteropathogenic strains of *Escherichia coli* during the manufacture and ripening of Camembert cheese, *J. Milk Food Technol.*, 63, 543, 1973.
46. **Frank, J. F., Marth, E. H., and Olson, N. F.,** Survival of enteropathogenic and non-pathogenic *E. coli* during the manufacture of Camembert cheese, *J. Food Prot.*, 40, 835, 1977.
47. **Frank, J. F., Marth, E. H., and Olson, N. F.,** Behavior of enteropathogenic *E. coli* during manufacture of brick cheese, *J. Food Prot.*, 41, 111, 1978.
48. **Wong, H. and Chen, Y.,** Effect of lactic acid bacteria and organic acids on growth and generation of *Bacillus cereus, Appl. Environ. Microbiol.*, 54, 2179, 1988.
49. **Mikolajcik, E. M., Kerney, J. W., and Kristofferson, T.,** Fate of *Bacillus cereus* in cultured and direct acidified skim milk and Cheddar cheese, J. *Milk Food Technol.*, 36, 317, 1973.
50. **Tuckey, S. L., Stiles, M. E., Ordal, Z .J., and Witter, L. D.,** Relation of cheese making operations to survival and growth of *Staphylococcus aureus* in different varieties of cheese, *J. Dairy Sci.*, 47, 604, 1964.
51. **Tatini, S. R., Wesala, W. D., JeJeski, J. J., and Marth, H. A.,** Production of staphylococcal enterotoxin A in Cheddar and Colby cheeses, *J. Dairy Sci.*, 54, 815, 1971.
52. **Kautter, D. A., Lilly, T., Jr., Lynt, R. K., and Solomon, H. M.,** Toxin production by *Clostridium botulinum* in shelf-stable pasteurized process cheese spreads, *J. Food Prot.*, 42, 784, 1979.
53. **Schaak, M. and Marth, E. M.,** Behavior of *Listeria monocytogenes* in skim milk and in yogurt mix during fermentation by thermophilic lactic acid bacteria, *J. Food Prot.*, 51, 607, 1988.
54. **Schaak, M. and Marth, E. M.,** Behavior of *Listeria monocytogenes* in skim milk during fermentation with mesophilic lactic starter cultures, *J. Food Prot.*, 51, 600, 1988.
55. **Ryser, E. T. and Marth, E. M.,** Fate of *Listeria monocytogenes* during the manufacture and ripening of Camembert cheese, *J. Food Prot.*, 50, 372, 1987.
56. **Ryser, E. T., Marth, E. H., and Doyle, M. P.,** Survival of *Listeria monocytogenes* during manufacture and storage of cottage cheese, *J. Food Prot.*, 48, 746, 1985.
57. **Ryser, E. T. and Marth, E. M.,** Behavior of *Listeria monocytogenes* during the manufacture and ripening of Cheddar cheese, *J. Food Prot.*, 7, 50, 1987.
58. **Cliver, D. O.,** Cheddar cheese as a vehicle for viruses, *J. Dairy Sci.*, 56, 1329, 1973.
58a. **Goel, M. C., Kulshrestha, D. C., Marth, E. C., Francis, D. W., Bradshaw, J. O., and Read, R. B.,** Fate of coliforms in yogurt, buttermilk, sour cream and cottage cheese during refrigerated storage, *J. Milk Food Technol.*, 34, 54, 1971.
59. **Lee, W. H., Reiman, H. P., and Al-Mashat, A. J.,** Controlled Fermentation and Prevention of Undesirable Bacterial Growth in Food, U.S. Patent #3794739, 1974.
60. **Bartholomew, D. T. and Blumer, T. N.,** The use of a commercial *Pediococcus cerevisiae* starter culture in the production of country style hams, *J. Food Sci.*, 42, 494, 1977.
61. **Tanaka, N., Meske, L., Doyle, M. P., Traisman, E., Thayer, D., and Johnston, R. W.,** Plant trials of bacon made with lactic acid bacteria, sucrose and lowered sodium nitrite, *J. Food Prot.*, 48, 679, 1985.
62. **Babel, F. J.,** Antibiotics by lactic culture bacteria, *J. Dairy Sci.*, 60, 815, 1977.
63. **Juffs, H. S. and Babel, F. J.,** Inhibition of psychrotrophic bacteria by lactic cultures in milk stored at low temperatures, *J. Dairy Sci.*, 58, 1612, 1975.
64. **White, C. H. and Shilotri, S. G.,** Inoculation of citric acid-fermenting bacteria in raw milk in farm bulk tanks, *J. Food Prot.*, 42, 51, 1979.
65. **Ross, G. D.,** The inhibition of growth of spoilage microorganisms in milk by *Streptococcus lactis* ssp. *diacetylactis, Leuconostoc cremoris* and *Leuconostoc dextranium, Aust. J. Dairy Technol.*, 36, 147, 1981.
66. **Gilliland, S. E. and Speck, M. L.,** Inhibition of psychrotrophic bacteria by lactobacilli and pediococci in nonfermented refrigerated foods, *J. Food Prot.*, 40, 903, 1975.

67. **Martin, D. R. and Gilliland, S. E.,** Inhibition of psychrotrophic bacteria in refrigerated milk by lactobacilli isolated from yogurt, *J. Food Prot.,* 43, 675, 1980.
68. **Honor, C.,** Raw milk psychrotrophs and the importance of their control, *Dairy Rec., 82,* 104D, 1981.
69. **Champagne, C. P., Girad, F., and Morin, N.,** Inhibition of the psychrotrophic bacteria of raw milk by addition of lactic acid bacteria, *J. Food Prot.,* 53, 400, 1990.
69a. **Kamau, D. N., Doores, S., and Pruitt, K. M.,** Antibacterial activity of the lactoperoxide system against *Listeria monocytogenes* and *Staphylococcus aureus* in milk, *J. Food Prot.,* 53, 1010, 1990.
70. **Cousin, M. A.,** Presence and activity of psychrotrophic microorganisms in milk and dairy products: a review, *J. Food Prot.,* 45, 172, 1982.
71. **Raccach, M. and Baker, R. C.,** Formation of hydrogen peroxide by meat starter cultures, *J. Food Prot.,* 41, 798, 1978.
72. **Reddy, S. G., Henrickson, R. L., and Olson, H. C.,** The influence of lactic culture on ground beef quality, *J. Food Sci.,* 35, 787, 1970.
73. **Reddy, S. G., Chen, M. L., and Patel, P. J.,** Influence of lactic starter cultures on the biochemical, bacterial and organoleptic changes in beef, *J. Food Sci.,* 40, 314, 1975.
74. **Raccach, M. and Baker, R. C.,** Lactic acid bacteria as an antispoilage and safety factor in cooked, mechanically deboned poultry meat, *J. Food Sci.,* 41, 703, 1978.
75. **Raccach, M., Baker, R. C., Regenstein, J. M., and Mulnix, E. J.,** Potential application of microbial antagonism to extend storage stability of a fresh type food, *J. Food Sci.,* 44, 43, 1979.
76. **Raccach, M. and Baker, R. C.,** Bacterial repressive action of meat starter culture in pasteurized liquid whole egg, *J. Food Sci.,* 44, 90, 1979.
77. **Speck, M. L.,** Control of foodborne pathogens by starter cultures, *J. Dairy Sci.,* 55, 1019, 1972.

Chapter 5

ACETIC, PROPIONIC, AND LACTIC ACIDS OF STARTER CULTURE BACTERIA AS BIOPRESERVATIVES

Bibek Ray and William E. Sandine

TABLE OF CONTENTS

The requirement for a high internal pH (pHi) for growth (of microorganisms) has been used for many years in the preservation of foods from microbial spoilage. The effect on pHi of weakly acidic food preservatives has been compared with that of other acids of similar structure and pK. From this study it was concluded that (weakly acidic) food preservatives are more effective at lowering pHi than other acids of comparable pK. This appears to be a consequence of inhibition of metabolism by the undissociated acid acting synergistically with the accumulation of the anion. Thus, although the (acidic) food preservatives have specific effects on metabolism, their inhibition on growth is exerted in part by their effect on pHi.[1]

I. INTRODUCTION

One of the earliest food biopreservation methods recognized by our ancestors was the ability of weak organic acids, produced by some desirable microorganisms, to reduce spoilage and extend the shelf life of many fermented foods. During the course of civilization it was recognized that different desirable bacteria naturally present in raw materials produce considerable amounts of organic acids, particularly lactic acid, during fermentation of many dairy foods, meats, and vegetables. These acids reduce the pH of the foods to the extent that undesirable microorganisms present in the raw materials, and introduced during subsequent handling, either fail to multiply or have a reduced growth rate. The longer shelf life of the fermented foods is thus achieved by suppression of the growth of undesirable microorganisms by the weak organic acids that are produced as byproducts of metabolism of the desirable bacteria. Rapid acid production is one of the desirable traits in selecting or developing strains of starter culture bacteria for use in controlled fermentation of many foods.

Among the weak organic acids, lactic acid is produced in relatively large amounts by most lactic acid bacteria. Also, considerable amounts of acetic acid can be produced from fermentable carbohydrates by some lactic acid bacteria and bifidobacteria, and propionic acid can be produced by propionic acid bacteria. As the antimicrobial concentrations and spectra of these three organic acids are greatly dependent upon the pH of the environment, and as they are more effective at lower pH levels (pH <5.0), their beneficial role in foods with a higher pH (>5.0) has been questioned. In contrast, several other organic acids such as fumaric, sorbic, benzoic, monolauric and their esters, have been found

TABLE 1
Production of Organic Acids by Lactic Acid and Proponic Acid Bacteria

Genera	% Acid Production	pH
Lactococcus	0.8	4.5
Streptococcus	1.0	4.0
Leuconostoc	0.6	4.8
Pediococcus	1.5	3.5
Lactobacillus	1.5	3.5
Bifidobacterium	1.0	4.0
Propionibacterium	1.0	5.0

to be effective at higher pH ranges; many are currently being used in food preservation. However, they do not fall in the category of biopreservatives of microbial origin. This chapter will concentrate on the uses of acetic, propionic, and lactic acid as food biopreservatives.

II. METABOLISM BY STARTER CULTURE BACTERIA

Lactic acid starter culture bacteria include species/strains from five genera: *Lactococcus, Streptococcus* (one species), *Leuconostoc, Pediococcus, Lactobacillus,* and *Bifidobacterium.* The latter are not usually included with lactic acid bacteria, LAB, but are so included in this section. These organisms produce fairly large amounts of lactic acid and, in some cases, acetic acid from fermentation of metabolizable carbohydrates (Table 1).[2–5] *Propionibacterium* species are capable of producing propionic and acetic acids from carbohydrates (or lactate)[5,6] The high amounts of weak acids produced by these bacterial genera reduce the pH of the environment considerably and inhibit or reduce the growth of many bacteria, including those that cause food spoilage and foodborne diseases. In addition, propionic acid inhibits growth of yeasts and molds. This is the basis of the longer shelf life of fermented foods. Starter culture bacteria metabolize fermentable carbohydrates through several different pathways, principally to generate energy, and in the process they produce weak acids as byproducts. As the weak acids are not utilized further by the producing cells, they are excreted into the environment. The different pathways used, and the byproducts produced, by these bacteria are presented schematically in Figures 1 through 5 and in Table 2. Emphasis has been given to the important enzymes catalyzing these reactions that lead to the production of the weak acids and related byproducts; energy generation has not been emphasized. In the homolactic fermentation, glucose and other usable hexoses are converted mainly to lactic acid, when the carbohydrate supply is unlimited, through the Embden-Meyerhoff-Prans (EMP) pathway; each molecule of glucose is potentially converted to two molecules of lactic acid (Figure 1). Species from *Lactococcus, Streptococcus (S. salivarious* ssp. *thermophilus), Pediococcus,*

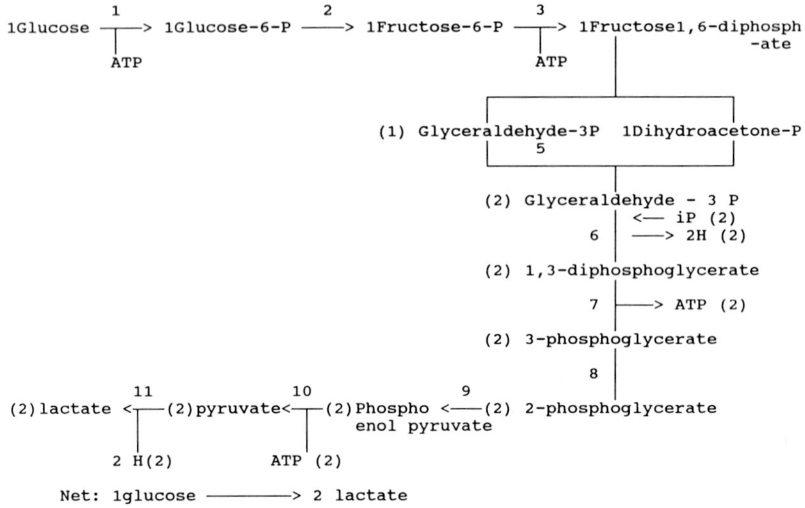

FIGURE 1. Metabolism of hexoses by homofermentative lactic acid bacteria by Embden-Meyerhoff-Pranas (EMP) pathway. Enzymes are: 1, hexokinase; 2, phosphoglucose isomerase; 3, 6 phosphofructokinase; 4, aldolase; 5, triosphosphate isomerase; 6, glyceraldehyde-3-phosphodehydrogenase; 7, 3-phosphoglycerate kinase; 8, phosphoglyceromutase; 9, enolase; 10, pyruvate kinase; 11, lactate dehydrogenase.

and group I and II *Lactobacillus* metabolize hexose by the EMP pathway (Table 2). However, in limiting concentrations of hexose, some strains of *L. lactis* ssp. *cremoris, S. salivarious* ssp. *thermophilus,* and some group II lactobacilli (*L. casei*), can produce acetate and CO_2 from pyruvate. Depending upon the type of lactate dehydrogenase a species harbors, the lactic acid produced can be L(+), D(–) or DL isomer(s). Leuconostocs and group III lactobacilli, however, metabolize hexose through the hexose monophosphate shunt (HMS) with production of equimolecular amounts of CO_2, acetate, and lactate (Figure 2). However, the production of acetate is reduced under a limited oxygen environment, since under these conditions acetyl-phosphate will be diverted to ethanol. In limiting hexose concentration, leuconostocs use a different pathway (mixed-acid type) to generate formate along with acetate and lactate (Figure 3). Those species capable of fermenting pentose sugars do so through the pentose-phosphate pathway with the production of equimolecular amounts of acetate and lactate (Figure 4); again the conversion of acetyl-phosphate to either acetate or ethanol is regulated by the availability of oxygen and the presence of the specific enzymes necessary (such as alcohol dehydrogenase). In all these pathways, in general, more lactate than acetate (on a molecular basis) is produced. In contrast, *Bifidobacterium* species produce more acetate than lactate (3:2) from hexose by the bifidus pathway (Figure 5).

TABLE 2
Summary of Production of Lactic, Acetic, and Proponic Acid by Food Grade Starter Culture Bacteria

Bacteria (genera)	Substrate(s)	Fermentative type[a]	Products[b]	Pathways[c]
Lactococcus	Hexose	Homo	Lactic, L(+)	EMP
Streptococcus	Hexose (nonlimiting)	Homo	Lactic, L(+)	EMP
	Hexose (limiting)	Hetero	Lactic, L(+), acetic, CO_2	EMP & PC
Leuconostoc	Hexose (nonlimiting)	Hetero	Lactic, D(–), acetic[d], CO_2 (1:1:1)	HMS
	Hexose (limiting)	Hetero	Lactic, D(–), CO_2/formic, acetic[d] (1:1:1)	MA
	Pentoses	Hetero	Lactic, D(–), acetic[d] (1:1)	PK
Pediococcus	Hexose	Homo	Lactic, DL	EMP
	Pentoses	Hetero	Lactic, DL, acetic (1:1)	PK
Lactobacillus				
Group I: *Thermobacterium*	Hexose	Homo	Lactic, L(+), D(–), DL	EMP
Group II: Streptobacterium	Hexose (nonlimiting)	Homo	Lactic, L(+), D(–), DL	EMP
	Hexose (limiting)	Hetero	Lactic, L(+), D(–), DL; acetic, CO_2 (1:1:1)	EMP & PC
	Pentoses	Hetero	Lactic, L(+), D(–), DL; acetic[d] (1:1)	PK
Group III: Betabacterium	Hexose	Hetero	Lactic, L(+), D(–), DL; acetic[d], CO_2 (1:1:1)	HMS
	Pentoses	Hetero	Lactic, L(+), D(–), DL, acetic[d] (1:1)	PK
Bifidobacterium	Hexose	Hetero	Lactic, L(+), acetic (2:3)	BP
	Hexose	Hetero	Lactic, DL, acetic, CO_2/ formic	BP & PC

TABLE 2 (continued)
Summary of Production of Lactic, Acetic, and Proponic Acid by Food Grade Starter Culture Bacteria

Bacteria (genera)	Substrate(s)	Fermentative type[a]	Products[b]	Pathways[c]
Propinibacterium	Pyruvate or lactic	—	Propionic, acetic, CO_2	SP

[a] Homolactic or heterolactic fermentation

[b] Products: L(+), levorotatory; D(–), dextrorotatory, DL, recemie lactic acid are produced by the respective lactose dehydrogenase present in a species/strain. Ratios in parentheses indicate approximate molar ratios.

[c] Pathways: EMP, Embden-Meyerhoff-Parnas; HMS, hexose monophosphate shunt (or phosphogluconate); PC, by phosphoroclastic cleavage (Pyruvate → Acetate + CO_2); PK, phosphoketolase (or pentosephosphate); BP, Bifido pathway; SP, succinate-propionate pathway.

[d] Depending upon oxidation-reduction potential of the environment ethanol, instead of acetate, can also form (under reduced state).

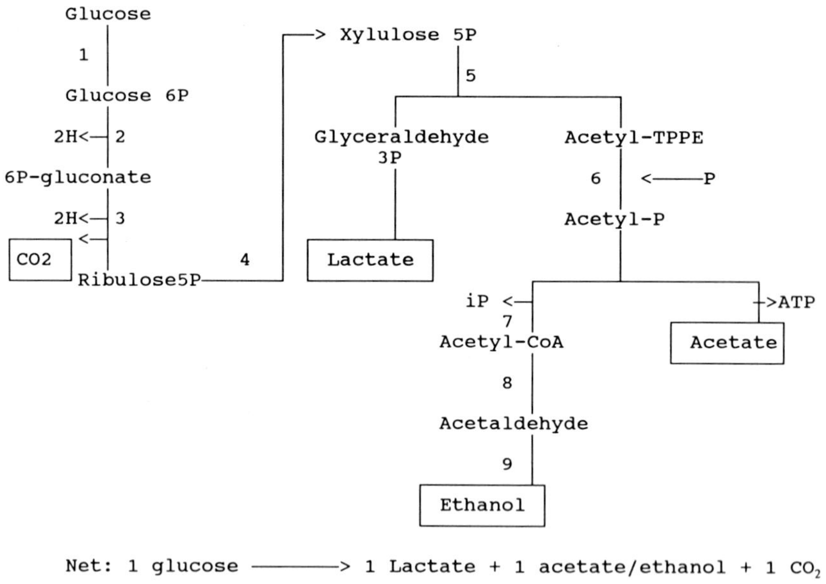

FIGURE 2. Heterolactic fermentation of hexoses by hexose monophosphate shunt (HMS). Enzymes are: 1, EMP pathway enzymes; 2, Glucose 6 phosphate dehydrogenase; 3, 6 phosphoglucono-dehydrogenase; 4, ribulosephosphate-3-epimerase; 5, phosphoketolase; 6, acetate kinase; 7, phosphotransacetylase; 8, acetaldehyde dehydrogenase; 9, alcohol dehydrogenase.

Propionibacterium species are capable of converting pyruvate, obtained either from glucose through EMP pathway or from dehydrogenation of lactate, to produce considerable amounts of propionic acid, along with some acetate and CO_2 through the succinate-propionate pathway (Figure 6).

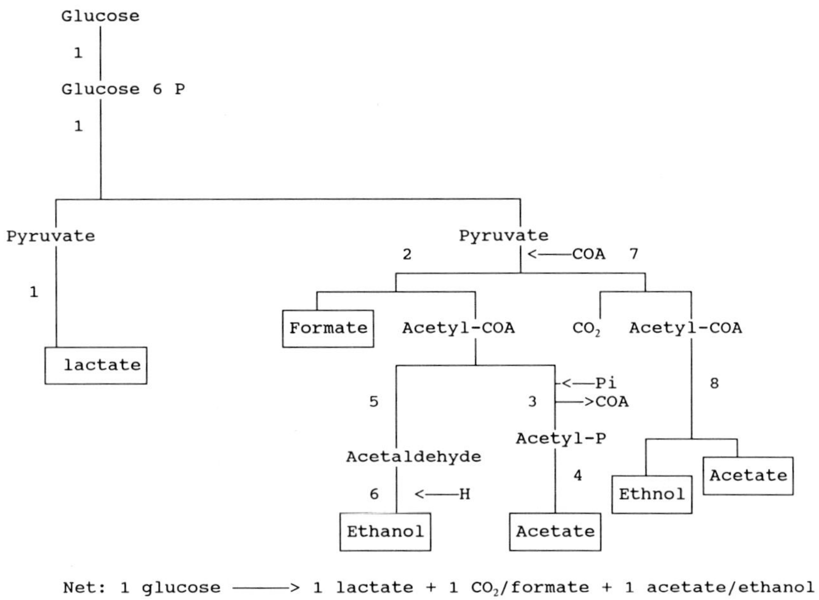

FIGURE 3. Mixed acid (MA) fermentation in limiting hexose concentration by some lactic acid bacteria. Enzymes are: 1, EMP pathway enzymes; 2, pyruvate formate lyase; 3, phosphotransacetylase; 4, acetatekinase; 5, acetaldehyde dehydrogenase; 6, alcohol dehydrogenase; 7, pyruvate kinase; 8, same as steps 3 to 6.

From the above discussion it is clear that many starter culture bacteria used in food fermentations have elaborate and diverse mechanisms to convert different carbohydrates to lactic, acetic, and propionic acids, along with several other byproducts. These byproducts not only may contribute to desirable flavor and body of these foods, but many of them, especially the weak organic acids, are capable of stabilizing the foods by inhibiting growth of undesirable microorganisms.

III. MECHANISM OF ANTIMICROBIAL ACTION

Two earlier observations, namely that the organic acids on an equimolar basis are more antimicrobial than the strong inorganic acids, and that the antimicrobial efficiency of a weak acid increases as the pH is reduced, lead to the assumption that the antimicrobial property of a weak organic acid is a function of its undissociated molecules.[7–9] For example, it has been shown that *Salmonella typhimurium* cells in milk died faster at pH 4.0 when the pH was adjusted with lactic acid rather than with hydrochloric acid. Similarly, at pH 5.5 the cells grew slower when lactic acid, instead of hydrochloric, was used to adjust the pH.[10] Relative antimicrobial efficiency of acetic, propionic, and lactic acids against several foodborne bacterial pathogens also has been reported. Sorrells et al.[11] observed that at any particular pH between 4.4 and

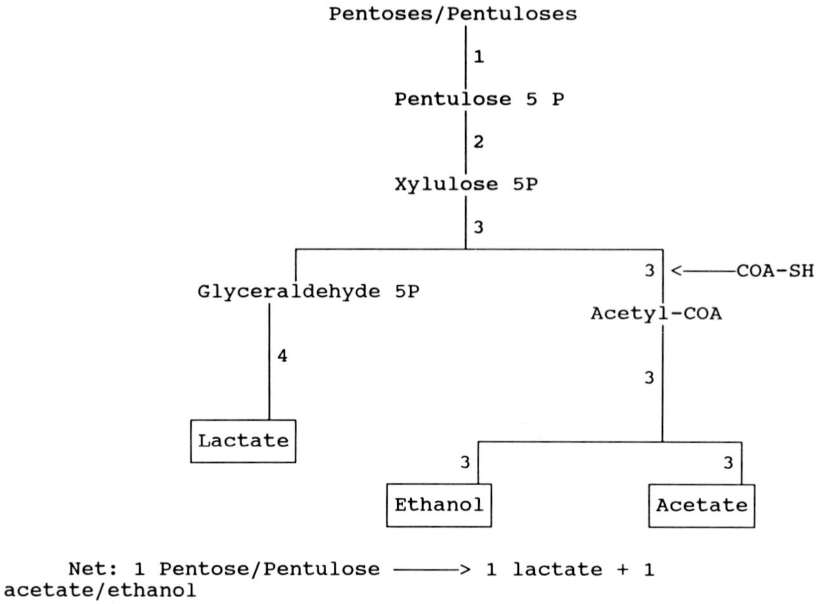

FIGURE 4. Metabolism of pentoses through pentose phosphate (PP) or phosphoketolase (pK) pathway by lactic acid bacteria through pentose phosphate pathway. Enzymes are: 1, specific kinases and dehydrogenases; 2, specific isomerases; 3, enzymes of HMS; 4, enzymes of EMP pathway.

5.2, the inhibitory effect of acids against *Listeria monocytogenes* was in the following order: acetic > lactic > HCl. The greater antibacterial effect of acetic acid over lactic acid against *L. monocytogenes* was also reported by Ahmed and Marth[12] and the greater effect against *Staphylococcus aureus* was reported by Miner and Marth,[13] and Nunheimer and Fahian.[14] Other studies also reported that at the same pH, the order of antibacterial efficiency of the three weak organic acids produced by starter cultures is propionic > acetic > lactic against *Salmonella* spp.,[15] *L. monocytogenes*,[16] and probably many other microorganisms.[17] The data presented in Table 3 provide justification as to why propionic acid is the most, and lactic acid the least, inhibitory to microorganisms at pH levels between 4 and 7. Their findings strengthen the earlier assumption that the antimicrobial efficiency and concentration of undissociated molecules are directly related. Therefore, at a lower pH, the proportion of undissociated molecules of an acid is higher, and its antimicrobial efficiency greater, than at a higher pH (Table 3).

A. Interference of Proton Motive Force and pH Homestasis

The mechanism of antimicrobial action of weak acids such as acetic, propionic, and lactic have been explained on the basis of chemiostatic theory[18–20] and

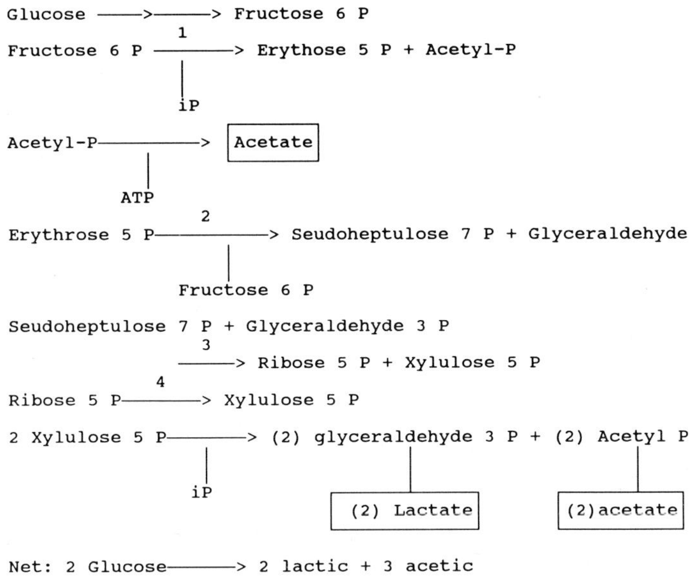

FIGURE 5. Metabolism of hexoses to lactate and acetate in bifidus pathway by Bif*idobacterium* species. Important enzymes are: 1, fructose 6 phosphatephosphoketolase; 2, transaldolase; 3, transketolase; 4, esomerase; 5, xylulose phosphate ketolase; See EMP and HMS pathway for enzymes in other steps.

pH homeostasis.[1,21] For normal survival and growth, microorganisms maintain a relatively constant internal cytoplasmic pH (pHi). In acidophiles and neutrophiles, which include most of the food spoilage and foodborne pathogenic microorganisms, the pH ranges from 6.5 to 7.0 and 7.5 to 8.0, respectively.[1] In some microorganisms, such as in *E. coli* and other respiratory organisms, pHi is tightly regulated so that a reduction in 1 pH outside (pHo) will lower the pHi by 0.1 unit or less. On the other hand, *Enterococcus faecalis* and other fermentative bacteria can withstand a greater variation in pHi. Similarly, for energy generation and active transport of nutrients, microbial cells maintain a transmembrane pH gradient of about 0.5 to 1.0 unit, with inside alkaline and outside acidic conditions. They also maintain a proton gradient of about 200 mV for respiratory microorganisms and a little less in fermentative bacteria. Together these gradients form the proton motive force (PMF), the driving force for the transport of protons and nutrients and for the generation of energy necessary for growth of microbial cells.[21–24]

When the weak organic acids such as acetic, propionic, and lactic are produced in the environment, some of the molecules dissociate to H^+ and anions, while others remain undissociated. The pH of the environment, pK of the acid, concentration, and several other factors will dictate the concentration of dissociated and undissociated molecules present (Table 3). This will cause an increase in transmembrane protons (greater than 200 mV) and pH gradients,

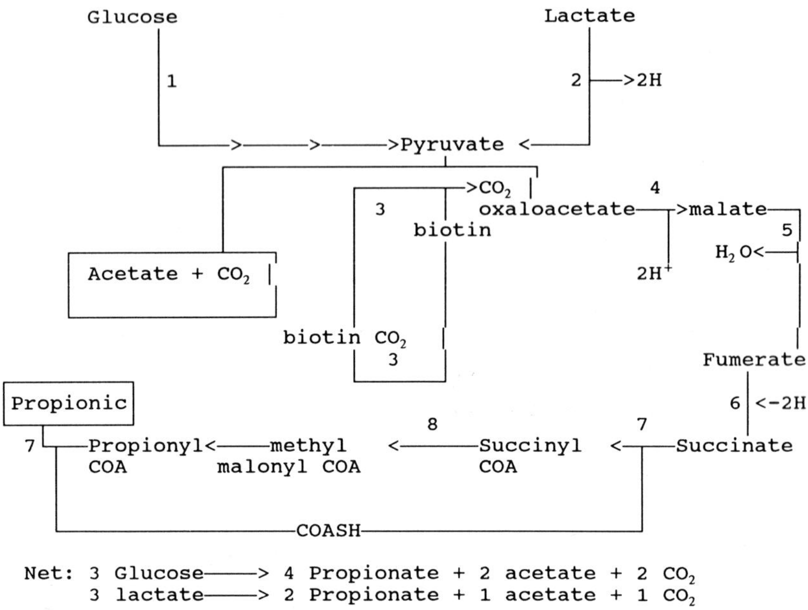

FIGURE 6. Succinate-propionate pathway of propionic acid production from glucose or lactate via pyruvate. Enzymes are: 1, EMP pathway enzymes; 2, lactate dehydrogenase; 3, methylmalonyl-COA-pyruvate transcarboxylase; 4, malate dehydrogenase; 5, fumerase; 6, fumerate reductase; 7, COA-transferase; 8, methylmalonyl-COA mutase + recemase.

TABLE 3
Influence of pH on the Percentage Amount of Undissociated Molecules Present in Aqueous Solutions of the Three Acids[a]

		% Undissociated of total at			
	pK[b]	pH 4	pH 5	pH 6	pH 7
Acetic	4.8	84.5	34.9	5.1	0.5
Propionic	4.9	87.6	41.7	6.7	0.7
Lactic	3.9	39.2	6.1	0.6	0.1

[a] Table adapted from References 7 and 32b.
[b] pK = pH at which concentrations of the undissociated molecules and the dissociated molecules are equal.

as well as a concentration gradient of undissociated cations (H^+) and anions.[1,17,25] The mechanism of antimicrobial effects of protons and undissociated molecules will be discussed separately.

To establish a normal gradient, cells tend to move protons inside. Although the lipid bilayer membrane is impermeable to protons, influx of protons can occur through proton pumps under PMF. A high concentration of H^+ outside will reduce the pHo considerably and expose the cell wall, cell membrane, periplasmic space, and outer surface of the cytoplasmic, or inner membrane to the detrimental effects of low pH. A very low pH (pHo) can cause denaturation of enzymes present in these structures and can also cause destabilization of the membrane permeability, which can inhibit growth and reduce cell viability.[1,21,23]

The lipid bilayer is quite permeable to lipophilic acetic, propionic, and lactic acids, and this facilitates their diffusion into the cytoplasm. The pHi being relatively alkaline, and the acids being highly soluble in the internal aqueous environment, the undissociated molecules will dissociate, producing protons and anions. The buffering capacity of the cytoplasmic materials will tend to prevent reduction of pHi by neutralizing some protons. When protons exceeding the buffering capacity are produced, they are transported out through the membrane-bound proton pumps, thus depleting the energy reserves. As the energy reserve is limited, proton extrusion through proton pumps stops, causing a lowering of pHi, which in turn denatures proteins and destabilizes other structural and functional components of the cells and interferes with viability and growth.[1,21,23]

The antimicrobial effect of the weak acids is thus principally produced by the undissociated molecules through the acidification of cytoplasm, destruction of the transmembrane proton motive force, and loss of active transport of nutrient through the membrane. In recent years these generalized effects, caused only by the undissociated molecules, have been questioned. Several studies have reported that growth inhibition of bacterial cells following treatment with a weak acid did not prevent the uptake of nutrients normally transported by the active transport systems. The total inhibitory action of weak acids is currently viewed as being the combined effects produced by undissociated molecules as well as the dissociated ions. Mathematical models based on the effects of both molecules have been developed. These models assume that the total effect of a weak acid under a certain condition is constant and equal to the combined effects produced by the undissociated and dissociated molecules separately. This effect could not be accounted for by assuming it was produced by the undissociated molecules alone.[17,23,25]

The specific mechanism of antimicrobial action of acetic, propionic, and lactic acids is probably produced in the same way as for other weak acids. Acetic acid is highly lipophilic and able to neutralize transmembrane pH gradients efficiently. Reduction of pHi and denaturation of cytoplasmic proteins by acetic acid have also been documented. Propionic acid, also highly lipophilic, has been found to neutralize PMF through the dissociation of undissociated molecules in the cytoplasm. This results in loss of active transport of some nutrient molecules, and in inhibition of cellular growth. Lactic acid, with its lower pK, has less effect on neutralizing the proton motive force as compared to acetic or propionic acids. Recent studies confirm that the total antimicrobial effects of these weak acids are produced by the combined effects of the undissociated and dissociated molecules. The effects are produced not only by lowering the internal pH and neutralizing the proton motive force, but also by causing sublethal injury and other unidentified harmful effects.[7,17,24,25]

B. Bacteriostatic and Bacteriocidal Action

It is known that many organic acids can inhibit growth (bacteriostatic) or cause viability loss (bacteriocidal) of microbial cells. The antimicrobial effects of the weak acids such as acetic, propionic, and lactic are produced through the destabilization of different functional and structural components of cells. The outcome of these multifunctional effects is dependent upon the particular acid, its concentration and pK, pH of the environment, time and temperature of exposure, and the target microbial species/strains. At a low acid concentration with a relatively high pH, the proton concentration that comes in contact with structures outside the membrane surface, the periplasmic space, and in the cell wall, will be low and may produce only a slight destabilization of some components. Similarly, the undissociated molecules may not cause an extreme lowering of pHi. Under these conditions a temporary growth inhibition, or a reduced rate of growth, of microorganisms can occur.

In contrast, at a higher acid concentration and especially at a lower pH, the high proton concentration can destabilize functions of different cell components, many of which can play key roles in cell viability. Similarly, a high concentration of the undissociated molecules will enter the cell, lower the pHi, and destabilize key components necessary for survival.[17,23]

C. Sublethal Injury

Exposure of suspensions *E. coli*,[26,27] *Listeria monocytogenes*,[29] *Salmonella bareilly*,[28] and yeast[30] cells to acetic and lactic acids were reported to cause viability loss and sublethal injury to some fractions of the population, while some cells retain normal characteristics. Thus, acid treatment of a population of microbial cells produces three types of cells: dead, injured, and uninjured or normal. Differentiation of these cell types can be accomplished by enumerating colony-forming units (cfu) of a cell suspension before and after the acid treatment, simultaneously in nonselective and selective agar media, and by measuring the differences in viable cfu. The fraction of the cells that loses the ability to form colonies in the nonselective medium is considered to be the dead cells, and those that form colonies equally well both in the nonselective and selective agar, are considered uninjured. Further, those that form colonies in the nonselective, but fail to do so in the selective agar, are considered injured or sublethally stressed cells. The injured cells have altered characteristics due to exposure to the acids, making them susceptible to ingredients in the selective medium — ingredients to which they are not normally resistant. In general, if the acid exposure is continued, there will be a shift in the relative fractions of the three groups: an increase in dead cells, due to conversion of injured cells to dead, and a reduction in uninjured cells, due to their conversion to injured state.[31]

In addition to their developed sensitivity to selective ingredients (inability to form colonies in a selective agar medium) and, with time, entering the death phase (inability to form colonies in a nonselective agar medium), the injured cells have two other important characteristics. First, the cells differ greatly in their level of injury, i.e., some are slightly injured, others relatively more, and others greatly injured. Second, the injury is reversible, i.e., if the injured cells are exposed in a suitable nonselective environment, they become insensitive to the selective pressure and gain back their ability to multiply and form colonies in a selective agar medium like the normal cells. The rate of repair can be instantaneous for some cells, while for others it requires a long incubation. The time required is directly related to the level and/or degree of injury a cell has sustained.[31]

Altered characteristics of injured cells are due to changes in structural and functional components of the cells occurring during exposure to the acid environment. Recovery from injury in a suitable environment results from the repair of those components. Repair can occur in several ways. In some states it can be in the form of reorganization of the existing components, while in other states it may involve synthesis of new macromolecules through normal metabolic processes.[31]

Roth and Keenan[26] reported that when *E. coli* cells were suspended in Tryptic Soy Broth acidified to pH 4.3 with acetic acid, or to pH 4.2 with lactic acid, >90% of the surviving cells failed to form colonies in violet-red bile (VRB) agar with 15 min exposure; but they formed colonies in nutrient agar. Similar differences in the ability to form colonies were observed when cells were treated with hydrochloric acid (pH 3.2). At these respective pHs, about 20% of acetic acid, 70% of lactic acid, and almost all of HCl are dissociated.[7] So the structural and functional components that were damaged, as reflected by the sensitivity of the injured cells to VRB agar, could not be due only to undissociated molecules of the acids. Rather, the dissociated molecules (H^+ and anions) are likely to be the ones primarily involved in this injury. Since acetic and lactic acids produced a greater percentage of injured cells than hydrochloric, as revealed by the differences in D values (about 232, 166, and 32 min, respectively, for acetic, lactic, and hydrochloric acids), the individual dissociated anions of the weak acids are likely to have the roles in this injury along with the H^+ common to all. The selectivity of VRB agar is principally due to bile salts, and the architecture of the lipopolysaccharide (LPS) molecules on the cell of *E. coli* outer membrane (OM) confers resistance to bile salts. Loss of resistance to bile salts, as reflected by the inability of the acid-injured cells to form colonies in VRB agar is then due, at least partially, to disruption of the architecture of LPS molecules. As a result, bile salts can pass through the LPS-barrier, come in contact with the cytoplasmic membrane and, by its detergent property, destabilize the membrane.[31] Ability of the *E. coli* cells injured by acetic acid to repair themselves and regain resistance to bile salts, and ability to form colonies in VRB agar, should thus be considered to result from repair of injury in the LPS molecules. Przbylski and Witter[27] observed that repair of acid-injury to LPS molecules and recovery of resistance to bile salts started almost instantaneously and could be accomplished in K-phosphate buffer at pH 7.0 to 8.0 (but not in Na-phosphate buffer at 7.0) as well as in a complex broth like Tryptic Soy Broth (TSB). They also reported that restoration of integrity in LPS molecules was not affected adversely by inhibitors of *de novo* synthesis of proteins, mucopeptide, DNA, RNA,energy (through coupling), or active transport. However, they were inactivated by streptomycin probably due to its role in K^+ efflux from cells. They concluded that *E. coli* cell injury, following exposure to acetic acid, results from an easily reversible equilibrium shift, and is more complex in nature than just acidification of cell interior. Blankenship[28] measured injury of *S. bareilly* cells following exposure to acetic acid by the ability of the injured cells to form colonies in nutrient agar (NA) and by their inability to grow in VRB agar containing 1% NaCl. In this study sensitivity of injured cells to two selective agents, bile salts and NaCl, was used.

In sublethal injury studies in Gram-negative bacteria, acquired sensitivity to bile salts has been used as an indicator of LPS injury in the OM, and NaCl has been used as an indicator of cytoplasmic membrane injury. Thus the method of combining bile salts and NaCl, as used by Blankenship,[28] actually measured injury in the OM as well as in the inner membrane of the acid treated cells. Using metabolic inhibitors, he reported that repair of acid injury, as reflected by the ability of injured cells to form colonies, was dependent upon *de novo* synthesis of RNA and proteins as well as energy production through the electron transport system. He also reported from the direct measurement of the activity of enzymes, extracted from acid-treated and untreated cells, that acetic acid-treated cells had lower activities of lactose dehydrogenase, glucose-6-phosphate dehydrogenase, lysine decarboxylase, and several other enzymes, but not of ornithine decarboxylase. Recently Ahmed and Marth[29] reported that cells of Gram-positive *L. monocytogenes*, following treatment with acetic and lactic acids, suffered injury as indicated by their inability to form colonies in tryptose agar (TA) containing 6% NaCl (TAS), but not in TA. As the incubation time (100 vs. 500 min) and temperature (13 vs. 35°C) were increased, there were increases in the proportion of dead and injured cells and a decrease in uninjured cells. The inability to form colonies by the injured cells in media containing salt reflected reversible injury to the cytoplasmic membrane and did not measure damage to other structural and functional components.

Exposure of microbial cells to weak acids can thus bring about changes in different structural and functional components of cells. These changes are brought about by the undissociated molecules as well as by the anions and H^+. Damages can occur in components present in the wall, membrane, and cytoplasm; some are reversible, while others are nonreversible, and a cumulative effect of these determines the ability of a cell to grow or remain viable without growth (bacteriostatic), or to die (bacteriocidal). The nature of an acid, concentration, pH, temperature, duration of exposure, and other factors play important roles in the final outcome.

D. Resistance

Death and reversible injury of microbial cells following exposure to weak acids are the cumulative outcome of the changes caused in different structural and functional components of the cells.[31] It is thus reasonable to assume that the antimicrobial effects of weak acids are multifactorial and probably not due to a single, or to a few, determinant(s). It is known that microorganisms differ in their sensitivity to weak acids, and a strain can be made acid tolerant to a limited degree by growing it in a medium containing increasing concentrations of an acid. These are phenotypic variants, since the developed characteristic to grow at high acid concentrations can be lost by growing the acid-tolerant cells in the absence of the acid. Temporary modifications of some structures, probably at the cell surface and membrane, may enable these cells to grow at lower pH.[23,24] Similar temporary ability of a bacterial strain to grow at a high concentration of an antibiotic has been known.

The ability of a microbial species to grow in high acid concentrations can also be a reflection of its ability to withstand reduction in pHi. In general, fermentative bacteria such as lactic acid bacteria, have the ability to grow and survive at a lower pH than the respiring bacteria, such as *E. coli*. Moreover, adaptation to cytoplasmic acidification by fermentative strains helps cells during the late exponential phase to grow even at a lower pH_o.[23,24]

In recent years one or more genetic determinants have been linked to the ability of certain strains to grow at higher concentrations of weak acids. Thus, the ability of a mutant of *Salmonella typhimurium* to tolerate acids was found to be due to over expression of a group of proteins that are induced by growing the mutant cells at a mild acid pH. Acid tolerance is thus linked to genetic determinants or genes. Identification of the location and the nucleotide sequences in the DNA coding for these proteins, and their relation to other stress proteins, will provide valuable information on genetic control of acid tolerance by microorganisms.[32,32a]

E. Factors Affecting Antimicrobial Properties

The antimicrobial effectiveness of acetic, propionic, and lactic acids are influenced by many different factors. For an effective use of these acids in foods to control growth of spoilage and pathogenic microorganisms, the following factors should be considered.[7,17,24,25]

1. pK, Concentration, and pH

As all these factors influence the proportion of undissociated molecules, an acid with higher pK, used at a high concentration and at a lower pH, will be the most effective. However, the maximum concentration of an acid that can be incorporated in a food is controlled by regulatory agencies according to the food in which it is used. In most food systems with a pH above 5.0, where spoilage and pathogenic bacteria are of concern, acetic and propionic acids will be better than lactic acid on an equimolar basis. The differences in results due to use of these acids in percent (%), but not in molar concentrations, could be due to differences in amounts of molecules and ions present.

2. Lipophilic Property and Solubility

An acid that is highly lipophilic, so that it can pass through the cell membrane easily and is more soluble in the aqueous phase, will have the greatest antimicrobial effect. Lactic acid is less lipophilic than acetic or propionic.

3. Temperature

At a higher temperature, the antimicrobial effect of an acid is increased as compared to its effect at lower temperatures. The protective effect of low temperatures should be considered in refrigerated foods for which long shelf life is required.

4. Resistance of Microorganisms

Microorganisms differ greatly in their ability to survive and/or grow at the pH of their environment. In general, yeasts and molds are able to grow at a lower pH than most bacteria, and among bacteria, Gram-negative bacteria are more sensitive to lower pH than Gram-positive bacteria. Fermentative bacteria are more resistant to lower pH_0 than respiratory bacteria. Spores and viruses are also more resistant to lower pH than vegetative bacterial cells.

Some microorganisms can metabolize lactic acid and thus can reduce its concentration to a level where other acid-sensitive microorganisms can grow. Propionic acid is more effective against molds than yeasts. Also, acid(s) present in foods can exert selective pressures resulting in enrichment of the microorganims that are able to grow in the presence of the acid(s).

5. Microbial Load

At a higher microbial load, a given concentration of an acid will be less effective.

6. Food Composition

Food components, either by acting as a physical barrier or by neutralizing or buffering, can reduce the antimicrobial effectiveness of an acid. They can also provide an environment favorable for repair of injury and thus facilitate cellular growth.

7. Other Factors

Presence of acid anions in the environment will increase the antimicrobial effect, probably by reducing the dissociation of an acid. Anaerobic conditions increase the antimicrobial effect of an acid over that of aerobic conditions. Thus propionic acid will have higher efficiency with CO_2. Acetic and lactic acids, when used together, have a synergistic effect. Weak acids can potentiate the antimicrobial effect of other inhibitors such as nitrite.

IV. PROPERTIES AND REGULATORY STATUS

A. Physical and Chemical Properties

Uses of acetic, propionic, and lactic acids in food, and their efficiency as antimicrobial agents, are dependent upon their physical and chemical properties. These will be described briefly in this section.

1. Acetic Acid[32b,33]

CH_3COOH; molecular weight 60.05; colorless liquid; pungent odor; weakly ionized in aqueous solution: $K_a = 1.8 \times 10^{-5}$; pK, 4.75; pH of aqueous solution: 1.0 M = 2.4; 0.1 M = 2.9, 0.01 M = 3.4. Excellent solvent for organic compounds; miscible with water, alcohol, glycerol, and lipids; incompatible with carbonate and phosphate; boiling point 39°C; melting point is –8.5°C; sour taste.

Prepared from conversion of ethanol by acetic acid bacteria to vinegar. Also obtained as glacial acetic acid from the destructive distillation of wood, or from acetylene and water by oxidation.

2. Propionic Acid[32b,33]

CH_3CH_2COOH; molecular weight 70.08; oily liquid; slightly pungent; $K_a = 1.34 \times 10^{-5}$; pK, 4.87; soluble in water, alcohol, and lipids; can be salted out from water by the addition of $CaCl_2$ or other salts; boiling point, 58°C (136°F); melting point, –21.5°C.

Present in some dairy products from the fermentation of dairy *Propionibacterium*. Can also be produced from wood-pulp waste liquor by fermentation with *Propionibacterium* or by several synthetic processes.

3. Lactic Acid[32b,33]

$CH_3CHOHCOOH$, 2-hydroxypropionic acid, occurs in three different forms: D(–) or levoratory, L(+) or dextrorotatory, and DL or recemic lactic acid; molecular weight, 90.08; hygroscopic, syrupy liquid; available in dry form also; K_a at 25°C is 1.38×10^{-4}; pK, 3.85; soluble in water, alcohol, and lipid; volatile in steam; melting point, 16.8°C; acrid taste.

D(–) lactic acid can be obtained from the fermentation of dextrose by *Lactobacillus delbrueckeii* ssp. *bulgaricus*, *L. delbrueckii* ssp. *leichmanni*, or by resolution of DL-lactic acid. It forms salts with Na^+, K^+, and Ca^{2+}; most salts are dextrorotatory. L(+) lactic acid can also be obtained by fermenting dextrose with *Lactobacillus* or by resolution from DL-lactic acid. This is the form that is produced in muscles. Most salts are levorotatory. DL-lactic, or ordinary lactic acid, can be produced by fermenting dextrose and other carbohydrates with suitable lactic acid bacteria such as *Pediococcus acidilactici*, *Lactobacillus acidophilus*, *L. helveticus*, and *L. plantarum*. The acid is also present in many fruits due to partial conversion of carbohydrates. Commerecially it is produced by the fermentation of whey, molasses, and starches.

B. Toxicoligical Information, Regulatory Status, and Uses in Food

1. Acetic Acid and Salts[7,32b–34]

LD_{50} for glacial acetic acid orally in rats is 3.53 g/kg. Feeding up to 200 mg/kg of body weight for 4 months in drinking water did not produce any adverse effect, as judged by food intake and weight gain. At higher concen-

TABLE 4
Regulatory Information for Use of Acetic, Propionic and Lactic Acids and Their Salts in Foods[7,32b,34,40]

Acid/Salt	LD_{50} oral in rats (g/kg BW[a])	ADI[b] (mg/kg BW[a])	Concentrations generally used (mg/kg food)	CFR[c]
Acetic	3.53	Not limited	No limit	182.1005
Acetate(Na^+)	NA[d]	Not limited		182.1721
Propionic	4.29	Not limited	Up to 3000	182.3081
Propionate (Ca^{2+})	3.34	Not limited		182.3221
DL-Lactic	3.73	Not limited	No limit	182.1061
Lactate (Na^+,K^+)	NA	Not limited		184.1639 (K^+) 184.1768 (Na+)

[a] BW, body weight.
[b] ADI, acceptable daily intake.
[c] CFR, Title 21 of the Code of Federal Registration.
[d] NA, not available.

trations, however, food intake and weight gain are reduced. The acid is regarded as safe for miscellaneous and general purpose uses under the Code of Federal Regulation (CFR), Title 21, 182.1005. No limit has been set on daily acceptable intake (Table 4). It is used in food for pickling meat, fish, and vegetables; in condiments such as mustard, salad dressing, mayonnaise, and catsup; and in some baby formulae. It is also used in water to spray beef carcasses.

Na-salt has been given GRAS (generally regarded as safe) status for use in foods. Limited toxicological studies have indicated that supplementing chicken diet with 5.4% Na-acetate resulted in reduced growth and increased mortality. However, this was attributed to higher Na^+ intake.

2. *Propionic Acid and Salts*[7,32b–34]

LD_{50} for propionic acid orally is 4.29 g/kg body weight. Feeding propionic acid to rats at a concentration of 50ml/kg rice for 110 d caused some lesions. However, salts of Na- and Ca-propionate did not produce any adverse effects on growth or mortality in rats. In an isolated study, 6 g daily oral dose of Na-propionate did not produce any toxic effect in a human male.

Both propionic acid and its salts (Na- and Ca- propionate) have been given GRAS status for general purpose uses (CFR 182.3081, 182.3221, and 182.3784 in Title 21, respectively, for acid, Na- and Ca-salts). They have been used in bread, other bakery products, and cheese as antimycotic agents. Salts have been used in packaging material for antimycotic properties (Table 4). They are generally used at 0.3% level. In Swiss cheese the *Propionibacterium* can produce as high as 1% propionic acid.

3. Lactic Acid and Salts[7,32b–34]

LD_{50} for DL-lactic acid orally in rats is 3.73g/kg. Long-term feeding studies in hamsters through drinking water at 40 mg/100 ml or in feed at 45.6 mg/100 g, did not produce any differences in growth as compared to controls. However, feeding premature human infants milk acidified with the three isomeric forms of lactic acid resulted in weight loss and dehydration with D- and DL-forms, but not with L(+) form. Thus, the L(+) isomer is recommended for use in foods for premature infants. Lactic acid has been approved as a GRAS substance for miscellaneous and general purpose uses (CFR 182.1061, Title 21). Salts of lactic acid (Na-, K-, and Ca- salts) have also been approved for use in foods. It is used in the manufacture of jam, jelly, beverages, confectionery products, and to adjust brine pH for pickles and olives (Table 4). The acid and salts are used to firm apple slices, to prevent discoloration of fruits, and in baking powder, salad dressing, mayonnaise, and processed meat products. The acid is also used with water to spray beef carcasses. In fermented foods, lactic acid bacteria can produce as high as 1.0 to 1.5% lactic acid.

C. Commercial Availability

1. Acetic Acid and Salts[33,35,36]

Acids and salts thereof for use in food are available commercially. Glacial acetic acid is not used in food. The acid is used in food in the form of vinegar that is produced through the conversion of ethanol to acetic acid by *Acetibacter aceti* and related species. Different forms of vinegar that differ in flavor and acid contents are available commercially. The acid content of vinegar can vary from 5 to 42%. In addition to acetic acid, vinegar contains as many as 50 other components; many are present in small amounts and give the characteristic aroma of the product.

2. Propionic Acid and Salts[33,37,38]

Food-grade propionic acid and its salts are available commercially from chemical suppliers.

A commercial product called Microgard™ for use in food as a preservative is produced by Wesman Food Inc., Beaverton, Oregon. Skim milk is fermented with a dairy strain of *Propionibacterium* and then heat treated (pasteurized) to kill the cells. It is available as an off-white liquid with a pH of 5.7 to 6.1 and is recommended for use at 1 to 2% level in dairy and nondairy products to increase shelf life by inhibiting Gram-negative pathogens and food-spoilage bacteria.[38a,38b]

3. Lactic Acid and Salt[33,39,40]

L(+), DL-lactic acid, and their salts are available commercially for use in foods from several different sources. Na- and K-salt of L(+) lactic acid at 60% is available from Purac, America, Inc., Lincolnshire, IL.[39] They are very hygroscopic with mild saline taste and strong humecant properties. Lactic acid may be produced by fermentation of sucrose by a *Lactobacillus* species. These bacteria are recommended for addition at 2% to enhance flavor and

TABLE 5
Inhibitory Concentrations of Three Weak Acids Against Microorganisms and t Percentage of Undissociated Molecules Present When Used at 1% Level[7,17,43]

Acids	%Undissociated acid required for inhibition of: Gm+[a] bact.	Gm–[b] bact.	yeasts	molds	Fraction undissociated acid present at 1% acid level at pH: 4	5	6
Acetic	0.1	0.05	0.5	0.1	0.84	0.35	0.05
Propionic	0.1	0.05	0.2	0.05	0.87	0.41	0.06
Lactic	>0.03[c]	>0.01[c]	>0.01[c]	>0.01[c]	0.39	0.06	<0.01

[a] Gm+, Gram-positive bacteria.
[b] Gm–, Gram-negative bacteria.
[c] Exact data not available.

stabilize microbial flora in meat products. They are permitted for use by the FDA.[40] DL-sodium lactate at 60% concentration is available from Pfanstiehl, Waukegan, IL.[41]

V. ANTIMICROBIAL EFFECTIVENESS

From experimental data and mathematical models, the antimicrobial effects of weak acids have been shown to be produced through the combined effects of the undissociated and dissociated molecules. Growth inhibition results primarily from the reduction of intercellular pH, both by the undissociated and dissociated acid forms, and relatively less from the inhibition of transport functions by the undissociated molecules. In general, undissociated molecules exert a more inhibitory effect. Concentration, pK, pH of the environment, and microbial types also play important roles in the antimicrobial efficiency of weak acids.

The effectiveness of acetic, propionic, and lactic acids and their salts that are approved for use in food against different spoilage and pathogenic microorganisms will be discussed in this section. The average effective concentrations of the undissociated molecule of the three weak acids necessary for the growth of Gram-positive and Gram-negative bacteria and yeasts and molds are listed in Table 5. In general, Gram-negative bacteria are most susceptible, and yeasts the least susceptible, to these three acids. If the acids are used in food systems at 1% level, at pH 4.0 all three acids have enough undissociated molecules to inhibit all four groups of microorganisms. At pH 5.0 only propionic acid will be effective against all four groups, and acetic acid will be effective against bacteria and molds but not against yeasts. For lactic acid, the only possible inhibition would be against the Gram-negative bacteria. At pH 6.0, acetic acid will inhibit some Gram-negative bacteria, and propionic acid will inhibit Gram-negative bacteria and molds. Lactic acid will not have any effect. However, if the inhibitory effect of the dissociated molecules and the ability of these acids to produce sublethal stress in microorganisms are con-

TABLE 6
Relative Inhibitory Effectiveness of Acetic Acid Against Different Groups of Microorganisms[12,16,17,43]

Bacterial species	Inhibitory pH	MIC Total acid (%)	Undissociated molecules (%)
Bacillus cereus	4.9	0.04	0.02
Staphylococcus aureus	5.0	0.03	0.01
Listeria monocytogenes	4.2	0.30	0.20
Listeria monocytogenes	4.5	0.07	0.05
Salmonella aertrykefi	4.9	0.04	0.02
Saccharomyces uvarum	4.5	2.40	1.60
Geotrichum condidum	4.5	2.40	1.60
Penicillium glaucum	3.5	1.00	0.90
Aspergillus parasiticus	4.5	1.00	0.70

sidered, some inhibition may be possible at 1% level of lactic acid at pH 6.0 or above.

A. Acetic Acid and Salts

The antimicrobial properties of acetic acid and several acetate salts have been studied in microbiological cultures and in food systems. They have been found to be quite effective against many bacteria, yeasts, and molds, although the effectiveness is much greater for bacteria (Tables 5 and 6). *Acetobacter* spp., and some lactic acid bacteria, are resistant to acetic acid[42] while most spore-forming and pathogenic bacteria are quite sensitive.[17,43] Pure culture studies showed acetic acid and acetate to be inhibitory for *Bacillus* spp., *Staphylococcus aureus*, *Salmonella* spp. *Pseudomonas aeruginosa*, *Clostridium* spp., including *C. botulinum*, *Saccharomyces* and other yeast genera, *Penicillium*, *Aspergillus* and other mold genera.[17,34,43] At a pH close to 5.0, both Gram-positive and Gram-negative bacteria were inhibited at a concentration below 0.05%. To inhibit both yeasts and molds, lower pH (~4.0) and a higher concentration were required. Bacteriostatic and bacteriocidal effects of acetate against two *L. monocytogenes* strains in tryptose broth were dependent upon the acid concentration (pH), incubation temperature, and incubation time. At 0.05% (pH 5.8 to 5.9) concentration, strains grew at temperatures from 7 to 35°C; bacteriostatic effect was evident at 0.2% (pH 4.4 to 4.6) while bacteriocidal effect was apparent at 0.3% (pH 4.2 to 4.3) and higher concentrations. Viability loss was faster at 35°C and slower as the temperature was reduced.[12] Four separate strains of *L. monocytogenes* in TSB acidified with acetic acid to pH

4.8 to 5.0 showed viability loss during incubation for 28 d at 10 to 35°C; greater viability loss was found at higher temperatures.[11] In another study, viability of four *L. monocytogenes* strains in TSB supplemented with 0.6% yeast extract and acidified to pH 4.5 (total acid 0.07%, undissociated acid about 0.05%) was reduced from 104 cfu/ml to <10/ml in less than two weeks at 10 to 30°C. Sublethal injury following exposure to acetic acid has been observed in *E. coli, Salmonella,* and *L. monocytogenes* strains. Exposure of *E. coli* cells in TSB adjusted to pH 4.3 with acetic acid for 30 min at 37°C resulted in greater than 90% injured cells among the survivors.[26] Similarly, *E. coli* cells exposed in buffer at pH 4.6 for 60 min at 32°C containing 0.1 and 0.4% Na acetate allowed 52% and 99% survivors, respectively.[27] High sublethal injury to *Salmonella* survivors in 0.2 N acetic acid and *L. monocytogenes*[29] in 0.3 and 0.5% acetic acid have also been observed.

Minor and Marth[13] observed that viability of *S. aureus* was reduced by 90% and 99% within 12 h in milk acidified with acetic acid to pH 5.2 and 5.0, respectively. Several studies have been conducted to control microbial load in fresh meat and meat products with acetic acid. Dipping pork chops in a 3% acetic acid solution prior to vacuum packaging and storage at 2 to 4°C for 6 weeks produced lower aerobic and psychrotrophic counts and effectively inhibited growth of *Enterobacteriaceae;* addition of Na-ascorbate helped to maintain the color of the meat.[44] Boneless pork loins that were spray washed with 2% acetic acid solution and then vacuum packaged and stored for 4 weeks at 4°C, had lower aerobic, anaerobic, and lactic acid bacterial counts than those either washed with water or with a 200 ppm hypochlorite solution; however, the acid treatment resulted in some discoloration of the meat. It was suggested that if the discoloration could be prevented, acetic acid wash would be useful in extending the storage life of fresh pork.[45] Thermal resistance of spores of *B. steathermophilus* at 121°C and pH 4.6, and *B. coagulans* at 110°C and pH 4.2 in a frankfurter emulsion slurry acidified with acetic acid, was reduced by over 50%.[46] Spraying beef carcass with 1% acetic acid, and storing under vacuum for 28 h at 2°C, resulted in significant reduction in aerobic plate counts, and lactobacilli were predominant among the microflora.[47] In another study, beef cubes inoculated with 5.2×10^6 *Salmonella typhimurium, Shigella sonnei, Yersinia enterocolitica, Escherichia coli, Pseudomonas aeruginosa, or Streptococcus faecalis* were dipped for 10 sec in 1.2% acetic acid and enumerated for the viability loss.[47a] In general, the treatment reduced average numbers of pathogens, except for *E. coli*, by 17 to 34% of that of the controls (Table 7). Kalchayanand[47b] investigated the effect of 1.5 or 3% acetic acid dip or spray at 25 to 55°C on lamb quarters during vacuum storage at 0°C for 16 weeks. Microbiological quality and softness of subcutaneous fat was observed. At 8 weeks total aerobic, psychrotrophic, coliform, and *B. thermosphacta* counts were generally lower in the treated samples than in the control. Treatment with 3% acid produced lower counts of coliforms than treatment with 1.5% acid (Table 8). At 16 weeks of storage, except for coliforms, counts of other groups were the same as the controls; coliforms were much lower in the treated

TABLE 7
Reduction in Viability of Pathogens in Beef Following Acetic Acid Dip

	% Recovery of pathogens of original[a]		
Bacterial species	**Treated**	**Control**	**%Reduction**
Salmonella typhimurium	73.3	42.1	31.2
Shigella sonnei	62.2	33.5	28.7
Yersinia enterocolitica	69.7	43.1	26.6
Escherichia coli	45.8	40.7	5.1
Pseudomonas aeruginosa	77.2	43.6	33.6
Streptococcus faecalis	62.1	44.9	17.2

[a] Beef cubes inoculated with 5.2×10^6 cfu of each pathogen were treated in 1.2% acetic acid solution (treated) or were put out in sterile water (control) for 10 sec, and enumerated for cfu in agar media. The recovery is calculated using the original numbers as 100%. Adapted from Reference 47a.

TABLE 8
Effect of Acetic Acid Treatment on Microbial Counts on the Surface of Vacuum-Packaged Lamb Quarters Stored at 0°C for 8 Weeks

	Colony forming units (cfu)/cm^2 [b]			
Treatment[a]	**APC**	**Psychrotrophic**	**Coliforms**	**Brochothrix**
Dip,1.5% acid, 25°C	4.9	3.9	2.5	1.7
Dip, 3% acid, 25°C	5.1	4.0	0.3	0.7
Dip, 1.5% acid, 55°C	5.8	5.4	2.6	2.2
Dip, 3% acid, 55°C	5.2	4.1	0.9	2.1
Spray,1.5% acid, 25°C	5.5	5.2	2.1	1.0
Spray, 3% acid, 25°C	4.8	3.0	1.9	1.1
Spray,1.5% acid, 55°C	5.8	4.3	1.7	1.3
Spray, 3% acid, 55°C	5.3	3.3	0.9	2.4
Control	6.3	5.5	3.6	2.3

[a] Lamb quarters were either dipped or sprayed (30 psi) for 10 sec with appropriate acid solutions, chilled overnight at 0°C, vacuum packaged and stored at 0°C.
[b] At selected intervals surface core samples from five places were collected in a sterile bag, mixed with 100 ml 0.1% peptone water, homogenized for 1 min with a stomacher, serially diluted and used for enumeration of the bacterial groups. The methods used were: aerobic plate count agar for APC (25°C for 3 d), and psychrotrophs (7°C for 10 d), violet-red bile agar (32°C for 2 d), Streptomycine-thallus-acetate agar for *Brochothrix thermosphacta* (25°C for 4 d). Adapted from Reference 47b.

samples. During storage of up to 8 weeks, the moisture content and melting point of the subcutaneous fat of the treated samples were not much different from the control samples. However, spraying methods resulted in higher content of residual acetic acid in the samples. Results indicated that acetic acid treatment can be used to increase shelf life of vacuum-packaged lamb up to at least 8 weeks. Acetic acid wash has also been found to be effective in increasing shelf life of poultry; however, this treatment caused the skin to be hard and leathery.[48] Addition of 0.1% acetic acid to the scald water drastically reduced the D52 values (time necessary for a one log decrease in bacterial numbers at 52°C) of two *Salmonella* species and one strain of *Campylobacter jejunii*[49] by about 10 to 15%.[49] Addition of 0.5% acetic acid to the scald water reduced the aerobic plate count; *Enterobacteriaceae* and *Salmonellae* were not detected.[50]

Mayonnaise samples acidified with vinegar to pH 5, 4.5, 4, and 3.6 for incubation at 4, 25, and 35°C for 5 d were inoculated with *Salmonella enteritidis* to determine the inhibitory effect of the vinegar. At pH 5, vinegar was bacteriostatic at all temperatures. However, at pH 4.5 it was bacteriostatic at 4 and 20°C, but bacteriocidal at 35°C. At pH 4.0 and 3.6, vinegar was bacteriocidal at all temperatures.[51]

B. Propionic Acid and Salts

The effect of propionic acid and its salts on growth inhibition of bacteria and fungi has been studied both in culture media and food systems. Eklund[52,53] determined the inhibitory concentrations of propionic acid against several species of microorganisms. *Bacillus subtalis* required higher concentrations than *E. coli* for 50% inhibition. This is in contradiction to data presented in Table 9 which indicate that Gram-negative bacteria are more sensitive than Gram-positive bacteria to propionic acid. However, Gram-negative bacteria are more sensitive than Gram-positive bacteria to the undissociated acid, and yeasts are much more resistant than bacteria.[54] Results also indicated that undissociated molecules are much more inhibitory than dissociated molecules. Propionic acid at pH 5.0 and 0.23% (about 0.1% undissociated) was reported to inhibit growth, but did not cause viability loss of *L. monocytogenes* during 12 weeks of storage at 10°C in TSB supplemented with 0.6% yeast extract.[16] Chung and Goepfert[54] observed that growth of three species of *Salmonella* was inhibited by propionic acid of pH 5.5. Moon[45] reported that 1% propionate in a broth containing glucose at pH 4.5 inhibited growth of acid-tolerant yeasts, such as *Saccharomyces uvarum, Goetrichum candidum, Endomycopsis burtonii,* and *Hansenula canadensis* during 2 weeks of incubation at 32°C. Synergistic growth inhibition against the yeasts by propionic acid with acetic acid and lactic acid was also reported.

Propionic acid and salts have been used for their inhibitory efficiency against molds in cheese, butter, and bakery products. At a concentration of 0.1%, they are not inhibitory to yeasts and molds unless the pH drops below 5.0.[17,55,56] However, at pH 5.0, even lower concentrations (0.1 to 0.2%) of propionate have been found to effectively inhibit growth of rope-forming bacilli in bread.[57–59] Sodium propionate at 0.1 to 5% has also been reported to inhibit growth of *S. aureus, Sarcina lutea, Proteus vulgaris, Lactobacillus plantarum,* and yeasts, and to delay spoilage in fresh figs, syrup, apple sauce, cherries, and berries.[34,60]

TABLE 9
Inhibitory Concentration of Propionic Acid to Several Strains of Microorganisms

	mM Propionic acid[a]		
	50% inhibition (pH 6.6)	MIC[b](undiss)	MIC(diss)
Escherichia coli	32 (0.237%)	70	800
Pseudomonas aeruginosa ATCC 9027	14 (0.0104%)	NA	NA
Bacillus subtilis ATCC 6633	10 (0.074%)	24	420
Bacillus subtilis W 23	14 (0.104%)	19	400
Bacillus cereus JMS 103	NA[c]	17	380
Staphylococcus aureus ATCC 25923	NA	19	830
Candida albicans JMS 101	NA	39	680

[a] 100 mM propionic acid = 0.74%
[b] MIC, minimum inhibitory concentrations of undisassociated and disassociated propionic acid molecules (mM).
[c] NA, not available. (Adapted from References 52 and 53.)

Microgard™, a milk-based pasteurized product fermented with a dairy *Propionibacterium* species containing propionic acid, has been reported to be inhibitory against Gram-negative bacteria and yeasts.[37,38] Salih et al.[61] studied the antimicrobial effect of freeze-dried Microgard™ by adding up to 10% in commercial yogurts and evaluating microbial quality during 28 d of storage at 5°C (Table 10). Total viable, Gram-negative bacterial and yeast counts increased in the untreated controls. In the treated samples, Gram-negative counts were <10/g as opposed to >104/g in the control. Treated samples had lower counts for viable cells and yeasts; the inhibitory effect increased as Microgard™ concentrations increased.

C. Lactic Acid and Salts

This acid is probably one of the earliest used in foods. It has been used for acidification, flavor enhancement, and microbial inhibition.[33] Due to its low pK (3.85), in most foods with pH above 5.0 it would be expected to have very little antimicrobial effect.[7] Other studies have shown that at pH 5.0 and above it is effective against many bacteria but not against yeasts or molds. This could be due to the undissociated ions, synergistic effect with other weak acids and food preservatives, and sublethal injury. Exposure of *E. coli* and *L. monocytogenes* strains to broth or buffer containing lactic acid has been

TABLE 10
Influence of Microgard™ Treatment on Microbial Population in Commercial Yogurt During Storage Up to 28 Days at 5°C

Microbial types[a] (cfu/g)	days at 5°C	% Freeze dried Microgard added to yogurt: 0	1	5	10
Viable counts	0	2.3×10^6	2.3×10^6	2.6×10^6	2.3×10^6
	14	3.5×10^9	4.3×10^8	5.2×10^7	3.9×10^8
	28	7.0×10^9	2.6×10^8	2.5×10^8	1.5×10^8
Gram-negative	0	<10	<10	<10	<10
	14	3×10^4	<10	<10	<10
	28	5×10^4	<10	<10	<10
Yeasts	0	$\sim 1 \times 10^5$	$\leq 1 \times 10^5$[b]	$\leq 1 \times 10^5$	$\leq 1 \times 10^5$
	14	$\sim 1 \times 10^7$	$\sim 2.5 \times 10^6$	$\sim 1.5 \times 10^6$	$\sim 1 \times 10^6$
	28	$\sim 4 \times 10^7$	$\sim 9 \times 10^6$	$\sim 8 \times 10^6$	$\sim 4 \times 10^6$

[a] Viable counts were determined with brain-heart infusion agar and Gram-negative bacteria enumerated in crystal violet tetrazolium agar after incubation of plates at 30°C for 48 h. Yeast counts were made with yeast malt extract agar containing 100 μg/ml oxytetracycline and incubating plates at 25°C for 5 d. pH at the product was 4.1 at 0 d, 5.3 at 14 days and 5.8 at 28 d.
[b] Adapted from Reference 61.

reported to produce reversible cell wall and cell membrane damage among surviving cells.[26,29]

In microbiological media, the presence of lactic acid or lactates, depending upon concentration and pH, have been reported to produce both bacteriostatic and bacteriocidal effects. Lactic acid showed bacteriostatic action above pH 4.5, but bacteriocidal action at pH 4.5, for *L. monocytogenes* strains in broth.[16] Ahmed and Marth[12] also reported that lactic acid in tryptose broth was bacteriostatic 0.2%, but bacteriocidal above 0.2%, for two strains of *L. monocytogenes* at temperatures from 7 to 35°C; however, the authors did not indicate the corresponding pH of the broth. Sorrells et al.[11] observed that *L. monocytogenes* strains grew in TSB containing lactic acid at pH 4.8 and above, but at pH 4.6, cells did not grow between 10 to 35°C during 4 weeks of storage. At this pH, the cells also lost viability during storage, with maximum loss at 35°C ($\log_{10}{}^{6.3}$) and minimum at 10°C ($\leq \log_{10}{}^{1}$)[11]. Motlagh[63] observed that addition of 3% Na lactate in TSB (pH 6.5) had bacteriostatic effects for up to 2 weeks for *L. monocytogenes* and *Y. enterocolitica* strains, and for up to 3 weeks for *E. coli* (Table 11). The bacteriostatic effectiveness of different concentrations of Na-lactate was studied by measuring the time required for a 90% increase in growth (from initial 10^3/ml to 10^4/ml) in a bacteriological medium during incubation at 25°C. All the Gram-positive bacteria tested, except *Leuconostoc* and *Enterococcus* spp., had extended generation times at 2.5% and higher lactate concentrations (Table 12). Among the three Gram-

TABLE 11
Bacteriostatic Effect of 3% Na-Lactate in Tryptic Soy Broth (pH 6.5) Against Pathogens During Storage at 10°C

	log$_{10}$ cfu/ml of							
Storage time (weeks)	LM[a] (Scott A)		LM (Ohio$_2$)		EC[b] (0157:H7)		YE[c] (Y7 P)	
	Ctrl[d]	Trted[e]	Ctrl	Trted	Ctrl	Trted	Ctrl	Trted
0	4.2	4.2	4.5	4.5	4.5	4.5	4.5	4.5
2	5.7	4.4	5.4	4.6	5.0	4.4	5.0	4.5
4	7.1	5.0	6.6	4.8	5.9	4.5	8.9	6.9
6	8.2	6.1	8.5	6.1	6.7	5.0	9.1	7.2

Note: Enumeration of colony forming units (cfu) by pour plating in tryptic soy agar.

[a] LM, *Listeria monocytogenes* strain.
[b] EC, *Escherichia coli.*
[c] YE, *Yersinia enterocolitica.*
[d] Ctrl, control, untreated control samples.
[e] Trtd, treated, Na-lactate treated samples.

negative bacteria, *Escherichia* and *Proteus* strains, but not *Salmonella*, showed some inhibition. Surprisingly, both yeast species showed inhibition at 2.5% lactate concentration. At 5% both Na- and K-lactate also reduced the growth rate of *L. monocytogenes* strains in TSB.[64]

Lactic acid and their salts (mainly Na- and K-lactate) have been used in food systems to study their antimicrobial effect. Woolthuis and Smulders[65] reported that a 1 min exposure of a 1.25% L-lactic acid spray on veal carcasses reduced APC by 0.8 to 1.3 log$_{10}$cfu/cm^2 and reduced *Enterobacteriaceae* below detection level. Also, the acid spray did not produce unacceptable discoloration at 1.25% level (Table 13). Treatment of porcine liver by dipping in 0.2% lactic acid for 5 min reduced the APC, *Enterobacteriaceae,* and *Lactobacillus* load considerably.[66] When the samples were vacuum-packaged and stored at 3°C, the level of these three groups were 1 to 3 magnitudes lower in the treated as compared to the control samples (Table 13). Addition of 2% Na-lactate (pH 6.5) in vacuum-packaged fresh beef was also found to keep the bacterial population lower during storage for up to 12 weeks at 2 to 3°C (Table 13). The meat samples had a slightly acid flavor and some discoloration. For the use of lactate in fresh meat, these flavor and color aspects need to be considered.[67] Incorporation of 4% Na– or K-lactate in sterile comminuted beef and chicken meat inoculated with *L. monocytogenes* cells extended the lag phase and reduced the growth rate during storage at 5°C; the inhibitory effect was more pronounced in beef than in chicken. At 20°C and above, this inhibition was not observed.[64] The use of Na-lactate at 2% in liver pate increased the shelf life at 6°C. The control samples had three magnitudes

TABLE 12
Effectiveness of Na-Lactate in Increasing Generation Time of Microorgansims

Microorganisms	Hours at 25°C for 1 log increase in numbers at Na-lactate concentrations of (%)[a]			
	0	**1**	**2.5**	**4.8**
Micrococcus spp.	26	25	53	>68
Lactobacillus (homo)	37	45	45	>96
Lactobacillus (hetero)	39	44	50	115
Leuconostoc ssp.	24	26	34	27
Brochothrix thermosphacta	103	142	140	149
Enterococcus faecalis	13	15	18	18
Staphylococcus aureus	31	32	49	55
Bacillus cereus	27	53	65	>116
Escherichia coli	30	49	31	93
Proteus morganii	48	34	36	55
Salmonella typhimurium	17	21	22	32
Saccharomyces cerevisiae	25	28	51	69
Saccharomyces uvarum	130	165	125	151

[a] Growth was measured from colony forming units in broth containing lactate at pH 7.0. (Provided by Dr. E. Bontenbal, Purac, America, Inc., Lincolnshire, IL.)

greater aerobic, anaerobic, and lactic acid bacterial counts than the controls at the end of 6 weeks of storage (Table 13); at 1% the inhibition was insignificant.[68] Addition of Na-lactate in frankfurter formulations produced bacteriocidal effects during storage at 4.5°C for over 36 d; during this time APC increased in samples with 2% lactate but not with 4% lactate.[69] At 21° C, *L. monocytogenes* initially showed viability loss, but after 6 d the surviving cells grew rapidly in products even with 4% lactate.[69] Mass et al.[41] studied the inhibitory effect of Na-lactate on toxin production by *C. botulinum* type A and type B spores in comminuted turkey heated to 71.1°C. They observed that even at 3.5%, lactate did not completely prevent toxin production but delayed it by about 7 to 8 d at 27°C; in the untreated controls, toxin was detected in 3 days. In a similar type of study, Anders et al.[70] claimed that the growth of *C. botulinum* in fish and poultry-type products could be delayed by adding lactate (Na, K, Ca, NH_4) at 1 to 7% after cooking the product below sterilization, vacuum packaging it and storing it at 26.5°C or a lower temperature. Lactic acid has

TABLE 13
Reduction in Bacterial Level in Lactic Acid or Lactate Treated Foods

Sample	Concentration & method used	ST[b]	Bacterial type[c]	Log $_{10}$ cfu/g[a] Control	Treated	Ref.
Veal	1.2%, 1 min		APC	3.5	2.2	65
carcas	spray	1 d	Ent	ND	1.5	65
Porcine	0.2%, 5 min dip	5 d	APC	5.2	2.4	66
liver			Ent	2.2	1.5	66
			LAB	4.1	2.8	66
Fresh	2%, Na-lactate	0 w	APC	4.2	1.1	67
beef			Psy	3.7	1.4	67
			LAB	3.3	0.7	67
		8 w	APC	6.6	4.4	67
			Psy	7.3	4.4	67
			LAB	6.4	3.9	67
		12 w	APC	7.2	5.2	67
			Psy	6.8	5.7	67
			LAB	7.0	5.2	67
Liver	2%, Na-lactate	0 w	APC	3.7	3.6	68
pate			An	3.6	3.4	68
			LAB	<10	2.8	68
		4 w	APC	3.6	2.7	68
			An	3.3	2.5	68
			LAB	3.2	2.3	68
		6 w	APC	6.7	2.7	68
			An	6.6	3.0	68
			LAB	6.8	3.2	68
Frankfurter	2%, Na-lactate	36 d	LM	3.0	2.7	69
	4%, Na-lactate	60 d	LM	3.0	1.7	69
	2%, Na-lactate	35 d	APC	2.7	5.0	69
	4%, Na-lactate	35 d	APC	2.7	2.0	69

[a] Control and treated.
[b] Enumerated after storage for days (d) or weeks (w).
[c] APC, aerobic plate counts; Ent, enterobacteriaceae; LAB, lactic acid bacteria; Psy, psychrotrophic counts; An, anaerobic counts; LM, *L. monocytogenes*

also been reported to reduce thermal inactivation time of *B. steathermophilus* and *B. coagulans* spores in frankfurter emulsion slurry; however, the efficiency is dependent upon the pH of the slurry.[46]

VI. CONCLUDING REMARKS

Among the weak organic acids produced by starter culture bacteria, lactic and propionic acid can be present at 1 to 1.5% in fermented products or mature starter cultures. Acetic acid can be produced by *Bifidobacterium* spp. in considerable amounts; however, acetic acid is also available as vinegar. At one time the antimicrobial effectiveness of these acids was considered only to be due to undissociated molecules. Thus, lactic acid with a lower pK was thought to be of less value as an antimicrobial agent than acetic or propionic acids in food systems with pH 5.0 and above. Recent studies, however, have shown that the combined effects of dissociated and undissociated molecules are responsible for the total antimicrobial effects of weak acids. These molecules cause reductions in cytoplasmic pH, transmembrane potential, and proton gradient. These changes adversely affect the ability of microbial cells to multiply.

Despite these findings, there are still unanswered questions about some other changes brought about by the weak acids in microbial cells. For example, why do cells become sublethally injured through reversible injuries sustained in the cell wall and cell membrane? How can similar sublethal injuries be produced by a strong, as well as a weak, acid? What is (are) the common characteristics of these two groups of acids that can cause similar effects in treated cells? Different studies have revealed that sublethal injury caused by other stresses bring about changes in the conformation of structural and functional macromolecules of biological systems, and viability loss (inability to multiply) is an extension of sublethal injury. How do sublethally injured cells, when exposed to a selective environment, undergo irreversible injury and suffer viability loss? Although these weak acids at a higher pH may not consist of many undissociated molecules, and therefore cannot bring about immediate bacteriocidal effects, at higher concentrations they can cause injury. Even if the cells are injured (such as in minimum heat treatment, freezing, and thawing) these acids can, by their selective nature, cause loss in viability of the injured cells over a long period. This combination can be used in some foods.

Other aspects to consider are why a weak acid is bacteriostatic and bacteriocidal at different concentrations, and why some strains are capable of growing rapidly in the presence of an acid after an initial lag. This is probably because the sensitivity and resistance of different microbial species manifested through multiple physiological factors, and not controlled by one or two genetic determinants. However, if genetic determinants for resistance/sensitivity to acids are found, they can provide valuable information.

Despite these unanswered questions, it is clear that the acids produced by starter culture bacteria can be incorporated in suitable foods as secondary barriers to control microbial spoilage and improve food safety.

REFERENCES

1. **Booth, J. R.,** Regulation of cytoplasmic pH in bacteria, *Microb. Rev.,* 49, 359, 1985.
2. **Konings, W. N. and Otto, R.,** Energy transduction and solute transport in *Streptococci, Antonie Van Leeuwenhoek,* 49, 247, 1983.
3. **Gilliland, S. E.,** Ed., *Bacterial Starter Cultures for Foods,* CRC Press, Boca Raton, FL, 1985.
4. **Kandler, O.,** Carbohydrate metabolism in lactic acid bacteria, *Antonie Van Leeuwenhoek,* 49, 209, 1983.
5. **Sneath, P. H. A.,** Ed., *Bergey's Manual of Systemic Bcteriology, Sections 12,* 14, 15, Vol. 2, Williams & Wilkins, Baltimore, 1986.
6. **Hettinga, D. H. and Reinbold, G. W.,** The propionic acid bacteria — a review, part II — metabolism, *J. Milk Food Technol.,* 35, 358, 1972.
7. **Baird-Parker, A. C.,** Organic acids, in *Microb. Ecology of Foods,* Vol. 1, Factors Affecting Life and Death of Microorganisms, Silliker, J. H. et al., Eds., Academic Press, London, 1980, chap. 7.
8. **Ingram, M., Ottoway, F. J. H., and Coppock, J. B. M.,** The preservative action of acid substances in food, *Chem. Ind. (London),* 42, 1154, 1956.
9. **Levine, A. S. and Feller, C. R.,** Action of acetic acid on food spoilage microorganisms, *J. Bacteriol.,* 39, 499, 1940.
10. **Ray, B.,** unpublished data, University of Wyoming, Laramie.
11. **Sorrells, K. M., Enigl, D. C., and Hatfield, J. R.,** The effect of pH, acidulant, time and temperature on the growth and survival of *Listeria monocytogenes, J. Food Prot.,* 52, 571, 1989.
12. **Ahamad, N. and Marth, E. H.,** Behavior of *Listeria monocytogenes* at 7, 13, 21 and 35°C in tryptic soy broth acidified with acetic, citric or lactic acid, *J. Food Prot.,* 52, 688, 1987.
13. **Minor, T. E. and Marth, E. H.,** Growth of *Staphylococcus aureus* in acidified pasteurized milk, *J. Milk Food Technol.,* 33, 516, 1961.
14. **Nunheimer, T .D. and Fabion, F. W.,** Influence of organic acids, sugars, and sodium chloride upon strains of food poisoning staphylococci, *Am. J. Public Health,* 30, 1040, 1940.
15. **Chung, K. C. and Goepfert, J. M.,** Growth of *Salmonella* at low pH, *J. Food Sci.,* 35, 326, 1970.
16. **Connar, D. E., Scott, V. N., and Bernard, D. T.,** Growth, inhibition and survival of *Listeria monocytogenes* as affected by acidic conditions, *J. Food Prot.,* 53, 652, 1990.
17. **Eklund, T.,** Organic acids and esters, in *Mechanisms of Action of Food Preservation Procedures,* Gould, G. W., Ed., Elsevier, London, 1989, chap. 7.
18. **Mitchell, P.,** Performance and conservation of osmotic work by proton-coupled solute porter systems, *J. Bioenerget.,* 4, 63, 1973.
19. **Harold, F. M., Pavlasova, E., and Baarda, J. R.,** A transmembrane pH gradient in *Streptococcus faecalis,* origin and dissipation of proton conductors and N, N-dicyclohexylcarbodiimide, *Biophys. Acta,* 196, 235, 1969.
20. **Maloney, P. C., Kashket, E. R., and Wilson, T. H.,** A proton motive force drives ATP synthesis in bacteria, *Proc. Natl. Acad. Sci. U.S.A.,* 71, 3896, 1974.
21. **Booth, I. R.,** Control of proton permeability: its implications for energy translation and pH homeostasis, in *FEMS Symposium No. 44: Homeostatic Mechanisms in Microorganisms,* Wittenburg, R., Gould, G. W., Banks, J. G., and Board, R. G., Eds., Bath University Press, Bath, England, 1988, chap. 1.
22. **Maloney, P. C.,** Microbes and membrane biology, *FEMS Microbiol. Rev., 87,* 91, 1990.
23. **Booth, I. R. and Kroll, R. G.,** The preservation of foods by low pH, in *Mechanisms of Action of Food Preservation Procedures,* Gould, G. W., Ed., Elsevier, London, 1989, chap. 6.

24. **Brown, M. H. and Booth, I. R.,** Acidulants and low pH, in *Food Preservatives,* Russell, N. J. and Gould, G. W., Eds., Van Nostrand Reinhold, New York, 1990, chap. 3.
25. **Kabara, J. J. and Eklund, T.,** Organic acids and esters, in *Food Preservatives,* Russell, N.J . and Gould, G. W., Eds., Van Nostrand Reinhold, New York, 1990, chap. 4.
26. **Roth, L. A. and Keenan, D.,** Acid injury in *Escherichia coli, Can. J. Microbiol., 17,* 1005, 1971.
27. **Przybylski, K. S. and Witter, L. D.,** Injury and recovery of *Escherichia coli* after sublethal acidification, *Appl. Environ. Microbiol.,* 37, 261, 1979.
28. **Blankenship, L. C.,** Some characteristics of acid injury and recovery of *Salmonella bareilly* in a model system, *J. Food Prot.,* 44, 73, 1981.
29. **Ahmed, N. and Marth, E. H.,** Acid injury of *Listeria monocytogenes, J. Food Prot.,* 53, 26, 1990.
30. **Neal, A. L., Weinstock, J. O., and Lampen, J. O.,** Mechanism of protein synthesis in *Escherichia coli* and *Bacillus subtalis, J. Appl. Bacteriol.,* 54, 237, 1983.
31. **Ray, B.,** Ed., *Injured Index and Pathogenic Bacteria,* CRC Press, Boca Raton, FL, 1989.
32. **Foster, J. W. and Hall, H. K.,** Adaptive acidification tolerance response of *Salmonella typhimurium, J. Bacteriol.,* 172, 771, 1990.

32a. **Foster, J. W.,** *Salmonella* acid shock proteins are required for adaptive acid tolerance response, *J. Bacteriol.,* 173, 6896, 1991.

32b. **Windholz, M., Ed.,** *The Merck Index, 10th ed.,* #44, Merck, Rahway, NJ, 1983, 5172.

33. **Dziezak, J. D.,** Acidulants: ingredients that do more than meet the acid test, *Food Technol.,* 44(1), 76, 1990.
34. **Doores, S.,** Organic acids, in *Antimicrobials in Foods,* Brannen, A. F. and Davidson, P. M., Eds., Marcel Dekker, New York, 1983, chap. 4.
35. **Anon.,** *Technically Speaking About Vinegar, a Food Technologist's Guide to Vinegar,* The Vinegar Institute, Atlanta, 1988.
36. **Ebner, H. and Follman, H.,** Vinegar, in Biotechnology, Vol. 5, Rehm, H. J. and Reed, G. Eds.,Verlag Chemie, Weinheim, West Germany, 1983, chap. 4.
37. **Weber, G. H. and Broich, W. A.,** Shelf life extension of cultured dairy foods, *Cult. Dairy Prod. J., 21(4), 19, 1986.*
38. **Anon.,** *Microgard #2436,* Wesman Food, Beaverton, OR.

38a. **Salih, M. A. and Sandine, W. E.,** Inhibitory effects of Microgard™ on yogurt and cottage cheese spoilage organisms, *J. Dairy Sci.,* 73, 887–893, 1990.

38b. **Al-Zoreky, N., Ayres, J. W., and Sandine, W. E.,** Antimicrobial activity of Microgard™ against food spoilage and pathogenic microorganisms, *J. Dairy Sci.,* 74, 758–763, 1991.

39. **Duxbury, D. D.,** Natural sodium lactate extend shelf life of whole and ground meats, *Food Process. (Chicago), 44(1), 91, 1988.*
40. **Anon., Pot**assium and sodium lactate: affirmation of GRAS status, *Fed. Reg., 52(85),* 10884, 1987.
41. **Mass, M. R., Glass, K. A., and Doyle, M. P.,** Sodium lactate delays toxin production by *Clostridium botulinum* in cook-in-bag turkey products, *Appl. Environ. Microbiol.,* 55, 2226, 1989.
42. **Dankin, J. C. and Radwell, J. Y.,** Lactobacilli causing spoilage of acetic acid preserves, *J. Appl. Bacteriol.,* 34, 541, 1971.
43. **Moon, N. J.,** Inhibition of the growth of acid tolerant yeasts by acetate, lactate and propionate and their synergistic mixtures, *J. Appl. Bacteriol.,* 55, 453, 1983.
44. **Mendonca, A. F., Moline, R. A., Kraft, A. A., and Walker, H. W.,** Microbiological, chemical and physical changes in fresh, vacuum-packaged pork treated with organic acids and salts, *J. Food Sci.,* 54, 18, 1989.
45. **Cacciarelli, M. A., Springer, W. C., Anderson, M. E., and Naumann, H. D.,** Effects of washing and sanitizing on the bacterial flora on vacuum-packaged pork loins, *J. Food Prot.,* 46, 231, 1983.

46. **Lynch, D. J. and Potter, N. N.,** Effects of organic acids on thermal inactivation of *Bacillus stearothermophilus* and *Bacillus coagulans* spores in frankfurter emulsion slurry, *J. Food Prot.,* 51, 475, 1988.
47. **Hamby, P. L., Savell, J. W., Acuff, G. R., Vanderzant, C., and Cross, H. R.,** Spray-chilling and carcass decontamination systems using lactic and acetic acid, *Meat Sci.,* 21, 1, 1987.
47a. **Bell, M. F., Marshall, R. T., and Anderson, M. E.,** Microbiological and sensory tests of beef treated with acetic acid and formic acids, *J. Food Prot.,* 49, 207, 1986.
47b. **Kalchayanand, N.,** Interaction of Microorganisms and Acetic Acid on the Subcutaneous Fat of Vacuum-Packaged Lamb Quarters, M.S. thesis, University of Missouri, Columbia, 1986.
48. **Mountney, G. J. and O'Malley, J.,** Acids as poultry meat preservatives, *Poultry Sci.,* 44, 582, 1965.
49. **Okrend, A. J., Johnston, R. W., and Moran, A. B.,** Effect of acetic acid on death rates at 52 C of *Salmonella newport, Salmonella typhimurium* and *Campylobacterium jejunii* in poultry scald water, *J. Food. Prot.,* 49, 500, 1986.
50. **Lillard, H. S., Blankenship, L. C., Dickens, J. A., Craven, S. E., and Shackelford, A. D.,** Effect of acetic acid on the microbiological quality of scalded picked and unpicked broiler carcasses, *J. Food Prot.,* 50, 112, 1987.
51. **Perales, I. and Garcia, M. I.,** The influence of pH and temperature on the behavior of *Salmonella enteritidis* phage type 4 in home-made mayonnaise, *Lett. Appl. Microbiol.,* 10, 19, 1990.
52. **Eklund, T.,** Inhibition of growth and uptake processes in bacteria by some chemical food preservatives, *J. Appl. Bacteriol.,* 48, 423, 1990.
53. **Eklund, T.,** Inhibition of microbial growth at different pH levels by benzoic and propionic acids and esters of p-hydroxybenzoic acid, *Int. J. Food Microbiol., 2,* 159, 1985.
54. **Salmond, C. V., Kroll, R. G. and Booth, I. R.,** The effect of food preservatives on pH homeostasis in *Escherichia coli, J. Gen. Microbiol.,* 130, 2845, 1984.
55. **Haseltine, W. W.,** Sodium propionate and its derivatives on bacteriostatics and fungistatics, *J. Pharm. Pharmacol.,* 4, 577, 1952.
56. **Olson, J. C. and Macy, H.,** Propionic acid and its calcium and sodium salts as inhibitors of mold growth, *J. Dairy Sci.,* 23, 509, 1945.
57. **Chichester, D. F. and Tonner, F. W., Jr.,** Antimicrobial food additives, in *Handbook of Food Additives,* Vol. 1, 2nd ed., Furia, T. E., Ed., CRC Press, Cleveland, OH, 1972, 115.
58. **Woolford, M. K.,** Microbiological screening of the straight chain fatty acids (C_1–C_{12}) as potential silage additives, *J. Sci. Food Agric.,* 28, 219, 1975.
59. **O'Leary, D.K. and Kralovec, R. D.,** Development of *Bacillus mesentericus* in bread and control with calcium acid phosphate or calcium propionate, *Cereal Chem.,* 18, 730, 1941.
60. **Wolford, E. R. and Anderson, A. A.,** Propionates control microbial growth in fruits and vegetables, *Food Ind.,* 17, 622, 1945.
61. **Salih, M. A., Sandine, W. E., and Ayres, J. W.,** Inhibitory effects of Microgard on yogurt and cottage cheese spoilage organisms, *J. Dairy Sci.,* 73, 887, 1990.
62. **Rice, A. C. and Pederson, C. S.,** Factors influencing growth of *Bacillus coagulans* in canned tomato juice. II. Acidic constituents of tomato juice and specific organic acids, *Food Res.,* 19, 124, 1954.
63. **Motlagh, A. M.,** Antimicrobial Efficiency of Pediocin AcH and Genetic studies of the Plasmid Encoded for Pediocin AcH, Ph.D. thesis, University of Wyoming, Laramie, 1991.
64. **Shelef, L. A. and Yang, Q.,** Growth suppression of *Listeria monocytogenes* by lactates in broth, chicken and beef, *J. Food Prot.,* 54, 283, 1991.
65. **Woolthuis, C. H. J. and Smulders, F. J. M.,** Microb. decontamination of calf carcasses by lactic acid spray, *J. Food Prot.,* 48, 832, 1985.

66. **Woolthuis, C. H. J., Mossel, D. A. A., Van Logtestijn, J. G., De Kruijf, J. M., and Smulders, F. J. M.,** Microb. decontamination of porcine liver with lactic acid and hot water, J. *Food Prot.,* 47, 220, 1984.
67. **Rozbeh, M.,** Extension of Shelf life in Vacuum-packaged Fresh Beef, M.S. thesis, University of Wyoming, Laramie, 1990.
68. **Debevere, J. M.,** The effect of Na-lactate on the shelf life of vacuum-packaged coarse liver pate, *Fleischwirtsch,* 69, 223, 1989.
69. **Bacus, J. and Bontenbal, E.,** Controlling *Listeria, Meat Poultry, 37(6)*, 64, 1991.
70. **Anders, R. J., Cerveny, J. G., and Milkowaski, A. L.,** Methods for Delaying *Clostridium botulinum* Growth in Fish and Poultry, U.S. Patent # 4, 798, 729, 1989.

Chapter 6

DIACETYL OF LACTIC ACID BACTERIA AS A FOOD BIOPRESERVATIVE

Bibek Ray

TABLE OF CONTENTS

Why do microorganisms produce diacetyl? Microorganisms produce diacetyl (and acetoin) primarily as a means of converting excess pyruvate into nontoxic neutral compounds. They (diacetyl and acetoin) are of little use to the microorganisms that produce them.[1] Diacetyl has been shown to be unique as an antimicrobial compound in that it is more effective against Gram-negative bacteria, yeasts and molds than against Gram-positive bacteria. As such it possesses excellent potential as an antimicrobial in the food industry.[2]

I. INTRODUCTION

Diacetyl (biacetyl; 2,3 butanedione; dimethyl diketone; 2, 3 diketobutane; $C_2H_6O_2$; molecular weight 86.1) is produced by some species and strains of the genera *Lactococcus, Leuconostoc, Streptococcus, Pediococcus,* and *Lactobacillus.* It is associated with the desirable characteristic flavor of many fermented dairy products, especially butter, buttermilk, and cottage cheese.[3] It is also produced by many other microorganisms, and is found in milk, bay oil, and other oils. It can also be made by converting methyl ethyl ketone, first to isonitroso compound, and finally hydrolyzing it with HCl. It is available commercially in different concentrations as a starter culture distillate that is a yellowish-green liquid with a strong odor. It is volatile, with a boiling point of 88°C and soluble in oil in about 4 parts of water. LD_{50} of diacetyl by oral administration in rats is 1.58 g/kg body weight.[4] In food, diacetyl is important not only because it is responsible for the desirable flavor in many foods, but also because it has antimicrobial property.[5] However, the antimicrobial property results only at relatively high concentrations; at lower concentrations it is ineffective and can even be metabolized or destroyed by some microorganisms.[6] Destruction of diacetyl at lower concentrations by *Bacillus* spp., *Pseudomonas* spp., *Escherichia coli, Enterobacter* spp., *Proteus* spp., and *Alkaligenes* spp. has also been reported.[7]

II. ANTIMICROBIAL PROPERTIES OF DIACETYL

A. Bacterial Sensitivity

The antimicrobial action of diacetyl was recognized as early as 1927 by Lemoigne,[7] who found only in high concentrations that it inhibited the growth of *Bacillus* spp. Its antimicrobial effects on some bacteria, associated with food spoilage and foodborne diseases, was studied by other earlier researchers that were cited by Jay.[8] These bacteria include *Enterobacter aerogenes, Escherichia coli, Pseudomonas* spp., *Salmonella* spp., and *Staphylococcus aureus.* These and other studies confirmed that, in general, the antimicrobial effect of diacetyl is produced only at high concentrations, and that Gram-negative bacteria and fungi are more susceptible than Gram-positive bacteria, with lactobacilli being the least sensitive.[5–13] Elliker[7] observed that 150 µg/ml of diacetyl produced bacteriostatic effects, while Hedgecock and Jones[5] reported that an antibacterial effect was produced at about 500 to 2,500 µg/ml (0.05 to 0.25%). Pinheiro et al.[9] found that discs dipped in 1 M diacetyl produced 30 mm zones around

TABLE 1
Determination of Bacterial Sensitivity to Diacetyl by Disc Assay Method[a]

Bacterial strain	Diacetyl concentration (disc size, mm)[c]	Inhibition (zone size, mm)[c]	Ref.
Pseudomonas fragi	0.1 M (13)	–	9
	1.0 M (13)	+ (30)	9
Escherichia coli	1000 ppm (13	–	10
	10,000 ppm (13)	Complete	10
Salmonella typhimurium	100 ppm (13)	–	10
	1000 ppm (13)	+ (17)	10
	10,000 ppm (13)	Complete	10
Staphylococcus aureus	1000 ppm (13)	–	10
	10,000 ppm (13)	Complete	10
Listeria monocytogenes[b]	100 ppm (6)	–	15
Listeria spp.[b]	100 ppm (6)	–	15
Salmonella spp.[b]	100 ppm (6)	–	15
Pathogenic E. coli spp.[b]	100 ppm (6)	–	15
Yersinia enterocolitica[b]	100 ppm (6)	–	15
Aeromonas hydrophila	100 ppm (6)	–	15

[a] Assay was done by placing diacetyl-soaked discs over an agar plate seeded with a bacterial strain, incubating the plates, and examining the zone of inhibition around the disc.

[b] *L. monocytogenes*, 9 strains: Scott A, CA, V7, $Camp^+Beta^+$, $Camp^+Beta^-$, $Ohio_2$, 101, 103, and 117. *Listeria* spp: *L. ivanovii* and *L. innocua*. *Salmonella* spp: *S. anatum* and *S. typhimurium*. *E. coli*: 0157: H7 and 018. *Y. enterocolitica*: 117 and Y7P. *A. hydrophila*: ATCC 1414 and 7915.

[c] Except for studies in Reference 15, discs were soaked in diacetyl solutions of different concentrations. For studies in reference 15, 30 μl of a diacetyl solution (equivalent to 100 ppm) was used in each disc.

[d] +, inhibition; –, no inhibition; complete inhibition indicates no growth in the plate of about 10^5 cells.

Pseudomonas fragi (Table 1). Kulshrestha and Marth[14] reported that diacetyl at 1000 ppm or more was required to show inhibition by disc assay against *E. coli*, *Salmonella* spp., and *S. aureus*. More recently, Motlagh et al.[15] tested nine strains of *Listeria monocytogenes*, and one strain each of *Listeria ivanovii* and *Listeria innocua*, and two strains each of *Salmonella* spp., pathogenic *Escherichia coli*, *Yersinia enterocolitica*, and *Aeromonas hydrophila*, with 100 μg diacetyl per disc, but did not observe any inhibitory zone. These differences could be due to the lower concentrations of diacetyl that were used by Motlagh et al.[15]

B. Mode of Antimicrobial Action

The exact mode of antimicrobial action(s) of diacetyl has not been determined, especially in live cells. *In vitro* studies have indicated that diacetyl deactivated enzymes from several microorganisms. Alcohol dehydrogenase and adenylate cyclase from yeasts,[16,17] and glutamate dehydrogenase[18] and transketolase in pentose cycle,[19] have been reported to be deactivated by diacetyl. It is postulated that the α, α-dicarbonyl group in diacetyl molecule (CH_3-CO-CO-CH_3) reacts with the guanido portions of amino acid arganine (H_2N-$C(NH)_2$-$(CH_2)_3$-$CH(NH_2)COOH$) in the microbial enzymes and deactivates the enzymes either by blocking or modifying the catalytic sites.[20] However, it is not clear if this mode of antimicrobial action is also produced in the viable cells. It is not also known if diacetyl can even enter into viable cells.[21]

The antimicrobial effectiveness of diacetyl is dependent upon several factors. It can exist in monomer, dimer, and trimer forms, with trimer being the most effective as an arginine-modifying agent. Also, the antimicrobial activity of diacetyl is maximum at the acid pH level (maximum at pH 5.0). Finally, diacetyl is more effective against Gram-negative bacteria. It is not known if these characteristics are all dependent on the ability of diacetyl to deactivate some enzymes in microbial cells by reacting with arginine or not.

C. Bacteriostatic and Bacteriocidal Efficiency

The effectiveness of diacetyl in inhibiting or reducing the growth of and/or killing of cells (viability loss) of several pathogenic and spoilage bacterial species and strains has been studied in broth cultures and in food systems. Kulshrestha et al.[12–14] inoculated cells of *E. coli, S. typhimurium,* and *S.aureus* strains in a nutrient broth with both low (100 ppm) and high (1,000 ppm) concentrations of diacetyl and incubated the broths at 37°C for 2 h and 24 h (Table 2). While at low levels diacetyl did not reduce the growth rate of treated samples as compared to the untreated controlled samples (3.9 to 19.7%), at high levels it resulted in over 99% viability loss within 2 h. Recently, Motlagh et al.[15] reported that 344 ppm of diacetyl treatment reduced the viability loss of strains of *Listeria monocytogenes,* pathogenic *E. coli, Yersinia enterocolitica, Salmonella* spp., and *Aeromonas Hydrophila* in TSB from 0 to >95% during a 24 h incubation at 4°C (Table 3). Seven of the ten strains studied had over 50% viability loss at this low diacetyl concentration. Jay and his co-researchers[8,21] found that incorporation of diacetyl at 250 ppm to 350 ppm (μg/ml) either in the plating media or in the broth media, at a pH range of five to seven under aerobic or anaerobic growth, reduced both cfu in agar media and OD at 600 nm in broths of Gram-positive and Gram-negative pathogenic and spoilage bacteria, as well as four species of yeasts (Table 4). However, the inhibitory effect of diacetyl was greatly influenced by the composition of the media. Thus, diacetyl incorporated in plate-count agar (PCA) inhibited growth of 10 out of 12 Gram-positive bacteria, 14 out of 14 Gram-negative bacteria, and 4 out of 4 yeasts; however in TSA it inhibited none out of 12 Gram-positive bacteria, only 3 out of 14 Gram-negative bacteria, and none out of 4 yeast strains.

TABLE 2
Growth Inhibition and Viability Loss of Three Bacteria Species and by Diacetyl in Nutrient Broth

Bacterial species	Incubation time(log_{10} cells/ml)[a]	Growth inhibition(%) in 100 ppm diacetyl	Viability loss in 1000 ppm diacetyl[c]	Ref.
Escherichia coli	2 h (3.7)	9.4	54.6	11
	14 h (8.2)	9.4	> 99.0	11
Salmonella	2 h (3.8)	19.7	> 99.0	12
typhimurium	14 h (8.2)	17.3	> 99.0	12
Staphylococcus	2 h (3.6)	3.9	> 99.0	13
ureus	14 h (6.9)	18.4	> 99.0	13

[a] About 10^3 cells per milliter were inoculated in nutrient broth, incubated at 37°C and enumerated for cfu/ml at 2 and 14 h (expressed as log_{10} cells per milliter).

[b] Growth inhibition was compared from log_{10} cfu/ml between control (100%) and treated samples for each incubation time.

[c] Cfu/ml after 2 and 14 h were less than 10.

TABLE 3
Bacteriocidal Effect of Diacetyl of Pathogens

Pathogens (strains)[a]	cfu/ml Untreated (control)	cfu/ml Treated	% Reduction
Listeria monocytogenes (Scott A)	8.2	7.6	5
Listeria monocytogenes (CA)	6.3	6.2	9
Listeria monocytogenes (V7)	8.9	8.6	27
Escherichia coli (018)	8.6	7.6	90
Escherichia coli (0157:H7)	8.6	8.0	54
Yersinia enterocolitica (Y7P)	8.2	6.3	98
Salmonella anatum	7.8	7.1	63
Salmonella typhimurium	7.1	6.1	90
Aeromonas hydrophila (ATCC 5914)	8.1	8.1	0
Aeromonas hydrophila (ATCC 1414)	7.9	6.5	94

[a] Cells for each strain from cultures grown overnight were transferred in duplicate into 10 ml Tryptic Soy Broth (Difco Lab) held at 4°C. Diacetyl was added only to the treated tubes at 344 ppm level. Both tubes were kept at 4°C and enumerated for colony forming units (cfu) by pourplating with Tryptic Soy Agar (Difco Lab) at 1 h and at 24 h. The data for untreated samples were from 1 h enumeration, while data for treated samples were from 24 h enumeration. Results are the average of two observations (Motlagh, A., 1991, unpublished data).

TABLE 4
Growth of Microorganisms in Agar or Broth Media Containing Diacetyl After 48 H at 30°C Under Different Conditions[a]

Test conditions		No. sensitive/no. tested of microbial type[b]			Study conditions
		Gram-positive	Gram-negative	Yeast	
pH effect:	5	3/4	13/13	3/3	Streaking on
	6	11/11	14/14	4/4	PCA containing
	7	12/12	14/14	4/4	350 μg/ml diacetyl
Plating medi:[b,c]	PCA	10/12	14/14	4/4	Streaking on agar
	TSA	0/12	3/14	0/4	plates containing
	NA	3/12	11/14	1/4	350 μg/ml diacetyl
	BHIA	0/12	0/14	0/4	
Additives:	no additives	3/12	11/14	1/4	Additives andd iacetyl
	1% glucose	3/12	11/14	2/4	(250 μg/ml) were
	0.5% gluconic acid	12/12	14/14	4/4	added to nutrient broth
Anaerobiosis:	PCA	1/5	4/4	Not tested	344 μg/ml diacetyl in
	Thioglycolate broth	1/5	1/4	Not tested	agar or broth media

[a] Adapted from Reference 8.
[b] Microorganisms include 12 Gram-positive nonlactic acid bacterial strains (including *Staphylococcus aureus*, *Bacillus* spp., *Micrococcus* spp., *Enterococcus* spp.), 14 Gram-negative strains (including *Pseudomonas* spp., *Salmonella* spp., *E. coli*, *Yersinia enterocolitica*, *Acinetobacter* spp.), 4 yeasts (including *Candida* spp., *Torulopsis* spp.).
[c] PCA, Plate Count Agar; TSA, Tryptic Soy Agar; NA, nutrient agar; BHIA, brain heart infusion agar.

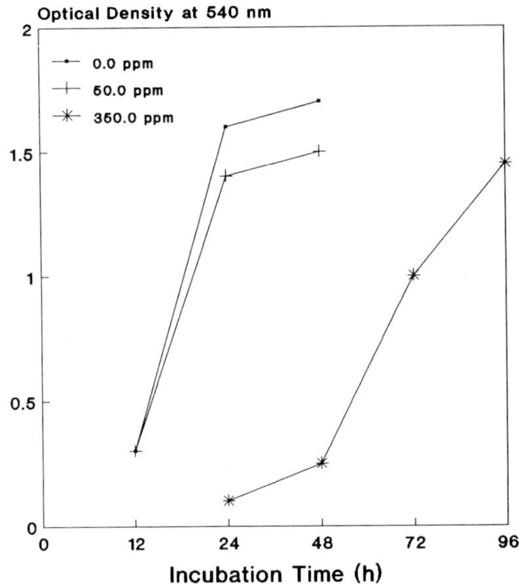

FIGURE 1. Influence of diacetyl on growth of *Staphylococcus aureus* under aerobic conditions. (From Farah, I. O., Food Science Department, University of Minnesota, 1989. With permission.)

In a recent study Farah[22] observed that diacetyl at concentrations of 50 to 250 ppm, inhibited the growth of *S. aureus* strains in NZ-amine broth at 37°C from 12 to 24 h. Although the cells resumed growth later, the final cell concentrations in all treated samples were less than the untreated control samples (Figure 1). The levels of enterotoxin produced by the strain were also less when the cells were grown in a broth containing diacetyl than in the control (Figure 2). The ability of the *S. aureus* cells to resume growth after an initial lag in media with diacetyl was due to the reduction in diacetyl content that volatilizes during incubation in broth at 37°C. The rate of loss from evaporation was much more rapid in shake cultures than in non-shake cultures (Figure 3); as a consequence, cells in the shake culture resumed growth earlier than the cells growing as static cultures.

D. Antibacterial Effect in Combination with Heat

Tatini and his associates[22–24] investigated the bacteriocidal efficiency of diacetyl following the exposure of several pathogens at different temperatures in broth and food systems. In a phosphate-saline buffer (pH 6.0) 1,000 ppm diacetyl reduced the cfu of *S. aureus* at 37°C by about 4.0 logs in 6 h, and by over 7 logs in 9 h; in contrast, at 45°C, a reduction of over 7 logs occurred in 6 h (Table 5).[22] The reduction in cfu of *Salmonella typhimurium* in buffer containing 1,000 ppm diacetyl was over 4 logs at 45°C in 1 h. In the presence of 30% sucrose in heated broth the reduction in cfu in the presence of diacetyl for both *S. aureus* and *S. typhimurium* was less. Heating cell suspensions to

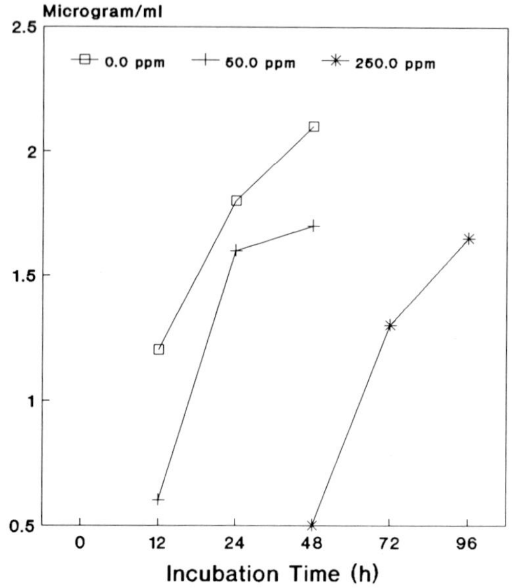

FIGURE 2. Influence of diacetyl on enterotoxin production by *Staphylococcus aureus* under aerobic conditions. (From Farah, I. O., Food Science Department, University of Minnesota, 1989. With permission.)

55°C for 1 h resulted in complete viability loss of 2×10^7 cells per milliliter of *S. aureus;* for *S. typhimurium* 2 h heating at 55°C was needed for complete viability loss of 2×10^7 cells per milliliter. Three pathogenic strains of *L. monocytogenes* heated to 60°C for 30 min showed a viability loss of over 6 logs in a broth containing 30% sugar and 10,000 ppm diacetyl as compared to about 4 logs in a broth without diacetyl. Diacetyl at 1,600 ppm also increased the viability loss of *E. coli, S. typhimurium,* and *S. aureus* in whole milk during heating at 54°C for 15 min (Table 6).[24] However, *S. typhimurium* strains showed a higher resistance to the combination of diacetyl and heating than either *E. coli* or *S. aureus* strains. Heating of spores of *Clostridium sporogenes* PA3679 at 110°C for 15 min in the presence of 1,600 ppm diacetyl also resulted in a greater level of destruction both in sirloin burger and cream-style corn. These results suggested that diacetyl can be used alone or in combination with other antimicrobial treatments to increase the destruction of microorganisms in foods. Diacetyl can also be used to increase the viability loss of microorganisms during subpasteurization heat treatment of foods.

III. DIACETYL PRODUCTION BY STARTER CULTURE BACTERIA

A. Metabolic Pathways in Lactic Acid Bacteria

Diacetyl at present is available primarily as a starter culture distillate. An understanding of the biochemical and genetic control systems in its production

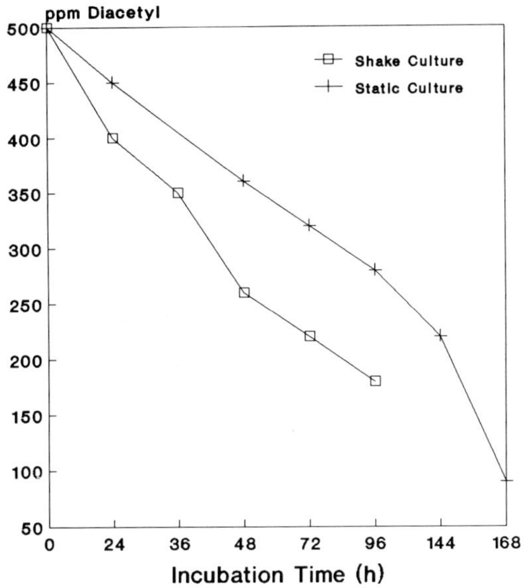

FIGURE 3. The residual levels of diacetyl during incubation under static and aerobic conditions at 37°C. (From Farah, I. O., Food Science Department, University of Minnesota, 1989. With permission.)

in starter culture bacteria will be helpful in increasing its economical production and commercial usage as a food biopreservative.

Many microorganisms, including lactic acid bacteria, can synthesize diacetyl by different metabolic pathways. The key metabolite involved in the synthesis of diacetyl is pyruvate. Normally, lactic acid bacterial strains capable of synthesizing diacetyl produce very little of it while growing actively in broths containing metabolizable carbohydrates such as glucose or lactose. However, as the cells enter into the stationary phase, diacetyl production from metabolizable carbohydrates is observed.[25] The biochemical pathways involving the use of pyruvate by the exponentially growing cells, and cells in the stationary phase, are presented in Figure 4. During active growth a metabolizable hexose such as glucose is metabolized by many lactic acid bacteria to pyruvate, and generates ATP by hexose diphosphate or Embden-Meyerhof-Parnas pathway.[25,26] For each molecule of glucose, the pathway also generates two molecules of NADH from two molecules of NAD. For the reactions to continue, and to supply the growing cells with ATP, the NADH molecules must be converted back to NAD, and this can be accomplished by converting pyruvate to lactic acid by lactate dehydrogenase; the lactate is then released into the environment. As the cells enter into the stationary phase, their need of ATP decreases and conversion of NADH to NAD becomes less critical. Pyruvate, which is highly toxic to cells, starts accumulating. Strains capable of producing, diacetyl can convert these pyruvate molecules to nontoxic diacetyl

TABLE 5
Synergistic Antibacterial Effect of Diacetyl and Heat in Broth

Bacterial species	Suspending media	Treatment	Log_{10} cfu/ml (growth)
Staphylococcus aureus	*Buffer*[a]	6 to 9 h at 37–45°C	9.3 to 9.4
	Buffer[a]+diacetyl[b]	6 h at 37°C	5.6
	Buffer[a]+diacetyl[b]	9 h at 37°C	<2.0
	Buffer[a]+diacetyl[b]	3 h at 45°C	7.0
	Buffer[a]+diacetyl[b]	6 h at 45°C	<2.0
	Sucrose[c]	3 h at 45°C	9.2
	Sucrose[c]+diacetyl[b]	3 h at 45°C	5.7
	Sucrose[c]	1 to 3 h at 55°C	(+)[f]
	Sucrose[c]+diacetyl[d]	1 h at 55°C	(-)[f]
Salmonella typhimurium	Buffer[a]	1 to 3 h at 45°C	9.0 to 9.3
	Buffer[a]+diacetyl[b]	1 h at 45°C	5.9
	Buffer[a]+diacetyl[b]	3 h at 45°C	5.1
	Sucrose[e]	3 h at 45°C	9.8
	Sucrose[e]+diacetyl[b]	3 h at 45°C	7.7
	Sucrose[e]	1 to 6 h at 55°C	(+)f
	Sucrose[c]+diacetyl[d]	1 h at 55 C	(+)[f]
	Sucrose[c]+diacetyl[d]	2 h at 55 C	(-)[f]
Listeria monocytogenes CA	Sucrose[e]	30 min at 60°C	3.40
	Sucrose[e]+diacetyl[b]	30 min at 60°C	1.48
Listeria monocytogenes V7	Sucrose[e]	30 min at 60°C	3.70
	Sucrose[e]+diacetyl[b]	30 min at 60°C	1.84
Listeria monocytogenes Scott A	Sucrose[e]	30 min at 60°C	3.00
	Sucrose[e]+diacetyl[b]	30 min at 60°C	1.60

[a] Na-phosphate, 10 mM + 2% NaCl buffer, pH 6.0.
[b] Diacetyl 1000 ppm.
[c] 30% sucrose in NZ-amine broth.
[d] Diacetyl 10,000 ppm.
[e] 30% sucrose in TS broth + 0.6% yeast extract with initial 6×10^6 cells/ml.
[f] Growth(+) or no growth(–) of initial 2×10^7 cells per milliter during incubation for 14 d at 37°C following heating.

From Farah, I. O., Food Science Department, University of Minnesota, 1980. With permission.

molecules without the need for first converting NADH to NAD. The molecules are then released from the cells into the environment.

Some lactic acid bacteria such as *Streptococcus diacetylactis* (*Lactococcus* lactis subsp. *lactis* var *diacetylactis), Leuconostoc* spp., *Lactobacillus brevis,* and several other species that are capable of metabolizing citrate,[25–32] can produce diacetyl even during the exponential phase. The biochemical steps involving the metabolization of citrate to pyruvate, without the production of NADH, are described in Figure 5. It involves the transportation of citrate from the media to the cell cytoplasm by citrate permease, followed by the conversion

TABLE 6
Enhancement of Bacterial Death by Diacetyl Due to Heat in Food

Organism	Food system (pH)	Heat treatment	Diacetyl conc. (ppm)	Log_{10} cfu/ml
Escherichia coli	Whole milk (6.6)	None	0	1.5×10^7
		54°C, 15 min	0	2.0×10^5
		54°C, 15 min	1600	$< 10^1$
Salmonella typhimurium	Whole milk (6.6)	None	0	4.4×10^7
		54°C, 15 min	0	3.5×10^6
		54°C, 15 min	1600	2.7×10^3
Staphylococcus aureus	Whole milk (6.6)	None	0	1.1×10^7
		54°C, 15 min	0	1.5×10^4
		54°C, 15 min	1600	3.0×10^1
Clostridium sporogenes (PA 3679) spores	Burger (5.6)	110°C, 15 min	0	7.7×10^3
	Burger (5.4)	110°C, 15 min	1600	1.6×10^2
	Cream-style corn (6.8)	110°C, 15 min	0	3.6×10^3
	Cream-style corn (6.0)	110°C, 15 min	1600	1.9×10^3

Adapted from Tatini, S. R., U.S. Patent 4871563, 1989. With permission.

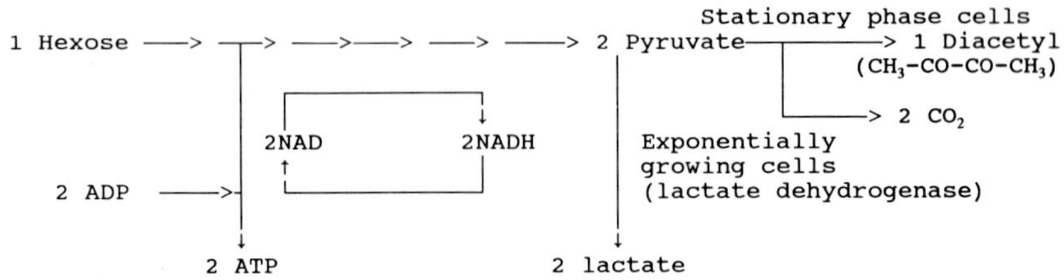

FIGURE 4. Biosynthesis of diacetyl via hexose-diphosphate pathway from metabolizable carbohydrates by cells growing exponentially and cells at stationary phase.

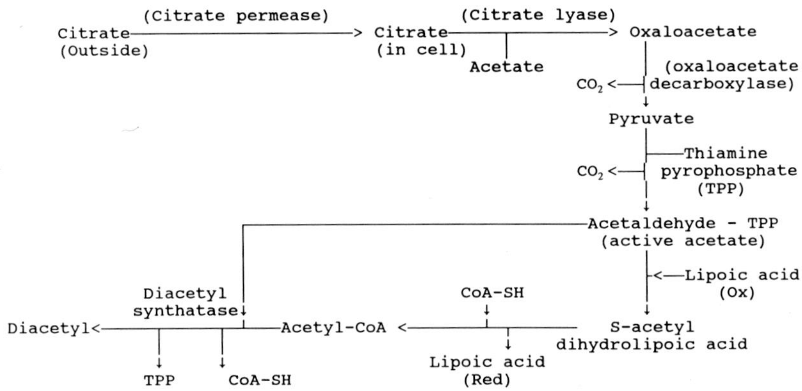

FIGURE 5. Biosynthesis of diacetyl from citrate.

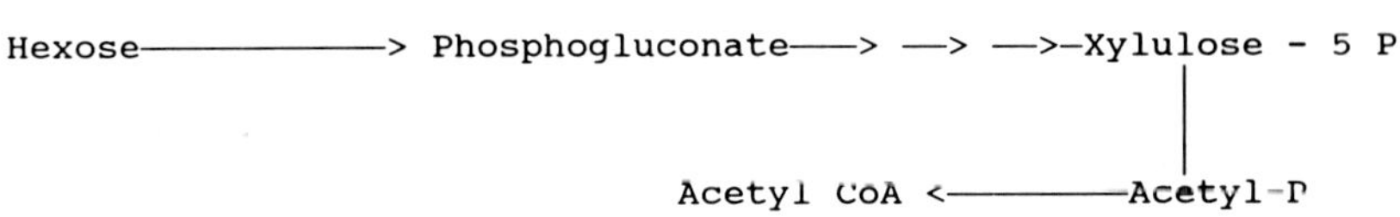

FIGURE 6. Synthesis of Acetyl-CoA via hexose monophosphate shunt by *Leuconostoc* spp.

of citrate by citrate lyase to acetate and oxaloacetate. Oxaloacetate is then converted to pyruvate by oxaloacetate dicarboxylase. As pyruvate production in this pathway does not require NAD-NADH system, the molecules can be channelized to produce diacetyl by converting pyruvate by pyruvate decarboxylase to acetate. Acetate molecules are then converted to active acetaldehyde (acetyl-TPP) and acetyl-CoA, which are then combined together by diacetyl synthatase to form diacetyl. Some *Leuconostoc* strains incapable of converting pyruvate to acetyl-CoA can generate this molecule through the metabolism of hexose via hexose monophosphate shunt which does not involve the production of pyruvate (Figure 6). Acetyl phosphate produced from xylulose 5-phosphate can be converted to acetyl-CoA, which then can combine with acetyl-TPP from pyruvate to produce diacetyl.[27]

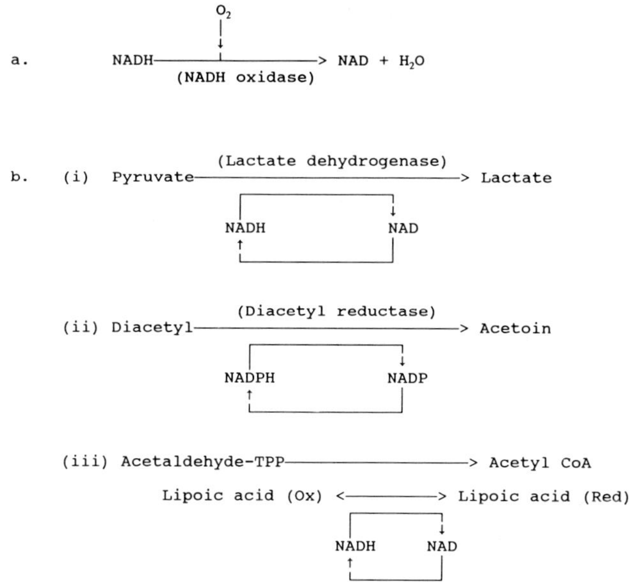

FIGURE 7. (a) Conversion of NADH to NAD due to oxygen stimulated production of NADH oxidase. (b) Regeneration of NAD by NADH oxidase (7a) reduces the reactions involved in the conversion of NADH to NAD (or NADP), and directly or indirectly increases the diacetyl concentration.

B. Factors Influencing Production

Different physiological, genetic, and environmental factors that control the metabolism of citrate and pyruvate can increase or decrease the production of diacetyl by lactic acid bacteria. The activity of citrate permease, which is an inducible enzyme, is pH dependent, being optimum at a pH of 5.4 and having a range of 4.3 to 6.5. Citrate lyase is constitutive in *L. lactis* subsp. *lactis* var *diacetylactis,* but inducible in *Leuconostoc* spp. It requires Mg^{2+} or Mn^{2+} and has an optimum pH of 7.0. Diacetyl synthatase is constitutive in *Leuconostoc* spp., but repressed by citrate in *L. lactis* subsp. *lactis* var *diacetylactis,* and has an optimum pH of 5.5. The pH of the growth media greatly influences the diacetyl production by these species. Normally, until the pH of the media drops to 5.5 very little diacetyl accumulation occurs.[25,27,33]

Aeration of culture media during growth of some bacteria such as *L. lactis* spp. *lactis* var *diacetylactis,* can increase diacetyl production.[25,34] It is postulated that oxygen, by stimulating production of NADH oxidase, increases the conversion of NADH to NAD without going through the steps involving the production of lactic acid from pyruvate (Figure 7a).

The unused pyruvate can then be used to generate diacetyl. Also, generation of NAD by aeration reduces the conversion of diacetyl to acetoin by diacetyl reductase. Finally, through the generation of NAD by NADH-oxidase, formation of acetyl-CoA is stimulated, which can then be used to produce diacetyl (Figure 7b).

Mutants of lactic acid bacteria have been isolated that differ greatly from their wild strains in diacetyl production. Kuila and Ranganathan[35] reported that a wild strain of *L. lactis* subsp. *lactis* var *diacetylactis*, which normally produces both lactic acid and diacetyl in moderate amounts, gave two types of mutants, one that produced 0.38% acid and 138 ppm diacetyl, while the second produced 0.93% lactic but only 42.8 ppm diacetyl. It was postulated that in the former group, high diacetyl production resulted from the reduction in lactic acid formation and concomitant accumulation of pyruvate. The excess pyruvate was converted to diacetyl by the mutants. McKay and Baldwin[36] observed that a mutant strain of *L. lactis* subsp. *lactis* C2 with defective lactate dehydrogenase converted pyruvate to diacetyl (and acetoin) in large amounts when grown in a broth. Mutants deficient in citrate permease have been isolated which could not transport citrate inside the cells, and thus did not use the citrate to produce diacetyl.

C. Genetic Control in Lactic Acid Bacteria

For a long time researchers working with *L. lactis* subsp. *lactis* var *diacetylactis* have observed that the cells would lose the ability to ferment citrate (changing phenotype from Cit^+ to Cit^-), and the mutants were unable to revert.[37–39] Further investigation by Collins and Harvey[38] showed that these mutants retained citrate-lyase activity but at a lower level. Later studies by Kempler and McKay[40] revealed that citrate fermentation ability in *L. lactis* subsp. *lactis* var *diacetylactis* strains is plasmid linked. They reported that the ability to utilize citrate in these strains could be cured by growing the cells in a broth containing acridine orange. The loss in the ability to ferment citrate was associated with the loss of a 5.5 MDal plasmid. They also observed that the mutants that were missing the 5.5 MDal plasmid were not deficient in citrate lyase activity, although the level of activity was lower than in the parent strains. This suggested that the 5.5 MDal plasmid encodes for citrate permease, and loss of this plasmid was associated with the loss in citrate permease activity. The lower activity of citrate lyase in these mutants was due to the repression of this enzyme from the absence of citrate molecules inside the cells due to the absence of citrate permease system. The inability to ferment citrate can also occur in the wild strains due to point mutation(s) in the gene encoding for citrate permease in the 5.5 MDal plasmid. Under this condition the mutants will have low citrate lyase activity, but they will carry the 5.5 MDal plasmid. Also, these mutants, as opposed to the mutants lacking the 5.5 MDal plasmid, can revert from Cit^- to Cit^+ phenotype. This correlation between a plasmid of about 5.5 MDal, and citrate-fermenting ability in these strains, has been reported by other researchers also.[41,42] Examination of the plasmids associated with citrate fermentation in different strains of *L. lactis* subsp. *lactis* var *diacetylactis* by restriction endonuclease digestion revealed that the 5.5 MDal plasmids are identical in all strains.[43]

Plasmid linkage of citrate permease in *L. lactis* subsp. *lactis* var *diacetylactis* provides the possibility of transferring the trait from a Cit^+ strain into a Cit^- strain. In addition, the gene encoding for the Cit^+ phenotype in the plasmid can be amplified to increase the production of diacetyl.[43] This can be achieved by specific mutations either to increase the copy number of the 5.5 MDal plasmid or to increase the copy number of the gene in the same plasmid.

IV. CONCLUDING REMARKS

There are advantages and disadvantages in using diacetyl as a food biopreservative. The advantages are that it has both bacteriocidal and bacteriostatic effects against many pathogenic and spoilage organisms, and it is economical. The disadvantages or limitations are that it has an intense flavor, and thus cannot be used in foods other than a few fermented dairy products. In addition, it is volatile and cannot be used in products that are to be stored even at refrigeration temperature for a long period of time. Finally, some bacteria can utilize it, thus reducing effective concentrations during the storage period.

The synergistic effects of diacetyl and heat treatment to enhance viability loss of pathogenic bacteria and spores has added a new dimension in the use of diacetyl as a food biopreservative. However, data from well-designed studies, either on the effectiveness of diacetyl alone or with heat treatment, on pathogens, spoilage bacteria, and spores, especially in different food systems, is nonexistent or limited. Future studies will be necessary to fill this gap.

REFERENCES

1. **Collin, E. B.,** Biosynthesis of flavor compounds by microorganisms, *J. Dairy Sci.,* 55, 1022, 1972.
2. **Jay, J. M.,** Effect of diacetyl on foodborne microorganisms, *J. Food Sci.,* 47, 1831, 1982.
3. **Vedamuthu, E. R.,** Fermented milks, in *Fermented Foods,* Rose, A. H., Ed., Academic Press, New York, 1982, 203.
4. **Windholz, M.,** Ed., *The Merck Index,* Merck, Rahway, NJ, 1984, 428.
5. **Hedgecock, L. W. and Jones, L. R.,** Antibacterial activity of diacetyl, in *Bacterial Proceedings,* American Society of Microbiology, Washington, D.C., 1950, 127.
6. **Lemoigne, M.,** Sur le metabolisme du diacetyle, *C.R. Soc. Biol.,* 97, 1479, 1927; cited by Jay, J. M., *Appl. Environ. Microbiol.,* 44, 525, 1982.
7. **Elliker, P. R.,** Effect of various bacteria on diacetyl content and flavor of butter, *J. Dairy Sci.,* 28, 93, 1945.
8. **Jay, J. M.,** Antimicrobial properties of diacetyl, *Appl. Environ. Microbiol.,* 44, 525, 1982.
9. **Pinheiro, A. J. R., Liska, B. J., and Parmelec, C. E.,** Inhibitory effect of selected organic chemicals on *Pseudomonas fragi, J. Dairy. Sci.,* 51, 223, 1968.

10. **Kulshrestha, D. C. and Marth, E. M.,** Inhibition of lactic streptococci and some pathogenic bacteria by certain milk-associated volatile compounds as measured by disc assay, *J. Milk Food Technol.*, 33, 305, 1970.
11. **Kulshrestha, D. C. and Marth, E. M.,** Inhibition of bacteria by some volatile and nonvolatile compounds associated with milk I. *Escherichia coli*, *J. Milk Food Technol.*, 37, 539, 1974.
12. **Kulshrestha, D. C. and Marth, E. M.,** Inhibition of bacteria by some volatile and nonvolatile compounds associated with milk II. *Salmonella typhimurium*, *J. Milk Food Technol.*, 37, 539, 1974.
13. **Kulshrestha, D. C. and Marth, E. M.,** Inhibition of bacteria by some volatile and nonvolatile compounds associated with milk III. *Staphylococcus aureus*, *J. Milk Food Technol.*, 37, 545, 1974.
14. **Kulshrestha, D. C. and Marth, E. M.,** Inhibition of lactic streptococci and some pathogenic bacteria by some milk-associated volatile compounds as measured by the disc assay, *J. Milk Food Technol.*, 33, 305, 1970.
15. **Motlagh, A. M., Johnson, M. C., and Ray, B.,** Viability loss of foodborne pathogens by starter culture metabolites, *J. Food Prot.*, 54, 873, 1991.
16. **Lange, L. G., III, Riordan, J. F., and Vallec, B. L.,** Functional arginyl residues as NADH binding sites of alcohol dehydrogenase, *Biochemistry*, 13, 4361, 1974.
17. **Varimo, K. and Londesborough, J.,** Evidence for essential arginine in yeast adenylate cyclase, *FEBS Lett.*, 106, 153, 1979.
18. **Hemmila, I. A. and Mantsala, P. I.,** Inactivation of glutamate dehydrogenase and glutamate synthatase from *Bacillus megaterium* by phenylglyoxal, butane 2,3-dione and pyridoxal 5-phosphate, *Biochem. J.*, 173, 53, 1978.
19. **Kremer, A. B., Egan, R. M., and Sable, H. Z.,** The active site of transketolase, *J. Biol. Chem.*, 255, 2405, 1980.
20. **Yankeelov, J. A. Jr.,** Modification of arginine by diketones, in *Methods in Enzymology*, Vol. 25 (Part B), Hirs, C. H. and Timasheff, S. N., Eds., Academic Press, New York, 1966.
21. **Jay, J. M., Rivers, G. M., and Boisvert, W. E.,** Antimicrobial properties of α-dicarbonyl and related compounds, *J. Food Prot.*, 46, 325, 1983.
22. **Farah, I. O.,** Control of Foodborne Pathogens in Low Water Activity System by Diacetyl, Ph.D. thesis, Food Science Department, University of Minnesota, St. Paul, MN, 1988.
23. **Tatini, S. R. and Daise, R.,** Enhancement of thermal destruction of foodborne bacterial pathogens and anaerobic bacterial spores by starter distillate and diacetyl, in *Proc. Second World Congress on Foodborne Infections and Intoxications*, Berlin, Germany, 1986, 617.
24. **Tatini, S. R.,** Process for Control of Food Pathogens, U.S. Patent, 4871563, 1989.
25. **Collins, E. B.,** Biosynthesis of flavor compounds by microorganisms, *J. Dairy Sci.*, 55, 1022, 1972.
26. **Marshall, V. M. E.,** Flavor development in fermented milks, in *Advances in the Microbiology and Biochemistry of Cheese and Fermented Milk*, Davies, F. L. and Law, B. A., Eds., Elsevier, New York, 1984, chap. 6.
27. **Marshall, V. M. E. and Law, B. A.,** The physiology and growth of dairy lactic acid bacteria, in *Advances in the Microbiology and Biochemistry of Cheese and Fermented Milk*, Davies, F. L. and Law, B. A., Eds., Elsevier, New York, 1984, chap. 3.
28. **Sandine, W. E.,** The *Streptococci:* milk products, in *Bacterial Starter Cultures for Foods*, Gilliland, S. E., Ed., CRC Press, Boca Raton, FL, 1985, chap. 1.
29. **Cogan, T. M.,** The *Leuconostocs:* milk products, in *Bacterial Starter Cultures for Foods*, Gilliland, S. E., Ed., CRC Press, Boca Raton, FL, 1985, chap. 3.
30. **Kempler, G. M. and McKay, L. L.,** Biochemistry and genetics of citrate utilization in *Streptococcus lactis* spp. *diacetylactis*, *J. Dairy Sci.*, 64, 1527, 1981.
31. **Harvey, R. J. and Collins, E. B.,** Role of citrate and acetoin in the metabolism of *Streptococcus diacetylactis*, *J. Bacteriol.*, 86, 1301, 1963.

32. **Speckman, R. A. and Collins, E. B.,** Diacetyl biosynthesis in *Streptococcus diacetylactis* and *Leuconostoc citrovorum, J. Bacteriol.,* 95, 174, 1968.
33. **Harvey, R. J. and Collins, E. B.,** Citrate transport system of *Streptococcus diacetylactis, J. Bacteriol.,* 86, 1301, 1963.
34. **Bruhn, J. C. and Collins, E. B.,** Reduced nicotinamide adenine dinucleotide oxidase of *Streptococcus diacetylactis, J. Dairy Sci.,* 53, 857, 1970.
35. **Kuila, R. K. and Ranganathan, B.,** Ultraviolet light-induced mutants of *Streptococcus lactis* subspecies *diacetylactis* with enhanced acid- or flavor-producing abilities, *J. Dairy Sci.,* 61, 379, 1978.
36. **McKay, L. L. and Baldwin, K. A.,** Altered metabolism in a *Streptococcus lactis* C2 mutant deficient in lactic dehydrogenase, *J. Dairy Sci.,* 57, 181, 1974.
37. **Kneteman, A.,** Enrichment and isolation of *Streptococcus citrophilus* Van Beynum et Pette, *Antonie van Leeuwenhock, J. Microbiol. Serol.,* 18, 275, 1952.
38. **Collins, E. B. and Harvey, R.J .,** Failure in the production of citrate-fermenting revertants by *Streptococcus diactylactis, J. Dairy Sci.,* 45, 32, 1962.
39. **Mostert, J. F.,** The ecology and taxonomy of *Streptococcus diacetylactis,* in *Agricultural Research,* J., Lowes, Ed., *Department of Technical Service,* Pretoria, South Africa, 118.
40. **Kempler, G. M. and McKay, L. L.,** Characterization of plasmid deoxyribonucleic acid in *Streptococcus lactis* spp. *diacetylactis*: evidence for plasmid-linked citrate utilization, *Appl. Environ. Microbiol.,* 37, 316, 1979.
41. **McKay, L. L.,** Functional properties of plasmids in lactic Streptococci, *Antonie van Leeuwenhock,* 49, 259, 1983.
42. **Davies, F. L. and Gasson, M. J.,** Reviews on the progress of dairy science: genetics of lactic acid bacteria, *J. Dairy Research,* 48, 363, 1981.
43. **Kempler, G. M. and McKay, L. L.,** Biochemistry and genetics of citrate utilization, in *Streptococcus lactis* spp. *diacetylactis, J. Dairy Sci.,* 64, 1527, 1981.

Chapter 7

HYDROGEN PEROXIDE, LACTOPEROXIDASE SYSTEMS, AND REUTERIN

Mark A. Daeschel and Michael H. Penner

TABLE OF CONTENTS

I. INTRODUCTION

Microorganisms are in constant competition with each other for nutrients and favorable environments. Mechanisms have evolved through mutation and exchange of genetic material that allow microorganisms to successfully grow and reproduce in changing environments. The diversity of the microbial world is proof of how change ensures survival. Apparent overt antagonism by production of molecules such as bacteriocins or environment modification by acid production are two examples of adaptation. Undoubtedly, microorganisms secrete compounds that influence the populations of microbial communities in ways scientists cannot discern. Many antagonistic products secreted are simply by products of detoxification processes of the producer microorganism, an example being hydrogen peroxide. Other metabolites secreted are in themselves very inhibitory, but seldom accumulate to concentrations that display discernable antagonistic effects, examples being higher alcohols and fatty acids. Carbon dioxide is a primary metabolite that we don't usually associate with antagonism, however, storage of fruits, vegetables, and beverages in a carbon dioxide environment is very effective in controlling the growth of spoilage microorganisms. In fact, it is likely that any microbial metabolite at a certain concentration will be antagonistic to other microorganisms. Applications in food protection mandate that antimicrobials be effective at low concentrations and not significantly impact the identity of the product. These requirements limit the number of microbial metabolites that may have application in food science. In this chapter we will discuss potential and realized food preservation applications of (1) hydrogen peroxide, as it relates to its production by lactic starters, (2) the lactoperoxidase system, and (3) reuterin, an inhibitory metabolite produced by *Lactobacillus reuterii*.

II. HYDROGEN PEROXIDE

A. Properties

Hydrogen peroxide (H_2O_2, molecular weight 34.02) is a colorless liquid; it has a density of 1.46 at 0°C, a boiling point of 152°C, and is miscible with water and insoluble in petroleum ether.[1] Hydrogen peroxide is a strong oxidizing agent and has several applications in the food industry, including use as a bleaching agent, a starch-modifying agent, a preservative, and an antimicrobial agent. The Food and Drug Administration [2] permits its use as a GRAS substance in milk (0.05%), whey (0.04%), starch (0.15%), and corn syrup (0.15%). The Bureau of Alcohol, Tobacco and Firearms allows a limit of 3 ppm in wine processing, however the final product must not contain a detectable residue[1]. The antimicrobial activity of hydrogen peroxide increases dramatically with rises in temperature. Because of this property, hydrogen peroxide has been used as a sterilant for container materials in aseptic packaging systems. Hydrogen peroxide is considered an effective sterilant in aseptic systems at concentrations of 30% if heated to 85 to 90°C for 3 to 5 sec.[3] With a view toward food applications, hydrogen peroxide has the desirable property of having innocuous degradation products ($H_2O + O_2$) and being active at relatively low concentrations, with 25 ppm or less needed to prevent growth of vegetative bacteria.[4] Residual hydrogen peroxide can be removed by the addition of catalase to foods (e.g., 0.1–0.5g/1000 lb milk[5]), or in aseptic packaging it is blown off with hot sterile air. Hydrogen peroxide reacts with organic materials which may limit its usefulness in certain foods. Food components such as ascorbic acid, cysteine, and glutathione have intentionally been considered for use in degrading residual hydrogen peroxide in raw milk[6]. Hydrogen peroxide may impart undesirable characteristics to food such as oxidized flavors, bleaching, and degradation of nutrients.[7]

Hydrogen peroxide is an effective antimicrobial by virtue of its oxidizing potential. Block[4] discussed and summarized the mechanism of action of hydrogen peroxide as the generation of a toxic specie (the hydroxy radical) by reaction of hydrogen peroxide with superoxide ion, the latter formed by reduction of molecular oxygen. The hydroxy radical, being very reactive, will immediately react and damage essential cellular components such as membrane lipids and DNA.

Many fermentative bacteria, including the lactic acid bacteria, produce hydrogen peroxide as a mechanism for protecting themselves against oxygen toxicity. Hydrogen peroxide can be generated by a variety of mechanisms, including nonenzymatic superoxide reduction, and oxidase and dehydrogenase-mediated reactions (reviewed by Fleming et al.[8]). Hydrogen peroxide-generating cells protect themselves from hydrogen peroxide either by enzymatically degrading it with catalase, NADH peroxidase, a manganese pseudocatalase, or by secretion and dilution into the bulk environment to nontoxic levels.

B. Hydrogen Peroxide and Starter Culture Bacteria

Shortly after the initial characterizations of nisin[9] and diplococcin[10] in 1944, a report[11] appeared on an antibiotic-like substance produced by a strain of *Lactobacillus lactis* isolated from Gruyere cheese. The substance, designated lactobacillin, was later concluded[12] to likely be hydrogen peroxide, although its physical presence was not documented. This report was the first observation that certain antagonistic activities by lactic acid bacteria could be due to hydrogen peroxide. Undoubtedly, during the earlier investigations of antagonistic activity of lactic acid bacteria, there were some uncertainties and misinterpretations as to the identities of the inhibitory substances produced by lactobacilli. In retrospect, after reviewing the literature, it appears that some of the uncertainties may stem from the probability that the observed antagonism was the result of several substances (i.e., bacteriocins, acids, and H_2O_2) which were not evaluated individually.

The observation of hydrogen peroxide production and accumulation was most often associated with the lactobacilli in contrast to the other lactic genera. Dahiya and Speck[13] investigated the inhibitory activity in culture filtrates of

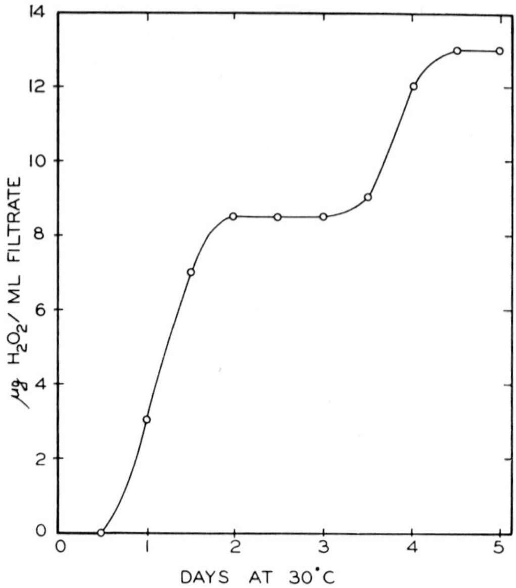

FIGURE 1. Accumulation of H_2O_2 in 1% peptone broth during growth of *Lactobacillus* L-3 at 30°C. (From Price, R. J. and Lee, J. S., *J. Milk Food Technol.*, 33, 13, 1970. With permission.)

Lactobacillus lactis and *Lactobacillus bulgaricus* that were antagonistic to *Staphylococcus aureus*. The antagonism was demonstrated to be due to hydrogen peroxide by the dissipation of inhibitory activity when catalase was added to filtrates, and by direct measurement of hydrogen peroxide produced. Moreover, hydrogen peroxide concentration was maximal (12.5 mg/ml) when producer cells were held in glucose (0.25%) containing buffer at pH 7.0 for 15 d at 5°C. Under their experimental conditions it was reported that 6 mg/ml was bacteriostatic and 20–22 μg/ml was bacteriocidal for *S. aureus*.

Price and Lee[14] identified strains of *Lactobacillus plantarum,* isolated from seafood, that were antagonistic to *Pseudomonas, Bacillus,* and *Proteus* species. The antagonistic component in *L. plantarum* culture filtrates was dialyzable, inactivated by heat, and completely inactivated when exposed to catalase (300 IU/ml). Furthermore, hydrogen peroxide concentrations were measured and found to reach levels of about 8 to 9 μg/ml after 2 d incubation at 30°C in 1% peptone broth (Figure 1). Moreover, between 2 to 4 d there was no increase in hydrogen peroxide levels, but at 5 d a concentration of 13 mg/ml was recorded that slowly decreased to 10 to 12 ug/ml after 10 d incubation. The inhibitory activity of the culture filtrates paralleled measured hydrogen peroxide, providing additional proof that *L. plantarum* produced hydrogen peroxide at levels inhibitory to certain bacteria. It was noted that *Pseudomonas* species were the most sensitive to the hydrogen peroxide containing *L. plantarum* culture filtrates. In addition, the minimum bacteriostatic concentrations for *Pseudomonas* species were determined to be from 2 to 8 mg/ml, which had

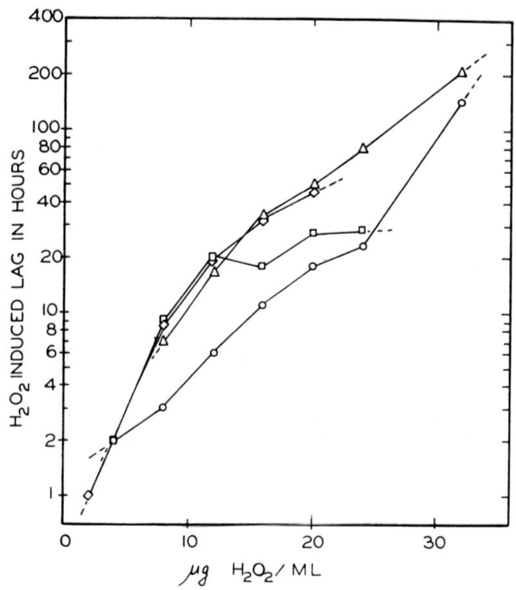

FIGURE 2. Inhibitory effect of hydrogen peroxide on *Pseudomonas* in 1% peptone broth at 30°C. Symbols: O, *Pseudomonas* PI-406; ◊, *Pseudomonas* PII-320; , *Pseudomonas* PIII-332; Δ, *Pseudomonas* PIII-985. (From Price, R. J. and Lee, J. S., *J. Milk Food Technol.*, 33, 13, 1970. With permission.)

the effect of increasing the bacterial growth lag time from 1 to 7 h. At hydrogen peroxide concentrations of 30 μg/ml the lag times were in excess of 200 h (Figure 2) and were predicted to become infinite at higher concentrations. The applied significance of these observations was discussed; the authors concluded that lactobacilli, by virtue of their inhibitory properties, may be beneficial in food products by preventing the growth of spoilage microorganisms such as *Pseudomonas*.

The aforementioned study, and that of Dahiya and Speck, clearly demonstrated that lactobacilli under defined conditions are bacteriostatic, though not necessarily bacteriocidal, to certain microorganisms that contaminate foods. Collins and Aramiki[15] evaluated the ability of 4 strains of *L. acidophilus* to produce hydrogen peroxide under different environments. They observed that one of the strains (C) when held at 37°C in reconstituted skim milk (10%) and continuously shaken, produced in excess of 40 mg/ml at 12 h of incubation (Figure 3), but less then 5 mg/ml when held under static conditions. The same strains were further evaluated for their inhibitory effect against *Pseudomonas fragii* when inoculated into milk (4°C) containing the *Pseudomonas* bacteria. The authors' data indicated that a growth-retarding effect on *P. fragii* was evident (Figure 4) with some strains of *L. acidophilus*, however they also presented evidence that *P. fragii* could degrade a portion of the hydrogen peroxide produced.

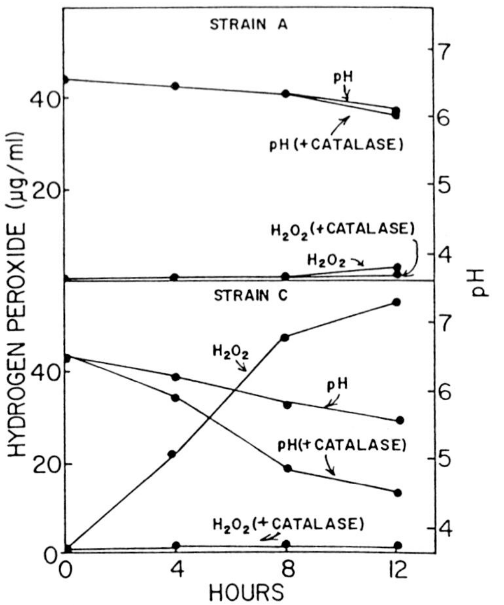

FIGURE 3. Production of hydrogen peroxide by two strains of *Lactobacillus acidophilus* continuously shaken at 37°C with and without the addition of catalase (42 μg/ml). (From Collins, E. B. and Aramak, K., J., *Dairy Sci.*, 63, 353, 1980. With permission.)

Sweet acidophilus milk is a product in which viable *L. acidophilus* strains are added (>1 × 10^6 cfu/ml) to milk that is held under refrigeration conditions until consumed. The aforementioned results of Collins and Aramiki[15] led them to conclude that "Our results indicate that an impractically large number of *L. acidophilus* would be required for the accumulation of enough hydrogen peroxide to be a factor in extending the shelf life of such products" (i.e., Sweet acidophilus milk). Gilliland and Speck[16] observed antagonistic activities by *L. acidophilus* strains when grown in associative cultures with the foodborne pathogens, S. *aureus* and *Salmonella typhimurium,* and with enteropathogenic *Escherichia coli.* To eliminate pH lowering of the culture medium by *L. acidophilus* as a source of inhibition, automatic pH controllers were employed. The authors concluded that acid production and pH changes were not sufficient to explain the observed inhibition of the pathogens. The inclusion of catalase into the associative cultures did reduce the antagonistic activities of *L. acidophilus,* but still greater than 50% of the antagonistic activity remained. They summarized their study with the statement that "The antibacterial action produced by *L. acidophilus* is probably due to a combination of factors including acid, hydrogen peroxide and other inhibitory factors including the antibiotics previously reported" (referring to Hamdan and Mikolajick,[17] Vakil and Shahani,[18] and Vincent et al.[19]).

The formation of hydrogen peroxide by the dairy lactococci has been studied,[20,21] and with some strains,[20] the concentration observed to accumulate

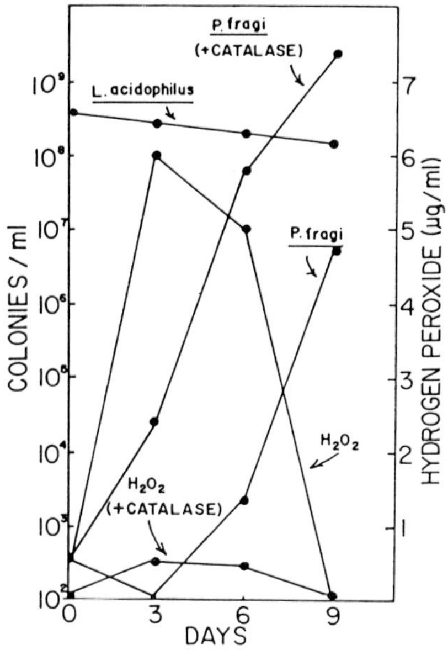

FIGURE 4. Influence of *Lactobacillus acidophilus* (strain C) on the growth of *Pseudomonas fragi* continuously shaken at 4°C with and without the addition of catalase (42 µg/ml). (From Collins, E. B. and Aramak, K., J., *Dairy Sci.*, 63, 353, 1980. With permission.)

was inhibitory to the producer strain. It appears, however, that the lactococci do not accumulate hydrogen peroxide, in concentrations observed with the lactobacilli, perhaps reflecting a greater sensitivity to hydrogen peroxide, or more efficient enzymatic degradation systems.[22] A recent investigation[23] indicated that hydrogen peroxide production in *Lactobacillus leichmanii* could be induced by saturated fatty acids, i.e., palmitic acid, with subsequent accumulations of up to 7 mM (238 mg/ml). The amounts of hydrogen peroxide produced by dairy lactococci have been reported[24] to be quite small (<.70 mg/ml) with strains of *L. lactis* subsp. *diacetylactis* and *L. lactis* subsp. *cremoris* after growth at 25 to 30°C under static conditions in milk. Likewise, the same small amounts of hydrogen peroxide were observed to be produced by *Leuconostoc dextranicum*. Similarly, hydrogen peroxide production and accumulation with *Pediococcus cerevisiae*,[25] a meat starter culture, was very low (<1 mg/ml).

The roles of starter culture bacteria in food preservation in regard to hydrogen peroxide production have been reviewed by Babel[26] and Gilliland.[27] Both reviews described several approaches for using antagonistic starters in dairy products and other foods. Preservation strategies that may be considered are (1) addition of viable starter cells to foods where they can synthesize and secrete antagonistic components, (2) addition of spent microbial media containing antagonistic metabolites to foods, and (3) addition of purified antagonistic metabolites, i.e., nisin and hydrogen peroxide, etc. directly to foods. Accordingly, the efficacy of the first two approaches depends on consistent strain performance, either in the food or in the medium to be added to the food. Furthermore, the antagonistic metabolites, all other metabolites, and the medium used would need to be compatible with the chemical and sensory identity standards of the food product. Some of these constraints also apply when using purified microbial metabolites.

The addition of dairy lactococci to refrigerated milk (3.5 and 7°C.) to inhibit psychrotrophic bacteria was investigated by Juffs and Babel.[28] They observed that mixed cultures were more inhibitory then pure cultures of either *L. lactis* subsp.*diaceytlactis* or *L. lactis* subsp. *cremoris,* in suppressing psychrotrophs. The observed inhibition effect in the milk was concluded to be bacteriostatic rather than bacteriocidal. Furthermore, at least a portion of the inhibitory activity could be ascribed to hydrogen peroxide. Moreover, inhibition was also apparent in the absence of any significant acid production. Gilliland and Speck[29] used cultures of lactobacilli and pediococci to inhibit psychrotrophic bacteria in nonfermented refrigerated foods. They observed with ground beef held at 5°C having an initial psychrotroph count of 6.7×10^2 cfu/ml, that *L. bulgaricus* and *L. lactis*, when added, allowed the psychrotrophs to increase by only one log cycle in 3 d. In contrast, the control sample without lactic culture added permitted the psychrotrophs to increase almost three log cycles. Similarly, *Pediococcus cerevisiae* was also able to retard psychrotroph growth, but not to the same extent as the lactobacilli. In conclusion, these investigators suggested that the major inhibitory effect of the lactobacilli used in their study was likely from hydrogen peroxide production. Another investigation[30] on the inhibitory effects of *L. bulgaricus* on psychrotrophs, this time in sterile refrigerated milk (5°C), showed an average inhibitory effect of 26 to 82% as compared to the control afterstorage for 6 d.

III. LACTOPEROXIDASE SYSTEMS

Peroxidase systems have been known to have antimicrobial properties since the early 1940s when Agner[31] demonstrated the antimicrobial activity of a peroxidase preparation derived from leukocytes. It was later demonstrated, by Wright and Tramer,[32] that the peroxidase system of bovine milk had a similar inhibitory effect on certain classes of bacteria. The peroxidase system of bovine milk had been previously purified by Theorell and Åkesson[33] and given the name lactoperoxidase. Immunochemical techniques were subsequently used to show the presence of lactoperoxidase in bovine saliva.[34] Further studies have demonstrated LP activity in the saliva of different animals and selected human secretions.[35]

The lactoperoxidase system (LPS) is one of several nonspecific antimicrobial systems present in bovine milk.[36] The active LPS includes three primary components: the lactoperoxidase enzyme, thiocyanate (SCN^-), and hydrogen peroxide (H_2O_2). The LPS is antimicrobial due to the production of toxic, short-lived, oxidation products when lactoperoxidase catalyzes the oxidation of thiocyanate by hydrogen peroxide.

Lactoperoxidase is a heme containing glycoprotein with a molecular weight of approximately 78,000.[37] Fractionation studies using conventional chromatography followed by electrophoretic analyses have identified at least 10 fractions of LP in bovine milk. The different fractions are at least partially due to differences in the carbohydrate moieties of the enzyme. Some of the forms may be an artifact of the purification protocol and, hence, the significance of this heterogeneity in vivo is not clear.

The classical peroxidase catalyzed reaction involves hydrogen peroxide as a hydrogen/electron acceptor, and another compound, designated AH2, as the hydrogen/electron donor.

$$H_2O_2 + AH_2 \rightarrow 2H_2O + A \tag{1}$$

The general mechanistic scheme for the above redox reaction (1), is as follows.[38]

$$ROOH + E_{native} \rightarrow \text{compound I - RO} + H_2O \tag{2}$$

$$AH_2 + \text{compound I - RO} \rightarrow \text{compound II} + AH \cdot + ROH \tag{3}$$

$$A'H_2 + \text{compound II} \rightarrow E_{native} + A'H \cdot \tag{4}$$

$$A'H \cdot + AH \cdot \rightarrow A'H_2 + A \tag{5}$$

$$\text{Total:} \quad ROOH + AH_2 \rightarrow H_2O + ROH + A \tag{6}$$

Note that equations (1) and (6) are equivalent when letting R = H, which is the case when hydrogen peroxide serves as the peroxide-oxidizing agent. The native resting enzyme, E_{native}, is the peroxidase hemoprotein with its iron in the formal oxidation state of FEIII. In reaction (2) the peroxide is split in a two-electron oxidation of the enzyme complex to form water and a complex including compound I and -RO. The formal oxidation state of iron in compound I is FeV. Compound I is then reduced by a single electron donation from AH_2 to compound II, with the subsequent release of ROH. The iron of compound II has a formal cxidation state of FeIV. A second single electron transfer then completes the cycle by reducing the enzyme to its resting FEIII state.

The single electron donors likely to participate in the above reaction scheme, AH2, are compounds which give stable radicals such as phenols and ascorbate. However, the primary compound undergoing oxidation in the LPS is thiocyanate, which is not believed to be oxidized by sequential one-electron transfers, but rather by a single two-electron transfer process.[39] Evidence supporting this contention is that no compound II can be detected in defined LPS reaction mixtures, and when compound II is produced by other means it reacts very slowly with SCN^-.[39] Therefore, it has been postulated that compound I directly oxidizes SCN^- by a 2 electron process similar to that observed for the peroxidase-catalyzed oxidation of halides. The initial product(s) of the enzymatically catalyzed oxidation of SCN^- has not been conclusively determined. It has been established that the principle product to build up in active aqueous LPS systems is hypothiocyanite ($OSCN^-$),[40,41] which is in acid-base equilibrium with hypothiocyanous acid (HOSCN), pKa of approximately 5.3.[42] Proposed mechanisms for the production of hypothiocyanite/hypothiocyanous acid include the following.[41]

$$H_2O \ + \ 2SCN^- \ + \ 2H + LPO \ 2H_2O \ + \ (SCN)_2 \qquad (7)$$

$$(SCN)_2 + H_2O \rightarrow HOSNC \ + \ H+ \ + SCN^- \qquad (8)$$

$$HOSCN \rightarrow H+ \ + \ OSCN^- \qquad (9)$$

or

$$H_2O_2 + \ SCN^- \quad LPO \quad OSCN^- \ + \ H_2O \qquad (10)$$

The differences in the postulated mechanisms for peroxidase reactions with thiocyanate and "AH_2"-type compounds provide a rationale for the limiting antimicrobial activity of the LPS at low SCN^- concentrations. The limiting activity would be the result of compound I preferentially reacting with AH_2-type compounds at low SCN^- concentrations, to yield compound II. This is in contrast to compound I directly oxidizing available SCN^-. This results in a build up of compound II due to the relatively slow step of reducing compound II to E_{native}. As mentioned above, compound II does not react at an appreciable rate with SCN^-. Hence, there may be a competition between substrates in which the oxidation of SCN^- leads to toxic products, while the oxidation of AH_2 does not. Oxidation products other than hypothiocyanite have been detected, in lesser amounts, in active LPS preparations.[41] It was suggested that these trace products may be higher oxy acids of SCN^-. The significance of these products in vivo has yet to be determined.

The mechanisms through which the LPS exerts its antimicrobial properties have not been conclusively determined.[43] The primary products of the LPS, hypothiocyanite ($OSCN^-$) and thiocyanegen $(SCN)_2$ have been shown to oxidize protein sulfhydrals to sulfenyl thiocyanates (-S-SCN), disulfides (-S-S) and sulfenic acids (-S-OH).[43] It is conceivable that these oxidations may result in the deactivation of key metabolic enzymes or structural proteins[44] resulting in the death or inhibition of the target microorganism. The proposed reaction schemes for sulfhydral oxidation (below) are consistent with the experimentally observed proportionality between sulfhydral oxidation and H_2O_2 concentration, and the apparent independence of sulfhydral oxidation with respect to thiocyanate concentrations, under certain conditions.[44]

$$OSCN^- + Protein-SH \rightarrow Protein-S-SCN + OH^- \quad (11)$$

$$Protein-S-SCN + RSH \rightarrow Protein-S-S-R + SCN^- + H+ \quad (12)$$

$$Protein-S-SCN + H_2O \Leftrightarrow Protein-S-OH + SCN^- + H+ \quad (13)$$

The proposed "turnover" of SCN^- as depicted above may account for its apparent ability to participate in more than one oxidative cycle. Under these conditions, it may be rationalized that the total moles of oxidized sulfhydral derivatives will be proportional to the H_2O_2 concentration, not thiocyanate. The reversal of LPS microbial inhibition by thiol supplementation supports the suggestion that sulfhydral oxidation is an important factor in the antimicrobial mechanism.[45] In vitro studies coincubating proteins and amino acids with the active LPS have demonstrated the potential for modification of aromatic amino acids.[46] They also observed that hypothiocyanous acid does not readily react with these amino acids, suggesting the formation of more reactive, possibly transient, intermediate oxidation products in the active LPS system. Further evidence for the production of products more toxic than hypothiocyanite in the LPS have come from studies comparing the toxicity of the active LPS in the presence of target microorganisms, versus the LPS products produced separately and then coincubated with the microbial population. In these experiments, exposure of the microorganisms to the active LPS was found to be more inhibitory than exposure to seemingly similar products produced separately.[45,47]

In recent years there has been increasing interest in the practical aspects of using the bovine LPS as a milk preservative. To be utilized in this capacity it is clear that all three of the components of the LPS must be present before its antimicrobial properties are manifested. The concentration of the lactoperoxidase enzyme in raw milk is sufficient for bactericidal activity. In fact, lactoperoxidase is the most abundant enzyme in bovine milk.[48] The concentration of thiocyanate in milk is variable depending on the cow's diet, but levels between 1 and 10 ppm are typical.[49] This level is a little below that suggested for maximum activity of the LPS, and hence the addition of small amounts (~8 ppm) of thiocyanate may be appropriate. However, milk normally contains sufficient thiocyanate to activate the LPS.[50] It should be noted that the concentration of thiocyanate in human gastric juices (~23 ppm) is greater than that which would be found in milk even after any possible supplementation of milk.[49] In contrast to the other two components, the concentration of hydrogen peroxide in bovine milk is assumed to be negligible. Consequently, a source of hydrogen peroxide must be added to the milk to activate the LPS. Only relatively small amounts of hydrogen peroxide (~8 ppm) need to be added for the activation of LPS. This is in contrast to "traditional" methods of hydrogen peroxide preservation which require the addition of up to 800 ppm hydrogen peroxide.

Hydrogen peroxide can be supplied to milk in three main ways: (1) by direct addition of hydrogen peroxide, (2) by the addition of an isolated enzyme system which generates hydrogen peroxide (such as glucose/glucose oxidase), or (3) by inoculation with microorganisms that generate hydrogen peroxide. It appears that the most feasible method of supplying hydrogen peroxide for large scale dairy applications is to inoculate the milk with a safe, hydrogen peroxide-generating bacteria. One precedent for this methodology is the incorporation into raw bovine milk of Raw Milk Inoculum (RMI) currently marketed by Chr. Hansen's Laboratory, Milwaukee, WI (R. Sellars, personal communication).

The lactoperoxidase enzyme of bovine milk is somewhat resistant to heat inactivation. This has led to the speculation that the LPS system may be utilized for milk preservation following pasteurization. The rate of heat inactivation of LP is strongly influenced by the composition of the milk.[37] In whole milk, ~75% of the LP activity is lost after 15 s at 70°C. In contrast, partially purified LP was stable for 15 min at the same temperature. Martinez et al.,[51] reported that approximately 75% of milk's original LP activity is present when assayed 1 h following pasteurization at 63°C for 30 min. The same investigators noted complete recovery of full enzyme activity with additional postpasteurization time beyond 1 h. In their study they further demonstrated the antimicrobial properties of the LPS reactivated by thiocyanate and hydrogen peroxide supplementation following this pasteurization protocol, 63°C/30 min. The significance of this will be geographically limited, since in many locations milk is primarily pasteurized in the continuous HTST process (72–75°C for 15 s) or sterilized in the UHT process (135–140°C for a few seconds).

The LPS has been shown to inhibit or kill specific Gram-negative and Ggram-positive bacteria.[43] Reiter et al. [52] demonstrated that the system was bacteriocidal for some Gram-negative pathogens, including *Salmonella typhimurium* and *Pseudomonas aeruginosa*. The LPS is also inhibitory toward specific psychrotrophic bacteria in milk.[53] This is of particular importance because the heat-stable enzymes produced by these organisms may cause highly undesirable lipolysis and proteolysis in milk and/or subsequent end products. For example, Bjorck[53] found the free fatty acid content in cheese produced from LPS-treated milk was five times lower than that in corresponding controls following four months of maturation. In Bjorck's study it was shown that activation of the LPS has no effect on the physicochemical properties of the milk.

Several recent studies have addressed the effectiveness of utilizing the LPS to inhibit the growth of *Listeria monocytogenes*. The LPS was shown to be bacteriostatic to *Listeria monocytogenes* Scott A cultured in broth, and the net lag periods varied inversely with temperature.[54] It was then reported the LPS exhibits bacteriocidal activity against *Listeria* in raw milk at refrigeration temperatures; the activity was dependent on temperature, incubation time, and strain of *Listeria monocytogenes*.[55] From a different perspective, Kamau et al.[56] have shown the LPS enhances the thermal destruction of *Listeria monocytogenes* and *Staphylococcus aureus* suspended in milk, and they have proposed activation of the LPS, followed by heating, as a method to control milkborne pathogens.

One approach to optimize the application of the LPS is to identify/develop compatible bacteria capable of generating adequate amounts of hydrogen peroxide at target temperatures. Bacteria that would generate optimum amounts of hydrogen peroxide (extracellular concentrations of ~8 ppm) at 2–5°C would be most beneficial due to the common practice of storing milk at these temperatures. Hydrogen peroxide is generated in all known bacterial systems that utilize oxygen as an oxidant. H_2O_2 is the product of several enzymatic reactions, including those catalyzed by flavoprotein oxidases and copper-containing oxidases.[57] Examples of H_2O_2 generating reactions occurring in lactic acid bacteria are as follows (taken from Fleming et al.[8]):

$$O_2^- + 2H + Mn^2 + \xrightarrow{\text{nonenzymatic superoxide reduction}} H_2O_2 + Mn^{2+} \quad (14)$$

$$\text{Pyruvate} + O_2 + PO^3{}_{4-} \xrightarrow{\text{pyruvate oxidase}} \text{acetyl phosphate} + CO_2 + H_2O_2 \quad (15)$$

$$\text{Lactate} + O_2 \xrightarrow{\text{L-lactate oxidase}} \text{pyruvate} + H_2O_2 \quad (16)$$

$$\text{Lactate} + O_2 \xrightarrow{\text{NAD-independent D-lactate dehydrogenase}} \text{pyruvate} + H_2O_2 \quad (17)$$

$$\text{NADH} + H+ \; + O_2 \xrightarrow{\text{NADH oxidase}} \text{NAD} + H_2O_2 \quad (18)$$

Once hydrogen peroxide is produced in cells it is usually eliminated rapidly by enzyme-catalyzed reactions. The major enzyme involved in hydrogen peroxide degradation in aerobic organisms is catalase.[58] Intracellular peroxidases (hemoprotein and nonhemoprotein types) are also capable of eliminating hydrogen peroxide. Streptococci and lactobacilli organisms both contain a nonhemoprotein peroxidase which catalyzes the reduction of hydrogen peroxide to water.[59] Attempts to modify bacterial H_2O_2 production could potentially focus on the activity of either H_2O_2-generating or H_2O_2-degrading enzyme systems. Recent advances in genetic engineering, coupled with an increasing knowledge of hydrogen peroxide metabolism, suggests that the development of improved LPS-activating bacteria may not be unreasonable.

IV. *LACTOBACILLUS REUTERII* AND REUTERIN

An antimicrobial substance termed reuterin, produced by the obligate heterofermentative *Lactobacillus reuterii,* was discovered and characterized through collaborative efforts of the Microbiology Departments of the Swedish University of Agricultural Sciences (Uppsala) and of North Carolina State University (Raleigh, NC). *Lactobacillus reuterii,* previously classified as a biotype of *Lactobacillus fermentum,*[60] is commonly isolated from the gastrointestinal tract of man and animals. Many strains have been characterized[61] as having the ability to attach to epithelial cells, giving further credence to the belief that the preferred environment of this bacterium is the gut.

The unique antimicrobial activity associated with *L. reuterii* was first recognized by Axelsson and co-workers.[62] Strains of *L. reuterii* isolated from pig intestine,[63] and the type strain ATCC 23273, were found to have a distinct inhibitory effect against *Escherichia coli.* A novel screening procedure[62] for antimicrobial activity was developed based on measuring a decrease in B-galactosidase synthesis in *E. coli* cells induced for B-galactosidase activity, and then cocultured with potential antagonistic lactobacilli species. Using this procedure it was discovered that resting cells of certain *L. reuterii* strains could inhibit B-galactosidase activity in *E. coli*.. The inhibitory effect was confirmed by testing the viability of *E. coli* cells when cocultured with *L. reuterii* and observing a decrease in viability from 10^7 cfu to 10^1 cfu after 8 h. A key observation was that the inhibitory effect was present only when either glycerol or glyceraldehyde was added as a carbon source to the coculture system. This study also described the inhibition as being active against a broad spectrum of Gram-positive and Gram-negative bacteria. The inhibitor (reuterin) was characterized[64,65] for composition and synthesis requirements.[66] Initial investigations provided evidence that the inhibitor was not one of the common metabolic end products associated with lactic acid bacteria, such as lactic acid, acetic acid, or hydrogen peroxide. However, it was apparent that the inhibitor was a water-soluble, nonproteinaceous, neutral metabolite directly associated with glycerol metabolism. Subsequently it was determined that the inhibitor was an equilibrium mixture of monomeric, hydrated monomeric, and cyclic dimeric forms of β–hydroxypropionaldeyde (Figure 5). A pathway for reuterin synthesis was proposed as a result of experiments using uniformly labeled [^{14}C] glycerol as the substrate. The pathway is given in Figure 6.[65]

A cell culture system was developed for producing sufficient quantities of reuterin for purification and chemical analysis. Pure reuterin was analyzed[65] by Fourier transform infrared spectroscopy, nuclear magnetic resonance spectroscopy, and mass spectroscopy. Confirmation as to the composition of reuterin was provided by the chemical synthesis of β-hydroxypropinoaldehyde. The synthesized compound exhibited HPLC elution patterns and antimicrobial activity essentially as those produced by reuterin isolated from *L. reuterii.* Subsequently a key enzyme in the reuterin pathway (glycerol dehydratase) was purified and characterized.[67] The mode of action of reuterin has not to date been fully elucidated. Unpublished data cited by Talarico and Dobrogosz[65] indicated that reuterin can inhibit ribonucleotide reductase activity, which may help explain its broad spectrum of activity including antibacterial, antimycotic, and antiprotozoal effects. Also cited by Talarico and Dobrogosz were experiments

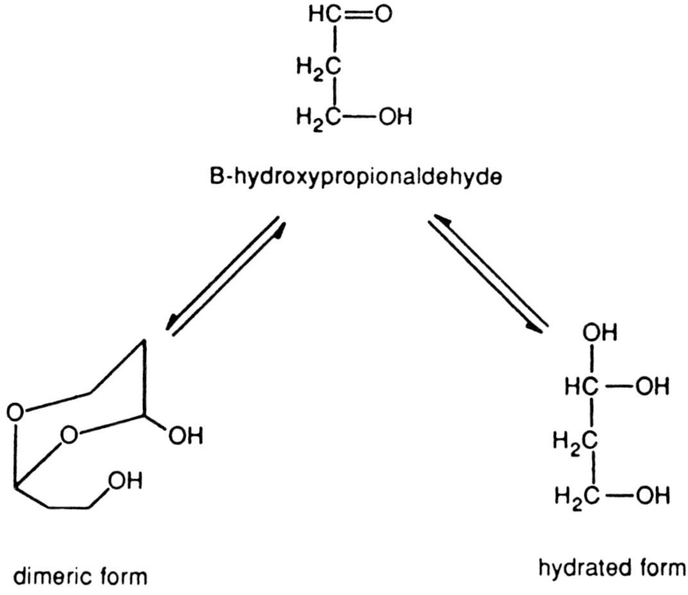

FIGURE 5. The three forms of reuterin present in aqueous solution. (From Talarico, T. and Dobrogosz, W. J., *Antimicrob. Agents Chemother.*, 33, 674, 1989. With permission.)

that demonstrated that the enzyme thiodoxin was inhibited by reuterin. More definitive research needs to be conducted to determine the mode of action of reuterin.

Reuterin has a broad spectrum of activity that makes it a candidate for an assortment of antimicrobial applications. However it is not evident from the existing literature what, if any, toxicity reuterin has toward mammalian organisms. This information would almost certainly be essential for reuterin to be approved for use in clinical settings or as a direct food additive.

Several general approaches can be considered for the application of reuterin in food processing and preservation. Reuterin-producing *L. reuterii* could be added with glycerol to a food or food system to suppress or contain naturally occurring spoilage microorganisms. An example of this type of approach was demonstrated by Lindgren and Dobrogosz[68] using herring filets stored at 5°C . The filets were dipped in a solution containing 1×10^9 cfu/ml of *L. reuterii* and 250 mM glycerol. Controls consisted of untreated herring and a non-reuterin producing strain of *L. reuterii* with glycerol. It was observed that initial levels of Gram-negative bacteria (~ 1 × 105 cfu/g) rose to approximately 1×10^8 cfu/g in 6 d with untreated fish, whereas with L. *reuterii*-treated fish, the numbers of Gram-negative bacteria rose only about 1 log within the same time period (Figure 7). The above-described experiment illustrates the potential of L. *reuterii* to extend the economic shelf life of perishable raw products such as refrigerated herring. What effects, if any, the addition of *L. reuterii* would have on sensory characteristics was not addressed.

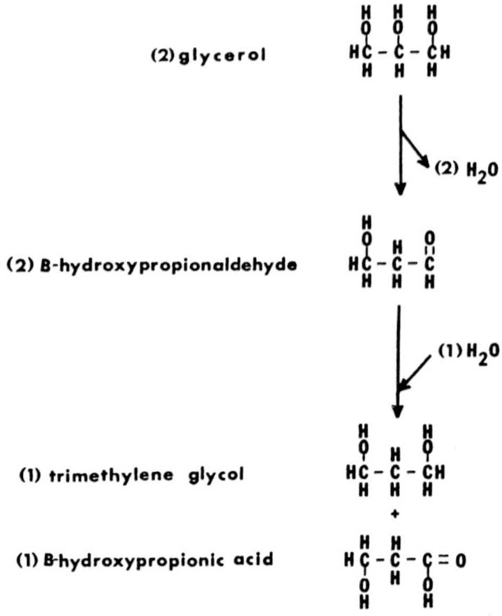

FIGURE 6. Proposed pathway of glycerol fermentation in *L. reuterii*. Glycerol is converted to β–hydroypropionialdehyde by a dehydrase. Two molecules of the aldehyde then undergo an aldehyde dismutation to form equimolar quantities of TMG and β-HPA. (From Talarico, T., Ca, I. A., Chung, T. C., and Dobrogosz, W. J., *Antimicrob. Agents Chemother.*, 32, 18754, 1988. With permission.)

A symbiotic relationship between *L. reutreii* and its host (animal digestive tract) was postulated and discussed by Chung et al.[66] It was proposed that *L. reuterii* may exist as an enteric symbiont and benefit the host by regulating the intestinal microflora and by preventing invasive pathogenic microorganisms from becoming established. A mechanism for this system was suggested to be the induction of reuterin synthesis, and excretion by physical contact by heterologous microbial cells. The process termed "heterologous induction" may have application in the promotion of animal health in commercial food species. Significant weight gains (20–30%) in commercially raised poultry have been observed (W.J., Dobrogosz, private communication) in birds that had reuterin-producing *L. reuterii* introduced into their digestive systems. The weight was believed to be the result of reuterins antagonizing or inhibiting intestinal microflora that can compromise avian health.

Lastly, another application approach, in systems where the intentional addition of microorganisms would not be desirable, would entail the direct addition of purified reuterin or reuterin-containing supernatants. An effective food preservative should be stable under conditions in which it is expected to be used. Studies on the stability of reuterin[64] indicated a half life of 14 d at 37°C and pH 6.5. Reuterin was also shown to be more stable under acidic conditions

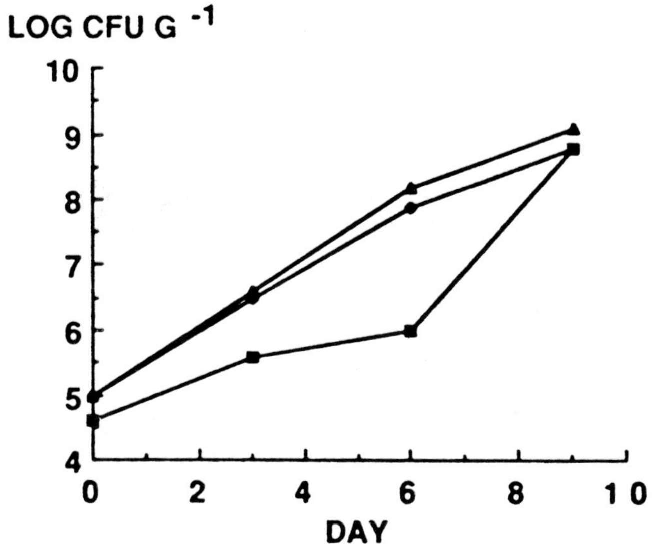

FIGURE 7. Levels of Gram-negative bacteria in herring filets stored in 100% N_2 at 5°C: Δ, nontreated control; ◊, treatment with L. re*uterii* 1068 (nonreuterin producer) and glycerol; ■, treatment with L. *reuterii* 1063 (reuterin producer) and glycerol (5 replicates). (From Lindgren, S. E. and Dobrogosz, W. J., *FEMS Microbiol. Rev.*, 87, 149, 1990. With permission.)

and irreversibly inactivated under alkaline conditions. However it was observed to be very stable (no decrease in activity within 6 months) when held at 5°C. Thus, the potential of reuterin as a food preservative would likely be with foods or beverages held under refrigeration temperatures. An example of such an application was reviewed by Daeschel,[69] in which reuterin was added to ground beef held at 4°C. Decreases in coliform populations were observed when 50 to 100 units of reuterin were added to the meat while control samples (no reuterin added) had 4 log increases in coliform populations (Figure 8).

V. CONCLUDING REMARKS

Studies addressing antagonistic activities in the lactic acid bacteria have clearly shown that antagonism can stem from a variety of secondary and primary metabolites. Low-molecular weight metabolites such as hydrogen peroxide and reuterin, in contrast to bacteriocins, are inhibitory to a broad spectrum of microorganisms, a desirable characteristic for food preservatives. However, compounds like hydrogen peroxide are highly reactive and interact with food components, hence limiting their usefulness. In situ generation of hydrogen peroxide by lactic acid bacteria in foods is an appealing preservation strategy, however the amounts generated by most lactic acid bacteria, with the

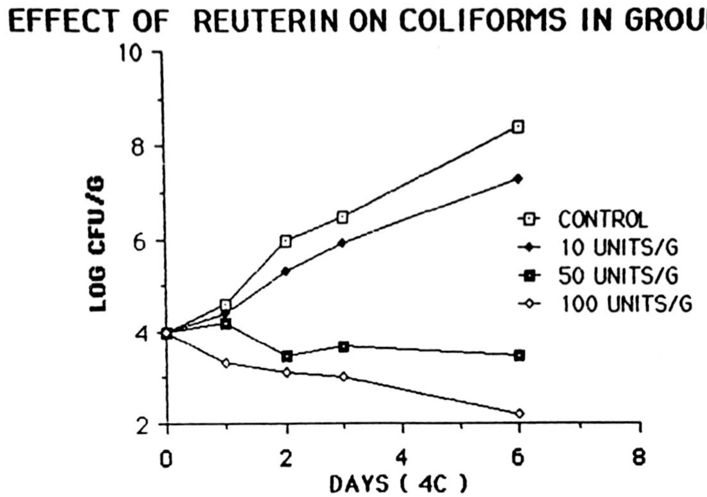

FIGURE 8. Effect of reuterin on coliforms in ground beef. (From Daeschel, M. A., *Food Technol.*, 43, 164, 1989. With permission.)

notable exception of some lactobacilli, are of insufficient quantity to produce significant preservation effects. Perhaps this limitation could be surmounted by genetic manipulation of the cellular regulatory systems governing hydrogen peroxide production and degradation, i.e., hyper hydrogen peroxide-producing starter cultures that are insensitive to amounts produced. An intriguing observation recently reported[23] was the production and accumulation of very large amounts of hydrogen peroxide by *Lactobacillus leichmanii* when it was cultured with saturated fatty acids. Additional research is warranted to further the understanding of how to manipulate hydrogen peroxide production. Furthermore, increased knowledge of starter culture hydrogen peroxide production would allow lactoperoxidase systems to function more efficiently by having optimum substrate amounts of hydrogen peroxide available. Also, in regard to the LPS, a better understanding of how the toxic primary products (hypothiocyanite and thiocyanegen) kill microorganisms would allow the development of more efficacious preservation approaches. New and continuing investigations are to be encouraged to understand the unique metabolite reuterin and how best to utilize its preservation properties. The study of microbial metabolites is an exciting endeavor with new and applicable information being published regularly. It is hoped the research momentum continues, and that highly efficacious, economic, and safe preservation strategies using microbial metabolites become viable alternatives or supplements to existing food preservatives.

REFERENCES

1. **Lewis, R. J.,** *Food Additives Handbook,* Van Nostrand Reinhold, New York, 1989, 252.
2. **von Bockelmann, B.,** Aseptic packaging, in *Disinfection, Sterilization, and Preservation,* 4th ed., Block, S. S., Ed., Lea & Feabiger, Philadelphia, 1991, 839.
3. **Anon.,** Substances generally recognized as safe, *Code of Federal Regulations,* Title 21, Part 182, Food and Drug Administration, Washington, D.C., 1981, 1366.
4. **Block, S. S.,** Peroxygen compounds, in *Disinfection, Sterilization, and Preservation,* Block, S. S., Ed., 4th ed., Lea & Feabiger, Philadelphia, 1991, 169.
5. **Davidson, P. M., Post, L. S., Branen, A. L., and McCurdy, A. R.,** Naturally occurring and miscellaneous food antimicrobials, in *Antimicrobials in Foods,* Marcel Dekker, New York, 1983, 385.
6. **Khan, T., Tahis, M.A., and Elahi, M.**, Decomposition with various food additives of residual hydrogen peroxide used for preservation of raw milk, *Aust. J. Dairy Technol.,* 30, 364, 1975.
7. **Foegeding, P. M., and Busta, F. F.,** Chemical food preservatives, in *Disinfection, Sterilization, and Preservation,* 4th ed., Block, S. S., Ed., Lea & Feabiger, Philadelphia, 1991, 820.
8. **Fleming, H. P., McFeeters, R. F., and Daeschel, M. A.,** The lactobacilli, pediococci, and leuconostocs: vegetable products in *Bacterial Starter Cultures for Foods,* Gilliland, S. E., Ed., CRC Press, Boca Raton, FL, 97, 1985, p. 97.
9. **Mattick, A. T. R. and Hirsch A.,** A powerful inhibitory substance produced by group N streptococci, *Nature (London),* 154, 551, 1944.
10. **Oxford, A. E.,** Diplococcin, an antibacterial protein elaborated by certain milk streptococci, *Biochem. J.,* 38, 178, 1944.
11. **Wheater, D. M., Hirsch, A., and Mattick, A. T. R.,** Latcobacillin, an antibiotic from lactobacilli, *Nature (London),* 168, 659, 1951.
12. **Wheater, D. M., Hirsch, A., and Mattick, A. T. R.,** Possible identity of "lactobacillin" with hydrogen peroxide produced by lactobacilli, *Nature (London),* 170, 623, 1952.
13. **Dahiya, R. S. and Speck, M. L.,** Hydrogen peroxide by lactobacilli, and its effects on *Staphylococcus aureus, J. Dairy Sci.,* 51, 1568, 1968.
14. **Price, R. J. and Lee, J. S.,** Inhibition of *Pseudomonas* species by hydrogen peroxide producing lactobacilli, *J. Milk Food Technol.,* 33, 13, 1970.
15. **Collins, E. B. and Aramaki, K.,** Production of hydrogen peroxide by *Lactobacillus acidophilus, J. Dairy Sci.,* 63, 353, 1980.
16. **Gilliland, S. E. and Speck, M. L.,** Antagonistic activity of *Lactobacillus acidophilus* toward intestinal and foodborne pathogens in associative cultures, *J. Food Prot.,* 40, 820, 1977.
17. **Hamdan, I. Y. and Mikolajick, E. M.,** Acidolin: an antibiotic produced by *Lactobacillus acidophilus, J. Antibiot.,* 27, 631, 1974.
18. **Vakil, J. R. and Shahani, K. M.,** Partial purification of antibacterial activity of *Lactobacillus acidophilus, Bacteriol. Proc.,* 9, 1965.
19. **Vincent, C. G., Veonett, R. C., and Riley, R. G.,** Antibacterial activity associated with *Lactobacillus acidophilus, J. Bacteriol.,* 78, 477, 1959.
20. **Anders, R. F., Hogg, D. M., and Jago, G. R.,** Formation of hydrogen peroxide by group N streptococci and its effect on their growth and metabolism, *Appl. Microbiol.,* 19, 608, 1970.
21. **Whittenbury, J. R.,** Hydrogen peroxide formation and catalase production in lactic acid bacteria, *J. Gen. Microbiol.,* 35, 13, 1964.
22. **Sharpe, E. L.,** Lactic acid bacteria in the dairy industry, *J. Soc. Dairy Technol.,* 32, 9, 1979.

23. **Nunez de Kairuz, M. S., Olazabal, M. E., Oliver, G., Pesce de Ruiz Holgado, A. A., Massa, E., and Farias, R. N.,** Fatty acid dependent hydrogen peroxide production in *Lactobacillus, Biochem. Biophys. Res. Commun.,* 152, 113, 1988.
24. **Ross, G. D.,** The inhibition of growth of spoilage microorganisms in milk by *Streptococcus lactis* subsp. *diacetylactis, Leuconostoc cremoris and L. dextranicum, Aust. J. Dairy Technol.,* 36, 147, 1981.
25. **Raccach, M. and Baker, R. C.,** Formation of hydrogen peroxide by meat starter cultures, *J. Food Prot.,* 41, 798, 1978.
26. **Babel, F. J.,** Antibiosis by lactic culture bacteria, *J. Dairy Sci.,* 60, 815, 1977.
27. **Gilliland, S. E.,** Role of starter culture bacteria in food preservation, in *Bacterial Starter Cultures for Foods,* Gilliland , S. E., Ed., CRC Press, Boca Raton, FL, 1985, 177.
28. **Juffs, H. S. and Babel, F. J.,** Inhibition of psychrotrophic bacteria by lactic cultures in milk stored at low temperature, *J. Dairy Sci.,* 58, 1612, 1975.
29. **Gilliland, S. E. and Speck, M. L.,** Inhibition of psychrotrophic bacteria by lactobacilli and pediococci in nonfermented refrigerated foods, *J. Food Sci.,* 40, 903, 1975.
30. **Martin, D. R. and Gilliland, S. E.,** Inhibition of psychrotrophic bacteria in refrigerated milk by lactobacilli isolated from yogurt, *J. Food Prot.,* 43, 675,1980.
31. **Agner, K.,** Verdoperoxidase, *Acta Physiol. Scand.,* 2, 1, 1941.
32. **Wright, R. C. and Tramer, J.,** Factors influencing the activity of cheese starters, the role of milk peroxidase, *J. Dairy Res.,* 25, 104, 1958.
33. **Theorell, H. and Åkesson, A.,** Purified milk peroxidase, *Arkiv. Kemi. Miner. Geol.,* 17B, 7, 1943.
34. **Morrison, M., Allen, P. Z., Bright, J., and Jayasinghe, W.,** Lactoperoxidase V. Identification and isolation of lactoperoxidase from salivary gland, Arch. *Biochem Biophys.,* 111, 126, 1965.
35. **Tenovuo, J. O. and Pruitt, K. M., Eds.,** The peroxidase system in human secretions in the lactoperoxidase system, in *The Lactoperoxidase System,* Marcel Dekker, New York, 1985, 101.
36. **Reiter, B.,** Review of the progress of dairy science: antimicrobial systems in milk, *J. Dairy Res.,* 45, 131, 1978.
37. **Paul, K. and Ohlsson, P.,** The chemical structure of lactoperoxidase, in *The Lactoperoxidase System,* Pruitt, K. M. and Tenovuo, J. O., Eds., Marcel Dekker, New York, 1985, 15.
38. **Walsh, C.,** Enzymatic *Reaction Mechanisms,* Bartlett, A. C. and McCombs, L. W., Eds., W. H. Freeman, San Francisco, 1979, 488.
39. **Thomas, E. L.,** Products of lactoperoxidase-catalyzed oxidation of thiocyanate and halides, in*The Lactoperoxidase System,* Pruitt, K. M. and Tenovuo, J. O., Eds., Marcel Dekker, New York, 1985, 31.
40. **Aune, T. M. and Thomas, E. L.,** Accumulation of hypothiocyanite ion during peroxidase-catalyzed oxidation of thiocyanate ion, E*ur. J. Biochem.,* 80, 209, 1977.
41. **Pruitt, K. M., Tenovuo, J., Andrews, R. W., and McKane, T.,** Lactoperoxidase-catalyzed oxidation of thiocyanate: polarographic study of the oxidation products, *Biochemistry,* 21, 3, 562, 1982.
42. **Thomas, E. L.,** Lactoperoxidase-catalyzed oxidation of thiocyanate: equilibria between oxidized forms of thiocyanate, *Biochemistry,* 20, 3273, 1981.
43. **Pruitt, K. M. and Reiter, B.,** Biochemistry of peroxidase system: antimicrobial effects, in *The Lactoperoxidase System,* Pruitt, K. M. and Tenovuo, J. O., Eds., Marcel Dekker, New York, 1985, 143.
44. **Aune, T. M. and Thomas, E. L.,** Oxidation of protein sulfhydryls by products of peroxidase-catalyzed oxidation of thiocyanate ion, *Biochemistry,* 17, 1005, 1978.
45. **Tenovuo, J., Mansson-Rahemtulla, B., Pruitt, K. M., and Arnold, R.,** Inhibition of dental plaque acid production by the salivary lactoperoxidase antimicrobial system, *Infect. Immun.,* 34, 1, 208, 1981.
46. **Aune, T. M. and Thomas, E. L.,** Lactoperoxidase-catalyzed incorporation of thiocyanate ion into a protein substrate, *Biochemistry,* 16, 4611, 1977.

47. **Bjoerck, L. and Claesson, O.,** Correlation between concentration of hypothiocyanate and antibacterial effect of the lactoperoxidase system against *Escherichia coli*, *J. Dairy Sci.*, 63, 919, 1980.
48. **Reiter, B.,** The lactoperoxidase system of bovine milk, in *The Lactoperoxidase System*, Pruitt, K. M. and Tenovuo, J. O., Eds., Marcel Dekker, New York, 1985, 123.
49. Reiter, B. and **Harnulv, G.,** Lactoperoxidase antibacterial system: natural occurrence, biological functions and practical application, *J. Food Prot.*, 47, 724, 1984.
50. **Reiter, B. and Harnulv, G.,** The preservation of refrigerated and uncooled milk by its natural lactoperoxidase system, *Dairy Ind. Int.*, 47, 13, 1982.
51. **Martinez, C. E., Mendoza, P. G., Alacron, F. J., and Garcia, H. S.,** Reactivation of the lactoperoxidase system during raw milk storage and its effect on the characteristics of pasteurized milk, *J. Food Prot.*, 51, 558, 1988.
52. **Reiter, B., Marshall, V. M. E., Bjorck, L., and Rosen, C.,** Nonspecific bactericidal activity of the lactoperoxidase-thiocyanate-hydrogen peroxide system of milk against *Escherichia coli* and some Gram-negative pathogens, *Infect. Immun.*, 13, 800, 1976.
53. **Bjorck, L.,** Antibacterial effect of lactoperoxidase system on psychrotrophic bacteria in milk, *J. Dairy Res.*, 45, 109, 1978.
54. **Siragusa, G. R. and Johnson, M. G.,** Inhibition of *Listeria monocytogenes* growth by the lactoperoxidase-thiocyanate-H_2O_2 antimicrobial system, *Appl. Environ. Microbiol.*, 55, 2802, 1989.
55. **Gaya, P., Medina, M. and Nuñez, M.,** Effect of lactoperoxidase system on *Listeria monocytogenes* — behavior in raw milk at refrigeration temperatures, *Appl. Environ. Microbiol.*, 57, 3355, 1991.
56. **Kamau, D. N., Doores, S., and Pruitt, K M.,** Enhanced thermal destruction of *Listeria monocytogenes* and *Staphylococcus aureus* by the lactoperoxidase system, *Appl. Environ. Microbiol.*, 56, 2711, 1990.
57. **Thomas, E. L.,** Bacterial hydrogen peroxide production, in *The Lactoperoxidase System*, Pruitt, K. M. and Tenovuo, J. O., Eds., Marcel Dekker, New York 1985b, 179.
58. **Schonbaum, G. R. and Chance, B.,** Catalase, in *The Enzymes*, Boyer, P., Ed., Academic Press, New York, 1976.
59. **Walker, G. A. and Kilgour, G. L.,** Pyridine nucleotide oxidizing enzymes of *Lactobacillus casei*: oxidase and peroxidase, *Arch. Biochem. Biophys.*, 111, 534, 1965.
60. **Lerch, M. and Reuter, G.,** Das Vorkommen aerob weachsender Gram-positiver Stabchen des genus *Lactobacillus* Beijerrinck im Darminhalt Erwaschsener Menschen, *Zentralbl. Bakteriol. Parasintenkd. Infektionskr. Hyg.* I (Abt. Orig.), 185, 446, 1962.
61. **Sherman, L. A. and Savage, D. C.,** Lipoteichoic acids in *Lactobacillus* strains that colonize the mouse epithelium, *Appl. Environ. Microbiol.*, 52, 302 , 1986.
62. **Axelsson, L., Chung, T. C., Dobrogosz, W. J., and Lindgren, S.,** Production of a broad spectrum antimicrobial substance from *Lactobacillus reuterii*, *Microb. Ecol. Health Dis.*, 2, 131, 1989.
63. **Axelsson, L. and Lindgren,** S., Characterization and DNA homology of *Lactobacillus* strains isolated from pig intestine. *J. Appl. Bacteriol.*, 62, 433, 1987.
64. **Talarico, T., Casas, I. A., Chung, T. C., and Dobrogosz, W. J.,** Production and isolation of reuterin, a growth inhibitor produced by *Lactobacillus reuterii*,*Antimicrob. Agents Chemother.*, 32, 1854, 1988.
65. **Talarico, T. and Dobrogosz, W. J.,** Chemical characterization of an antimicrobial substance produced by *Lactobacillus reuterii*, *Antimicrob. Agents Chemother.*, 33, 674, 1989.
66. **Chung, T. C., Axelsson, L., Lindgren, S. E., and Dobrogosz, W. J.,** In vitro studies on reuterin synthesis *Lactobacillus reuterii*, *Microb. Ecol. Health Dis.*, 2, 137, 1989.
67. **Talarico, T. and Dobrogosz, W. J.,** Purification and characterization of glycerol dehydratase from *Lactobacillus reuterii*, *Appl. Environ. Microbiol.*, 56, 1195, 1990.
68. **Lindgren, S. E. and Dobrogosz, W. J.,** Antagonistic activities of lactic acid bacteria in food and feed fermentations, *FEMS Microbiol. Rev.*, 87, 149, 1990.
69. **Daeschel, M. A.,** Antimicrobial substances from lactic acid bacteria for use as food preservatives, *Food Technol.*, 43, 164, 1989.

Chapter 8

BACTERIOCINS OF STARTER CULTURE BACTERIA AS FOOD BIOPRESERVATIVES: AN OVERVIEW

Bibek Ray

TABLE OF CONTENTS

Teleologically, it is tempting to assume some major role for bacteriocins and bacteriocin-like substances in the regulation of population dynamics in various bacterial ecosystems. One possible basis for this assumption is the seemingly universal occurrence of bacteriocins in the bacterial world and the recognition that to have so consistently survived the pressure of evolution they must fulfil some vital functions.[1] Recently, there has been a remarkable renaissance of research interest on the subject of the production by bacteria of antibiotic peptides and proteins,...there was an air of uncertainty concerning how best to categorize some of these peptides...produced by certain Gram-positive bacteria. On the one hand, there was some real aversion to users of the label antibiotic since it was considered that this may impact negatively on any future attempts to exploit these substances commercially. for example as food preservatives. On the other hand, could these small peptides justifiably be referred to as bacteriocins?...those inhibitors which to some extent are similar to colicins could be referred to as bacteriocin-like inhibitory substances-BIIS.[1a]

I. INTRODUCTION

Bacteriocins, a term designated originally to proteins of colicin type from *Escherichia coli* that are characterized by lethal biosynthesis. They are produced by many strains of Gram-positive and Gram-negative bacteria. The characteristics of bacteriocins from the two bacterial groups differ to a certain degree. Tagg et al.[1] suggested that the following three criteria could be used to consider an antimicrobial compound from a Gram-positive bacterium as a bacteriocin: that it is a biologically active protein moiety, that it has a narrow inhibitory spectrum, and that it is bacteriocidal to sensitive strains. However, the authors recognized that some well-defined bacteriocins from Gram-positive bacteria have wide spectrums of activity. They have also suggested that the term "bacteriocin" should be used for an antimicrobial compound only when all the above criteria have been met, and that antagonistic substances that are incompletely studied should be designated as "bacteriocin-like compounds." Many bacteriocins of Gram-positive bacteria contain one or more unusual amino acids, one of which is lanthionine (alanine-S-alanine), and recently a term "lantibiotics" has been coined for these antimicrobial peptides.[2–5]

Bacteriocins of Gram-positive bacteria vary greatly in their spectrum of activity. Some inhibit only a single or a few strains of the same species, or a few species from the same genus, while others have activities against many strains from different genera of Gram-positive bacteria. There are reports that some bacteriocins (or bacteriocin-like compounds) from Gram-positive bacteria are inhibitory to Gram-negative bacteria.[1]

Among Gram-positive starter culture bacteria many strains produce bacteriocin-like compounds that also differ greatly in their antagonistic properties. Some have only a narrow host spectrum, while others have wide spectrums that include not only strains of Gram-positive bacteria from different genera, but also include Gram-negative bacteria, and some even include yeasts and molds. However, Klaenhammer[6] has indicated that inhibition of Gram-negative bacteria has not been demonstrated with purified bacteriocins from Gram-positive bacteria in well-designed studies. One recent study reported that the receptor of a bacteriocin, pediocin AcH, is in the lipteichoic acid on the host cell wall.[7] This finding is in agreement with the suggestion that bacteriocins of Gram-positive bacteria may not be inhibitory to Gram-negative bacteria under normal physiological and environmental conditions as they do not have teichoic acid on cell walls. However, different bacteriocins can have different receptors, and well-designed studies with purified bacteriocins will be necessary to resolve this controversy.

Among the many bacteriocins of starter culture bacteria only a few have been studied as biopreservatives in food systems. These include nisin of *Lactococcus lactis* subsp. *lactis* and pediocins of *Pediococcus acidilactici.* Along with this introductory chapter on bacteriocins of starter culture bacteria, three separate chapters on bacteriocins are included. These include one on nisin, one on pediocins of *P. acidilactici,* and one on those that have been studied for chemical properties and genetic controls but have not been used in food systems. The purpose of this introductory chapter is to provide an overview of the properties of the many different bacteriocins of starter culture bacteria that are now being studied.

II. BACTERIOCINS OF STARTER CULTURE BACTERIA

Many strains of starter culture bacteria have been reported to produce metabolites that are inhibitory to other bacteria, and the antagonistic effect was not detected following treatment of the metabolic products with proteolytic enzymes. A comprehensive list of different strains of food-grade starter culture bacteria reported to produce such inhibitory compounds is presented in Table 1. It includes strains from *Lactococcus lactis* subsp. *lactis,*[8–13b] *Lactococcus lactis* subsp. *cremoris,*[11,14–18] *Lactococcus lactis* subsp. *lactis* var *diacetylactis,*[11,19–22] *Streptococcus salivarious* subsp. *thermophilus,*[23] *Leuconostoc* species,[24–27] *Pediococcus acidilactici,*[27–38], *Pediococcus pentosaceus,*[38–44] *Lactobacillus acidophilus,*[45–51] *Lactobacillus plantarum,*[34,52–54], *Lactobacillus delbrueckii* subsp. *lactis,*[55,56] *Lactobacillus helveticus,*[57–60] *Lactobacillus casei,*[61,62] *Lactobacillus brevis,*[62] *Lactobacillus fermentum,*[63], *Lactobacillus sake,*[64–66], *Lactobacillus gasseri,*[67] *Bifidobacterium* spp.,[68,69] and *Propionibacterium thoenii.*[70] Also included are several strains that are not used as starter cultures, but are found in food systems or have some importance in foods. These include *Leuconostoc gelidum,*[25,26] some unknown *Leuconostoc* spp.,[25,27], *Lactobacillus gasseri,*[67], *Lactobacillus viridescens,*[27], *Carnobacteria piscicola,*[27,71] and *Enterococcus faecium..*[71a,72] Some strains of starter culture bacteria that have been reported to produce antagonistic compounds, but have not been shown to be proteinaceous in nature, at least by the loss of activity following treatment with a proteolytic enzyme or by other methods, have not been included in this table. Some of these include unknown compounds designated as acidolin and acidophilin of *L. acidophilus,*[73–75] bulgarican by *Lactobacillus delbrueckii* subsp. *bulgaricus,*[76] and others.[77] It will be apparent even from the limited information listed in this table that variation exists among the researchers in their test procedures, especially in the material(s) that were tested to show the production of inhibitory compound(s) by a strain. The

TABLE 1
Bacteriocin Producing Starter Culture Bacterial Species/Strains

Species/Strain[a]	Specific name (Proteolysis)[b]	Testing materials[c]	Host range[d]	Ref.
***Lactococcus lactis* subsp. *lactis*:**				
ATCC 11454 ATCC 7962 DL 16, 354/07	Nisin (trypsin resistant)	Colony, culture broth, precipitate, gel filtrate	Wide: only Gram-positive bacteria	8–10
5D8, 6F3, 6F5	None (trypsin resistant)	Colony, culture broth, precipitate	Wide: only Gram-positive[b] bacteria	11
2F6, 7C1	None (trypsin resistant)	Colony, culture broth, precipitate	Wide: only Gram-positive[b] bacteria	11
DRC1	Dricin (trypsin and heat sensitive)	Colony, cross–streaking Culture broth	Narrow: related strains	12
10, 300	Lactostrepcins (trypsin sensitive)		Medium: Gram-positive bacteria	13
ADRIA85L030	Lactococcin	Culture broth purified		13a, 13b
***Lactococcus lactis* subsp. *cremoris*:**				
346	Diplococcin (trypsin sensitive)	Colony, culture broth, precipitate, gel filtrate	Narrow: related species//strains	14, 15
202	Lactostrepcin 5 (Protease and Phospholipase A sensitive)	Colony, precipitate	Narrow:related species//strains	16

AC1, 3A6, 4C6, 9B4, 3C6	None (trypsin sensitive)	Colony, culture broth, precipitate	Medium: Gram-positive	11, 17, 18
1A1, 3E9, KC3, W3, 4E9	None (trypsin and heat sensitive)	Colony, culture broth, precipitate	Medium: Gram-positive bacteria	11
Lactococcus lactis* subsp. *lactis* var *diacetylactis				
DRC1, 26–2	None (not tested)	Culture broth, gel filtration	Wide: Gram-positive and Gram-negative bacteria	19
S50	Bacteriocin S50 (trypsin sensitive)	Culture broth	Narrow: related species/ strains	20
WM4	None (trypsin sensitive)	Colony	Narrow: related species//strains	21, 22
6F7	None (trypsin sensitive)	Colony, culture broth, precipitate	Medium: Gram-positive bacteria	11

TABLE 1 (continued)
Bacteriocin Producing Starter Culture Bacterial Species/Strains

Species/Strain[a]	Specific name (Proteolysis)[b]	Testing materials[c]	Host range[d]	Ref.
		Streptococcus salivarius* subsp. *thermophilus		
ATCC 14485	None (pepsin resistant)	Culture broth, gel filtrate	Wide: Gram–positive and Gram-negative bacteria	23
		Leuconostoc mesenteroides* subsp. *mesenteroides		
C9, 24X, H	Leuconocin Lcm1 (trypsin sensitive)	Colony, culture broth, precipitate, gel filtration	Medium: Gram-positive bacteria	24
UL 5	Mesenteroicin 5 (Protease sensitive)	Broth, precipitate	Medium	24a
Leuconostoc spp. many strains	None (trypsin sensitive)	Colony, culture broth	Narrow: related species/strains	24
		Leuconostoc gelidum		
IN 139	None (protease sensitive)	Colony, culture broth	Medium: Gram-positive bacteria	25

UAL 187	None (trypsin sensitive)	Colony, culture broth, precipitate	Medium: Gram-positive bacteria	26
		***Leuconostoc* spp.**		
OX, VX	None (trypsin sensitive)	Colony	Wide: Gram-positive and Gram-negative bacteria	27
		Pediococcus acidilactici		
H, E, F, M	Pediocin AcH (trypsin sensitive)	Colony, culture broth, precipitate, gel filtrate	Wide: Gram-positive bacteria	7,27–35
		Pediococcus acidilactici		
PAC1.0	Pediocin PA1 (chymotrypsin sensitive)	Colony, culture broth, precipitate, gel filtrate	Wide: Gram-positive bacteria	36, 37

TABLE 1 (continued)
Bacteriocin Producing Starter Culture Bacterial Species/Strains

Species/Strain[a]	Specific name (Proteolysis)[b]	Testing materials[c]	Host range[d]	Ref.
		Pediococcus acidilactici		
PO2, B5627, PC	None	Colony	Wide: Gram-positive bacteria	38
		Pediococcus pentosaceus		
FBB–61, L7230	Pediocin A (pronase sensitive)	Colonies, culture broth	Wide: Gram-positive bacteria	39–42
FBB–63 and its derivatives	None	Colonies	Wide: Gram-positive bacteria	43, 44
		Lactobacillus acidophilus		
IFO 3205	None	Gel filtration	Wide: Gram-negative bacteria	45

AC1	None (trypsin sensitive)	Culture broth, precipitate, gel filtration	Wide: Gram-positive and Gram-negative bacteria	46
N2	Lactacin B (Proteinase K sensitive)	Colony, culture broth, gel filtration	Medium: Gram-positive bacteria	47, 48
88	Lactacin F (trypsin sensitive)	Colony, culture broth, precipitate, gel filtration	Medium: Gram-positive bacteria	49, 50
LAPT 1060	Acidophilucin A (trypsin sensitive)	Culture broth	Narrow: related species/ strains	51
		Lactobacillus plantarum		
83	None (trypsin sensitive)	Associative growth	Medium: Gram-positive bacteria	52
NCDO 1193	Plantacin B (trypsin sensitive)	Culture broth	Medium: Gram-positive bacteria	53

TABLE 1 (continued)
Bacteriocin Producing Starter Culture Bacterial Species/Strains

Species/Strain[a]	Specific name (Proteolysis)[b]	Testing materials[c]	Host range[d]	Ref.
		***Lactobacillus plantarum* (contd.)**		
C11	Plantaricin A (protease sensitive)	Colony, culture broth	Medium: Gram-positive bacteria	54
27	None (trypsin sensitive)	Colony, culture broth, precipitate	Medium: Gram-positive bacteria	34
		Lactobacillus delbrueckii* subsp. *lactis		
G4	None	Culture broth	Narrow: related species/strains	55
JCM 1106, JCM 1107	Lacticin (trypsin sensitive)	Culture broth	Narrow: species/strain	56
		Lactobacillus helveticus		
LP27	Lactocin 27 (trypsin sensitive)	Culture broth, gel filtration	Narrow: species/strain	57, 58
481	Helveticin J (trypsin sensitive)	Colony, culture broth, precipitate, gel purified	Narrow: species/strain	59, 60

B80	Caseicin 80 (trypsin sensitive an heat sensitive)	Culture broth, gel filtrate	Narrow: species/strain	61,62
		Lactobacillus brevis		
B 37	Brevicin 37 (trypsin sensitive)	Culture broth	Narrow: species/strain	62
		Lactobacillus fermentum		
466	None (trypsin sensitive)	Culture broth, gel filtration	Narrow: species/strain	63
		***Lactobacillus* sake**		
Lb706	Sakacin A (trypsin sensitive)	Culture broth	Medium: Gram-positive bacteria	64
L45	Lactocin S (trypsin sensitive)	Colony, culture broth, precipitate, gel filtrate	Medium: Gram-positive bacteria	65, 66

TABLE 1 (continued)
Bacteriocin Producing Starter Culture Bacterial Species/Strains

Species/Strain[a]	Specific name (Proteolysis)[b]	Testing materials[c]	Host range[d]	Ref.
		Lactobacillus gasseri		
LA 12, 14, 33, 39, 48, 50	Gassericin (trypsin sensitive)	Culture broth	Narrow: species/strains	67
		Lactobacillus viridescens		
PX, QX	None (trypsin sensitive)	Culture broth	Wide: Gram-positive and Gram-negative bacteria	27
		Bifidobacterium bifidum		
1452	Bifidin	Precipitate	Medium: Gram-positive and Gram-negative bacteria	68
		***Bifidobacterium* spp.**		
1	None	Colonies, culture broth	Wide: Gram-positive bacteria	69

P127	Propionicin PLG–1 (trypsin and heat sensitive)	Colonies, precipitate	Wide: Gram-positive and Gram-negative bacteria, yeasts and molds	70
		Carnobacterium piscicola		
LV17	None (trypsin sensitive)	Colonies, culture broth, precipitate	Wide: Gram-positive bacteria	71
GN, DX	None (trypsin sensitive)	Colonies	Medium: Gram-positive bacteria	27
		Enterococcus faecium		
NCIB2699 NCIB2702	None	Colonies, culture broth	Medium: Gram-positive bacteria	72

TABLE 1 (continued)
Bacteriocin Producing Starter Culture Bacterial Species/Strains

[a] Species used as starter cultures are included with the exception of *Leuconostoc gelidum*[25,26], unknown species of *Leuconostoc*,[24,27] *Lactobacillus gasseri*,[67] *Carobacterium pescicola*,[71] and *Enterobacter faecium*.[71a,72] Species/strains that have been reported to produce antimicrobial compounds, but that are not known if they are protein, are not included in this list.[73–77]

[b] Bacteriocins that have been tested either for sensitivity to proteolytic enzyme(s), or other characteristics to indicate that the active component(s) is (are) protein, are included. To indicate sensitivity to proteolytic enzyme (where available only), sensitivity to trypsin has been included (to compare against nisin which is resistant to trypsin, but sensitive to α–chymotrypsin). Where a bacteriocin was not tested against trypsin, information on sensitivity to another proteolytic enzyme is included to show the protein–nature of a bacteriocin.[9,10,16,23,25,36,40,42,54] If a bacteriocin is sensitive to heat, as opposed to resistance to heat, it is also indicated.[61,70]

[c] Materials used to show bacteriocin activity include colony (which also include growing cells, such as a stab culture), broth culture (cell-free supernatant, including dialyzed broth), precipitate (precipitated with ammonium sulfate or ethanol), gel filtrate (include ion exchange, shephedex).

[d] Host range: narrow includes several strains and species of same genus; medium includes strains from one or two other genera (excluding own genus); wide includes strains from three or more other genera, all from Gram-positive bacteria. Some bacteriocins were listed as active against Gram-negative bacteria[9,19,23,27,44,46,68,70] and are included in the wide-range category (although currently bacteriocins of Gram-positive bacteria are considered to be effective normally only against Gram-positive bacteria). On the basis of an antimicrobial spectrum, the compounds listed in Table 1 have been grouped arbitrarily as narrow, medium, and wide. In Table 2 this aspect has been elaborated further. While bacteriocins in the narrow range are inhibitory to one or few strains or species in the same genus, the one in the medium range are inhibitory to species/strains from two to three genera. If they are tested properly, many of these may show inhibitory activity against species/strains in other genera also and could be included as a wide-range bacteriocin. Only nisin and pediocin AcH and pediocin PA1 from lactic acid bacteria have been included in this group (wide range) as they have been tested with purified or partially purified materials. Propionicin PLG 1, a heat-labile bacteriocin of *P. thoenii*, showed a wide host range including Gram-positive and Gram-negative bacteria as well as yeasts and molds. However, the potency of this compound is quite low; a large volume (as much as 0.2 to 0.5 ml) gave only 1- to 4-mm zone of inhibition.

TABLE 2
Antimicrobial Spectrum of Some Partially Purified Bacteriocins from Starter Culture Bacteria

Partially purified bacteriocins[a] (producer species)	Sensitive bacteria[b]	Ref.
A. Narrow range		
Diplococcin (*L. lactis* subsp. *cremoris*)	*L. lactis* subsp. *cremoris* and subsp. *lactis*	14
None (*L. fermenti*)	*L. fermenti*	63
Caseicin 80 (*L. casei*)	*L. casei*	62
Helveticin J (*L. helveticus*)	*L. helveticus*	59
Lactacin 27 (*L. helveticus*)	*L. helveticus*	57
B. Medium range		
Lactocin S (*L. sake*)	*Lactobacillus, Leuconostoc,* and *Pediococcus* sp.	65, 66
Lactacin B (*L. acidophilus*)	*Lactobacillus* and *Enterococcus faecalis*	49
C. Wide range		
Nisin (*L. lactis* subsp. *lactis*) Heat stable	*Lactococcus, Leuconostoc, Lactobacillus, Enterococcus, Bacillus, Pediococcus, Staphylococcus aureus, Micrococcus, Clostridium, Listeria monocytogenes,* other *Listeria* spp. Spores: *Bacillus* spp. and *Clostridium* spp. (including *C. botulinum*)	9, 10

table includes mostly test of the growth inhibition of indicator strain(s) by the cells (colony, stab) or the cell-free culture broth. Only limited studies have been conducted after some purification of the materials. If the suggestions of Tagg et al.[1] are used as a model that includes "biologically active protein moiety" then the proof of bacteriocin production by a strain should include a purification step. Klaenhammer[6] has also indicated that reports on the inhibition of Gram-negative bacteria by a bacteriocin of lactic acid bacteria under normal conditions should be confirmed with a purified preparation. In the absence of such information, the proteolytic enzyme-sensitive inhibitory material in a broth culture or in a growing culture may be designated as a "bacteriocin-like compound."[1]

The spectrum of activity of bacteriocins vary greatly. In this review they have been grouped as narrow, medium, and wide (Table 2). Those with wide range will have better application as food biopreservatives.

Some characteristics of the bacteriocins of starter culture bacteria that will be important in their use in foods have been presented in Table 3. Diverse methods were used to purify the bacteriocins that varied in molecular weight ranging from 100 Da to 40,000 Da. Most showed loss of activity following hydrolysis by one or more of the three proteolytic enzymes of the gastrointestinal (GI) tract. Except for a few, most were heat stable, and thus can be advantageously included in the formulation of a food product prior to heat

TABLE 3
Some Characterisitics of Partially Purified Bacteriocins From Starter Culture Bacteria

Bacteriocin (Producer)	Purification Method	MW (Dal)	Amino Acid		Inactivated by			Heat Effect	Toxic to Animals (Method)	Host Range	Ref.
			Number	Unusual	Pepsin	Trypsin	α-Chymotrypsin				
Nisin[a] (*L. lactis* subsp. *lactis)*	Sephadex G-25	3500	34	4	NT[b]	–	+	Stable	None (feeding)	Wide	9
Diplococcin (*L. lactis* subsp. *cremoris)*	Carboxy methyl cellulose	5300	51	0	NT	+	+	Stable	NT	Narrow	14
None[c] (*L. lactis* subsp. *lactis* var *diacetylactis)*	Sephadex G-10	100 to 300	NT	NT	NT	NT	NT	NT	NT	Wide	19
None[d] (*S. salivarius* subsp. *thermophilus)*	Sephadex G-10	700	NT	NT	NT	NT	NT	Stable	NT	Wide	23
Pediocin AcH[e] (*P. acidilactici)*	SDS-PAGE and C18 column (FPLC)	2700	NT	3	NT	+	+	Stable	None (injection)	Wide	31, 32

Pediocin PA1 *(P. acidilactici)*	Sephacryl S-200	16,500	NT	NT	NT	NT	+	Stable	NT	Wide	36
None[f] *(Lactobacillus acidophilus)*	Sephadex G-25 and TLC	3500	25	0	NT	NT	NT	NT	NT	Wide	45
None[f] *(Lactobacillus acidophilus)*	Sephadex G-100	5400	NT	NT	NT	+	+	NT	NT	Wide	46
Lactacin B *(L. acidophilus)*	Sephacryl S-200	6500	NT	NT	NT	NT	NT	Sensitive	NT	Medium	48
Lactacin F[g] *(L. acidophilus)*	C18 column (HPLC)	2500	56	0	NT	+	NT	Stable	NT	Medium	50
Lactacin 27[h] *(L. helveticus)*	Sephadex G-25	12,400	123	0	NT	+	NT	Sensitiive	NT	Narrow	57

TABLE 3 (continued)
Some Characterisitics of Partially Purified Bacteriocins From Starter Culture Bacteria

Bacteriocin (Producer)	Purification Method	MW (Dal)	Amino Acid		Inactivated by			Heat Effect	Toxic to Animals (Method)	Host Range	Ref.
			Number	Unusual	Pepsin	Trypsin	α-Chymotrypsin				
Helveticin J *(L. helveticus)*	Sephadex G200-50	37,000	NT	NT	NT	+	NT	Sensitive	NT	Narrow	59
Caseicin 80[h] *(L. casei)*	Superose HR10130 (FPLC)	40,000	NT	NT	NT	+	NT	Sensitive	NT	Narrow	62
None *(L. fermenti)*[i,j]	Sephadex G-100	NT	16	0	NT	+	NT	Stable	NT	Narrow	63
Lactocin S[j] *(L. sake)*	Sephacryl S-100	NT	33	1	NT	+	NT	Stable	NT	Medium	66
Propionicin PLG 1[k] *(P. thoenii)*	Sephadex G-200	10,000	NT	NT	+	+	+	Sensitive	NT	Wide	70

[a] Sensitivity of nisin to α-chymotrypsin is relatively low. 200 μg/ml of α-chymotrypsin reduced its activity to about 50% of original in 2 h at 37°C, pH 7.0. Under the same conditions the same amount of activity units of pediocin AcH was completely lost (Ray, unpublished data). Nisin contains four unusual amino acids, viz., lanthionine, ß-methyllanthionine, dehydroalanine, and dehydrobutyrine, which are synthesized through posttranslation modification.

[b] NT, not tested.

[c] With molecular weight 100–300, it may not be a peptide or at the most a di- or a tri- peptide. As it is effective against Gram-negative bacteria, it may not be a bacteriocin.

[d] May not be a bacteriocin (see comments in footnote c).

[e] Initial amino acid analysis of the preparation indicated the presence of three amino acids different from the regular standards (see Reference 31).

[f] Effective against Gram-negative bacteria; may not fit the current definition of bacteriocins.

[g] Seemed to have too many amino acids for the molecular weight.

[h] Heat sensitive.

[i] A lipo-protein-carbohydrate complex; the active moiety is a glycoprotein.

[j] Molecular weight was not determined.

[k] Sensitive to heat. Effective against Gram-positive and Gram-negative bacteria, and against yeasts and molds. Does not fit with the current definition of bacteriocin of Gram-positive starter culture bacteria.

TABLE 4
General Characteristics of Bacteriocins of Starter Culture Bacteria

1. Chemical nature:	hydrophobic cationic peptides; form aggregates; some have unusual amino acids, especially lanthionine
2. Synthesis:	Primary or secondary metabolite; ribosomally synthesized; may undergo posttranslation modification
3. Genetic determinants:	Encoded either in plasmid or in chromosome (transposon)
4. Mode of action:	Bacteriocidal; very rapid; binds to specific receptor(s) on cell wall, then destabilizes membrane functions; producer and many resistant Gram-positive bacteria also adsorb, but are immune
5. Host range:	Specific strains of Gram-positive bacteria; strains from within the species or genus, or strains from different genera; under normal conditions Gram-negative bacteria are resistant
6. Stability:	Bacteriocidal property for most is retained after high heat treatment; active at wide pH range, more in acidic pH; most are stable over several months during frozen and refrigeration storage and after drying
7. Proteolytic enzyme:	Activity is lost by one or more proteolytic enzymes, including gastric proteinases
8. Toxicity:	Those examined are nontoxic to animals

processing. None of these, except for nisin and pediocin AcH, have been tested for toxicity to animals either by feeding or by injection.

Characteristics that are common to bacteriocins, at least for which the information is available, are listed in Table 4. These peptides are generally hydrophobic in nature and form aggregates, a property which must be considered during their purification. While nisin[9] and pediocin AcH[78] are produced in large amounts after the cell growth has reached a stationary phase, i.e., as secondary metabolites, several others are produced as primary metabolites during the exponential growth phase. Location of genetic determinates for bacteriocin production in these strains vary greatly. The gene for the pediocin AcH production phenotype resides in an 8.9 Kb plasmid,[33,79] and the gene for diplococcin[15] phenotype is encoded in an 81 Kb plasmid. In contrast, the gene for nisin is located in a transposon,[80] and for lactacin B,[47] it is chromosomally linked. Other characteristics listed in Table 4 show that there are many similarities that exist between the bacteriocins. In studying the properties of a new bacteriocin these characteristics may be included to make comparisons with others. It will also be important to check if the strains of a species are really different or if they are the same with respect to a specific bacteriocin. For example, four bacteriocin-producing *P. acidilactici* strains H, E, F, and M, isolated from fermented meat and obtained from commercial sources, were originally thought to be different in some properties.[33,34] However, comparisons of the molecular weight of the bacteriocins, and restriction digestion profiles of the plasmids encoding for bacteriocin phenotype for these strains, revealed that the plasmids are the same in all four strains.[32,79]

TABLE 5
Food Systems in Which Bacteriocins of Starter Culture Bacteria Have Been Tested for Antimicrobial Efficiency

Nisin[9,10,82–84]:	Dairy Products: pasteurized milk, evaporated milk, dairy desserts, cheese spread, different cheeses. Meat products: canned ham, pork slurry, bacon, fresh beef, beef byproducts. Fish products: fresh fish. Others: canned vegetables, canned soup, wine, beer.
Pediocin AcH[35,83–88]:	Dairy: milk, ice cream, cottage cheese. Meat: refrigerated fresh beef and beef byproducts, ground beef, sausage mix, cooked beef and pork products.
Pediocin PA137:	Dairy: cottage cheese, cheese spread, milk. Meat: beef.

III. SUITABILITY OF BACTERIOCINS OF STARTER CULTURE BACTERIA AS FOOD BIOPRESERVATIVES

The antibacterial properties of bacteriocins and bacteriocin-like compounds have mostly been studied with pure cultures in bacteriological media. Only three bacteriocins, nisin from *L. lactis* subsp. *lactis,* pediocin AcH from *P. acidilactici* H, and pediocin PA1 from *P. acidilactici* PAC 1.0 have been tested for their ability to cause a bacteriocidal effect in different food systems (Table 5). Nisin, recognized for over 50 years for its antibacterial property, has been extensively studied in many different food systems that include dairy, meat, fish, vegetables, wine, and beer, and has been found to be effective in preserving these foods against many Gram-positive spoilage and pathogenic bacteria with some limitations.[9,10,81–84] It has been approved for use in many countries, including the U.S.,[81] for use in specific foods. The efficiency of pediocin AcH, which was first identified in 1984, in controlling bacterial population in food systems, is currently being extensively studied.[35,83–87] It has been found to effectively control spoilage and pathogenic Gram-positive bacteria in liquid milk, ice cream, cottage cheese, fresh and cooked beef, ground beef, beef byproducts, and cooked pork products. Similarly pediocin PA1 has been found to control Gram-positive bacterial population in cottage cheese, cheese spread, and milk.[37]

There are several limitations one should recognize in using bacteriocins as food preservatives. They are effective, at least according to current understanding, against some Gram-positive bacterial strains. They are bacteriocidal to the sensitive strains; however, in a sensitive population there are cells that are resistant. As bacteriocins are not bacteriostatic, the resistant survivors will multiply and can reach a high population with time even in the presence of residual bacteriocins. However, a food produced under proper

TABLE 6
Some Criteria of a Bacteriocin From Starter Culture Bacteria for Use as Food Biopreservatives

1. Produced by a food-grade bacteria
2. Preferably have a wide host range
3. Preferably activity is heat stable
4. Effective under wide pH and temperature range
5. Retained activity during storage under different conditions
6. Activity is not lost in the presence of food additives
7. Effective in food systems during storage period
8. Effective in low concentrations
9. Easy to apply in food system
10. Safe for human consumption
11. Does not alter acceptance quality of a food
12. Economical

sanitary conditions and not subsequently grossly abused, is expected to have a low level of contamination of Gram-positive spoilage and pathogenic bacteria. Incorporation of a suitable bacteriocin as one of the several hurdles used to extend shelf life and enhance safety of this food (as a secondary barrier) will reduce the population of the Gram-positive spoilage and pathogenic bacteria to such a low level that, under proper conditions of storage, the survivors will take a long time to reach a population level to either spoil the food or make it unsafe (especially for pathogens that require high population levels to cause health hazards). The other limitation is that the bacteriocins of starter culture bacteria are not normally effective against Gram-negative bacteria, yeasts, and molds, and thus cannot be used to control their growth in foods in which they can grow.

In certain foods, such as fresh and minimally processed vacuum-packaged refrigerated foods with a need for extended shelf life, growth of Gram-positive anaerobic and facultative anaerobic psychrotrophic spoilage bacteria is of major concern.[32,87–90] These products can provide the necessary environment for growth of some psychrotrophic Gram-positive pathogens such as nonproteolytic *Clostridium botulinum* B and E, and *Listeria monocytogenes*. Suitable bacteriocin preparations can be used in these foods to reduce growth of these Gram-positive bacteria to extend their shelf life and enhance safety. However, suitability of a bacteriocin preparation as a food biopreservative will depend upon several factors. These include factors that are normally considered for food preservatives and additives (Table 6). In general, their safety to consumers, their bacteriocidal efficiency to a diverse group or Gram-positive bacteria in different food systems, their stability and uniform distribution in food systems, and their cost will be of major concern.

IV. CONCLUDING REMARKS

Current need for suitable food biopreservatives, especially in the developed countries, has renewed interest in the search for bacteriocins of starter culture bacteria as possible candidates. This is evident from the explosion in the publication of research articles and reviews and scientific symposiums in this area. However, most information is far from complete, and in many instances it is confusing and contradictory. It will be important to set up guidelines to generate information with an ultimate aim to test the suitability of a bacteriocin as a food biopreservative. A minimum need for this will probably be to test the antimicrobial effects not only with colonies and culture broths of producer strains, but also with partially purified preparations. Similarly, examination of the compound in food systems will be necessary before its usefulness as a food biopreservative can be claimed. Most currently published data are providing very little information on the effectiveness of many bacteriocins in food systems.

Bacteriocins of lactic acid bacteria such as nisin have often been connected with the term "antibiotics." One of the major reasons for the current search for biopreservatives is the consumer perception that many currently used food preservatives are unsafe and unnatural. Use of the term "antibiotic" for these bacteriocins may create confusion among consumers, may give cause to the consumer lobbyist groups to be vocal, may force the regulatory agencies to be overly cautious in allowing their use in foods, and finally, may create hesitation amongst food processors and even force them to stay away from using something that, in reality, is safe and effective. There is a definite need for a biopreservative that includes bacteriocins of starter culture bacteria to ensure the safety of foods such as the new generation refrigerated foods. Use of the term "antibiotic" for the bacteriocins could therefore turn out to be a poor choice. The concern in using the term "antibiotic" for nisin and other bacteriocins of lactic acid bacteria (LAB) is reflected in the following statement made in 1983 by Hurst[91] who has studied the different aspects of nisin since 1944:[92]

> In previous publications nisin has been called an antibiotic, but the reader may have noted that this term was carefully avoided when dealing with inhibitors formed by LAB. At present there is considerable and probably justifiable resistance to the use of antibiotics as food preservatives. Although nisin has been thought of as an antibiotic, it is suggested that it would be more appropriate to classify it as a bacteriocin and therefore it should be placed among the other biological inhibitors produced by the LAB and not among the antibiotics. ...nisin producing streptociocci occur in milk. ...Therefore it is likely that nisin or substances similar to it have been consumed for a long time without apparent ill effects. Concern that the use of nisin might hide sloppy manufacturing practices is unjustified because of its narrow antibacterial spectrum. ...As the number of organisms increase the inhibitory effectiveness of nisin decreases ...nisin is ineffective when poor quality starting materials were used.

There are reasons not to designate bacteriocins as antibiotics. Some of the differences between bacteriocins of starter culture bacteria and antibiotics used

TABLE 7
Some Differences Between Bacteriocins of Starter Cultures and Therapeutic Antibiotics[a,b]

	Bacteriocins of starter culture[c]	Therapeutic antibiotics
1.	Produced by safe food-grade bacteria	Produced by nonfood-grade microorganisms
2.	Ribisomally synthesized proteins; many undergo posttranslation modification	Nonribosomally synthesized and catalyzed by multienzyme complexes by microorganisms or synthesized in laboratory
3.	Relatively large peptide	Many are nonproteins; those which are proteins are small peptides; can be synthetic
4.	Destroyed by the proteolytic enzymes of GI tract	Not affected by these enzymes of the GI tract
5.	No adverse effect on microflora of the GI tract[d]	Upset the GI tract microflora
6.	Bind to specific receptors on cell wall prior to causing viability loss of sensitive bacteria	Inhibit cellular functions not through binding to any cell wall receptors

[a] Presented in a symposium at the joint annual meeting of the American Society of Animal Science, Lexington, KY, 1989.
[b] Antibiotics: extract of certain microorganisms (or their synthetic equivalents) employed against infections caused by other microorganisms (New Gould Medical Dictionary, 2nd ed., McGraw-Hill, 1956.)
[c] Based on information on bacteriocins of starter culture bacteria.
[d] Nisin is the only bacteriocin tested and found to have no effect on GI microflora.

for therapeutic purposes are listed in Table 7. Just the term "bacteriocin" or even the term "lantibiotic," at least for those that contain lanthionine, may prove to be a better choice. Alternatively, the recent suggestion of Tagg[1a] to designate the antibacterial peptide and proteins that are different from colicins in many respects as "Blis" may be accepted to include the antibacterial proteins of starter culture bacteria. In addition to this, a system in naming the specific Blis will be important to reduce the current confusion. Probably a good approach will be to purify and determine the amino acid sequence of a specific Blis from a species/strain before giving it a name. Also, the genus, species, and strain can be combined together to name a Blis. We used this system to designate an antibacterial peptide from *Pediococcus acidilactici* H as pediocin AcH.[28]

Finally, one should recognize that a suitable bacteriocin of starter culture bacteria should be used as one of the barriers. Use of bacteriocin in products that are produced under proper sanitation with a low initial level of Gram-positive bacteria, and handled and stored under the proper environment, will provide the most desirable results. In contrast, including a bacteriocin in a product produced under unsanitary conditions, and abused during subsequent handling prior to consumption, will not do any good.

REFERENCES

1. **Tagg, J. R., Dajani, A. S., and Wannamaker, L. W.,** Bacteriocins of Gram-positive bacteria, *Bacteriol. Rev., 40,* 722, 1976.

1a. **Tagg, J. R.,** Bacterial BLIS, *ASM News, Am. Soc. Microbiol.,* 57, 611, 1991.

2. **Schnell, N., Entian, K. D., Schneider, U., Gotz, F., Zahner, H., Kellner, R., and Jung, G.,** Prepeptide sequence of epidermin, a ribosomally synthesized antibiotic with four sulphide rings, *Nature* (London), 333, 276, 1988.

3. **Kaletta, C. and Entian, K. D.,** Nisin, a peptide antibiotic: cloning and sequencing of the *nis* A gene and posttranslation processing of its peptide product, *J. Bacteriol.,* 171, 1597, 1989.

4. **Weil, H.P., Beck-Sickinger, A. G., Metzger, J., Stevanovic, S., Jung, G., Josten, M., and Sahl, H. G.,** Biosynthesis of lantiobiotic Pep 5., *Int. J. Biochem.,* 194, 217, 1990.

5. **Horner, T., Ungerman, V., Zahner, H., Fiedler, H. P., Utz, R., Kellner, R., and Jung, G.,** Comparative studies on the fermentative production of lantibiotics by staphylococci, *Appl. Microbiol. Biotechnol.,* 32, 511, 1990.

6. **Klaenhammer, T. R.,** Bacteriocins of lactic acid bacteria, *Biochimie,* 70, 337, 1988.

7. **Bhunia, A. K., Johnson, M. C., Ray, B., and Kalchayanand, N.,** Mode of action of pediocin AcH from *Pediococcus acidilactici* H on sensitive bacterial strains, *J. Appl. Bacteriol.,* 70, 25, 1991.

8. **Gherna, R. and Pienta, P., Eds.,** American type culture collection, in *Catalogue of Bacteria and Bacteriophages,* 17th ed., American Type Culture Collection, Rockville, MD, 1989.

9. **Hurst, A., Nisin,** *Adv. Appl. Microbiol.,* 27, 85, 1981.

10. **Fowler, G. G. and Gasson, M. J.,** Antibiotics – nisin, in *Food Preservatives,* Russell, N. J. and Gould, G. W., Eds., AVI Publishing, New York, 1990, chap. 8.

11. **Geis, A., Singh, J., and Teuber, M,** Potential of lactic streptococci to produce bacteriocin, *Appl. Environ. Microbiol.,* 45, 205, 1983.

12. **Powell, I. B., Ward, A. C., Hiller, A. J., and Davidson, B. E.,** Simultaneous conjugal transfer in *Lactococcus* two genes involved in bacteriocin production and reduced susceptibility to bacteriophages, *FEMS Microbiol. Lett.,* 72, 209, 1990.

13. **Kozak, W., Bardowski, J., and Dobrzanski, W. T.,** Lactostrepcins-acid bacteriocins produced by lactic streptococci, *J. Dairy Res.,* 45, 247, 1978.

13a. **Thuault, D., Beliard, E., LeGuern, J., and Bourgeois, E. M.,** Inhibition of *Clostridium tyrobutyricum* by bacteriocin-like substance produced by lactic acid bacteria, *J. Dairy Sci.,* 74, 1145, 1991.

13b. **Dufour, A., Thuault, D., Boulliou, A., Bourglois, C. M., and LePennee, J. P.,** Plasmid-encoded determinants for bacteriocin production and immunity in *Lactococcus lactis* strain and purification of the inhibitory peptide, *J. Gen. Microbiol.,* 137, 2423, 1991.

14. **Davey, G. P. and Richardson, B. C.,** Purification and some properties of diplococcin from *Streptococcus cremoris* 346, *Appl. Environ. Microbiol.,* 41, 84, 1981.

15. **Davey, G.,** Plasmid associated with diplococcin production in *Streptococcus cremoris, Appl. Environ. Microbiol.,* 48, 895, 1984.

16. **Zajdel, J. K., Ceglowski, P., and Dobrzanski, W. T.,** Mechanism of action of lactostrepcin 5, a bacteriocin produced by *Streptococcus cremoris* 202, *Appl. Environ. Microbiol.,* 49, 969, 1985.

17. **van Belkum, M., Hayema, B. J., Geis, A., Kok, J., and Venema, G.,** Cloning of two bacteriocin genes from a lactococcal bacteriocin plasmid, *Appl. Environ. Microbiol.,* 55, 1187, 1989.

18. **van Belkum, M. J., Hayema, B. J., Jeeninga, R. E., Kok, J., and Venema, G.,** Organization and nucleotide sequences of two lactococcal bacteriocin operons, *Appl. Environ. Microbiol.,* 57, 492, 1991.

19. **Branen, A. L., Go, H. C., and Genske, R. P.,** Purification and properties of antimicrobial substances produced by *Streptococcus diacetylactis* and *Leuconostoc citrovorum, J. Food Sci.,* 40, 446, 1975.

20. **Kojic, M., Svircevec, J., Banina, A., and Topisirovic, L.,** Bacteriocin producing strain of *Lactococcus lactis* subsp. *diacetylactis* S50, *Appl. Environ. Microbiol.*, 57, 1835, 1991.
21. **Scherwitz, K. M., Baldwin, K. A., and McKay, L. L.,** Plasmid linkage of bacteriocin-like substances in *Streptococcus lactis* subsp. *diacetylactis, Appl. Environ. Microbiol.*, 45, 1506, 1983.
22. **Harmon, K. C. and McKay, L. L.,** Restriction enzyme analysis of lactose and bacteriocin plasmids from *Streptococcus lactis* subsp. *diacetylactis* WM4 and cloning of *Bcl* I fragments coding for bacteriocin production, *Appl. Environ. Microbiol.*, 53, 1171, 1987.
23. **Pulusani, S. R., Rao, D.R., and Sunki, G. R.,** Antimicrobial activity of lactic cultures: partial purification and characterization of antimicrobial compounds produced by *Streptococcus thermophilus, J. Food Sci.*, 44, 575, 1979.
24. **Ray, B.,** Bacteriocins of Psychrotrophic Leuconostocs Isolated from Refrigerated Meat Products, University of Wyoming, Laramie, 1990, unpublished data.

24a. **Daba, H., Pondian, B., Gosselin, J. F., Simard, R. E., Huang, J., and Lacroix, C.,** Detection and activity of bacteriocin produced by *Leuconostoc mesenteroides, Appl. Environ. Microbiol.*, 57, 3450, 1991.

25. **Charmaigne, D., Harding, C. D., and Shaw, B. G.,** Antimicrobial activity of *Leuconostoc gelidum* against closely related species and *Listeria monocytogenes, J. Appl. Bacteriol.*, 69, 648, 1990.
26. **Hastings, J. W. and Stiles, M. E.,** Antibiotics of *Leuconostoc gelidum* isolated from meat, *J. Appl. Bacteriol.*, 70, 127, 1991.
27. **Lewus, C. B., Kaiser, A., and Montville, T.,** Inhibition of food-borne bacterial pathogens by bacteriocins from lactic acid bacteria isolated from meat, *Appl. Environ. Microbiol.*, 57, 1683, 1991.
28. **Bhunia, A. K., Kim, W. J., Johnson, M. C., and Ray, B.,** Partial purification and chracterization of antimicrobial substances from *Pediococcus acidilactici* and *Lactobacillus plantarum*.Institute of Food Technologists Annual Meeting, Las Vegas, 143, (Abstr.), 1987.
29. **Bhunia, A. K., Johnson, M. C., and Ray, B.,** Direct detection of an antimicrobial peptide of *Pediococcus acidilactici* in sodium dodecyl sulfate-polyacrylamide gel electrophoresis, *J. Ind. Microbiol.*, 2, 319, 1987.
30. **Bhunia, A. K., Johnson, M. C., and Ray, B.,** Purification, characterization and antimicrobial spectrum of a bacteriocin produced by *Pediococcus acidilactici, J. Appl. Bacteriol.*, 65, 261, 1988.
31. **Bhunia, A. K.,** Characteristics and Mode of Action of an Antimicrobial Peptide Produced by *Pediococcus acidilactici* Strain H, Ph.D. thesis, University of Wyoming, Laramie, 1989.
32. **Ray, B., Kalchayanand, N., and Field, R. A.,** Isolation of a *Clostridium* spp. from spoiled vacuum-packaged refrigerated beef and its susceptibility to bacteriocins from *Pediococcus acidilactici*, in Proc. International Conference on Meat Science and Technology, Copenhagen, 285, 1989.
33. **Ray, S. K., Johnson, M. C., and Ray, B.,** Bacteriocin plasmids of *Pediococcus acidilactici, J. Ind. Microbiol.*, 4, 163, 1989.
34. **Kim, W. J.,** Gene Transfer System in Lactic Acid Bacteria used in Meat Fermentation, Ph.D. thesis, University of Wyoming, Laramie, 1990.
35. **Motlagh, A.,** Antimicrobial efficiency of pediocin AcH and genetic studies of the plasmid encoding for pediocin AcH, Ph.D. thesis, University of Wyoming, Laramie, 1991.
36. **Gonzalez, C. F. and Kunka, B. S.,** Plasmid associated bacteriocin production and sucrose fermentation in *Pediococcus acidilactici, Appl. Environ. Microbiol.*, 53, 2534, 1987.
37. **Pucci, M. J., Vedamuthu, E. R., Kunka, B. S., and Vandenbergh, P. A.,** Inhibition of *Listeria monocytogenes* by using bacteriocin PA-1 produced by *Pediococcus acidilactici* PAC 1.0, *Appl. Environ. Microbiol.*, 54, 2349, 1988.
38. **Hoover, D. G., Walsh, P. M., Kolaetis, K. M., and Daly, M. M.,** A bacteriocin produced by *Pediococcus* species associated with a 5.5 Mdal plasmid, *J. Food Prot.*, 51, 29, 1988.

39. **Fleming, H. P., Etchells, J. L., and Costilow, R. N.,** Microbial inhibition by an isolate of *Pediococcus* from cucumber brine, *Appl. Microbiol.,* 30, 1040, 1975.
40. **Daeschel, M. and Klaenhammer, T. R.,** Association of a 13.6 Mdal plasmid in *Pediococcus pentosaceus* with bacteriocin activity, *Appl. Environ. Microbiol.,* 50, 1538, 1985.
41. **Harris, L. J., Daeschel, M. A., Stiles, M. E., and Klaenhammer, T. R.,** Antimicrobial activity of lactic acid bacteria against *Listeria monocytogenes, J. Food Prot.,* 52, 3784, 1989.
42. **Okereke, A. and Montville, T. J.,** Bacteriocin inhibition of *Clostridium* spores by lactic acid bacteria, *J. Food Prot.,* 54, 349, 1991.
43. **Graham, D. C. and McKay, L .L.,** Plasmid DNA in strains of *Pediococcus cerevisiae* and *Pediococcus pentosaceus, Appl. Environ. Microbiol.,* 50, 532, 1985.
44. **Spelhaug, S. R. and Harlander, S. K.,** Inhibition of food-borne bacterial pathogens by bacteriocins from *Lactococcus lactis* and *Pediococcus pentosaceus, J. Food Prot.,* 52, 856, 1989.
45. **Hosono, A., Yastuki, K., and Tokita, F.,** Isolation and characterization of an inhibitory substance against *Escherichia coli* produced by *Lactobacillus acidophilus, Milchwissenschaft,* 32, 727, 1977.
46. **Mehta, A. M., Patel, K. A., and Dave, P. J.,** Isolation and purification of an inhibitory protein from *Lactobacillus acidophilus* AC1, *Microbios,* 37, 37, 1983.
47. **Barefoot, S. F. and Klaenhammer, T. R.,** Detection and activity of Lactacin B, a bacteriocin produced by *Lactobacillus acidophilus, Appl. Environ. Microbiol.,* 45, 1808, 1983.
48. **Barefoot, S. F. and Klaenhammer, T. R.,** Purification and characterization of the *Lactobacillus acidophilus* bacteriocin Lactacin B, *Antimicrobiol. Agents Chemother.,* 26, 328, 1984.
49. **Muriana, P. M. and Klaenhammer, T. R.,** Conjugal transfer of plasmid-encoded determinants for bacteriocin production and immunity in *Lactobacillus acidophilus* 88, *Appl. Environ. Microbiol.,* 53, 553, 1987.
50. **Muriana, P. M. and Klaenhammer, T. R.,** Purification and partial characterization of Lactacin F, a bacteriocin produced by *Lactobacillus acidophilus* 11088, *Appl. Environ. Microbiol.,* 57, 114, 1991.
51. **Toba, T., Yoshiko, E., and Itoh, T.,** Acidophilucin A, a new heat labile bacteriocin produced by *Lactobacillus acidophilus* LAPT 1060, *Lett. Appl. Microbiol.,* 12, 106, 1991.
52. **Anderson, R.,** Inhibition of *Staphylococcus aureus* and spheroplasts of Gram-negative bacteria by an antagonistic compound produced by a strain of *Lactobacillus plantarum, Int. J. Food Microbiol.,* 3, 149, 1986.
53. **West, C. A. and Warner, P.J .,** Plantacin B, a bacteriocin produced by *Lactobacillus plantarum* NCDO 1193, *FEMS Microbiol. Lett.,* 49, 163, 1988.
54. **Daeschel, M. A., McKenney, M. C., and McDonald, L. C.,** Bacteriocin activity of *Lactobacillus plantarum* C11, *Food Microbiol.,* 7, 91, 1990.
55. **Giraffa, G., Neviani, E., and Veneroni, A.,** Use of conductance to detect bacteriocin activity, *J. Food Prot.,* 53, 772, 1990.
56. **Toba, T., Yoshioka, E., and Itoh, T.,** Lactacin, a bacteriocin produced by *Lactobacillus delbrueckii* subsp. *lactis, Lett. Appl. Microbiol., 12,* 43, 1991.
57. **Upreti, G. C. and Hinsdel, R. D.,** Isolation and characterization of a bacteriocin from a homofermentative *Lactobacillus, Antimicrobiol. Agents Chemother.,* 4, 487, 1973.
58. **Upreti, G. C. and Hinsdel, R. D.,** Production and mode of action of Lactocin 27: bacteriocin for homofermentative *Lactobacillus, Antimicrobiol. Agents Chemother., 7, 1*39, 1975.
59. **Joerger, M. C. and Klaenhammer, T. R.,** Characterization and purification of helveticin J and evidence for a chromosomally determined bacteriocin produced by *Lactobacillus helveticus* 481, *J. Bacteriol.,* 167, 439, 1986.

60. **Joerger, M. C. and Klaenhammer, T. R.,** Cloning, expression and nucleotide sequence of the *Lactobacillus helveticus* 481 gene encoding the bacteriocin helveticia J, *J. Bacteriol.,* 172, 6339, 1990.
61. **Rammelsberg, M., Muller, E., and Radler, F.,** Caseicin 80: purification and characterization of a new bacteriocin from *Lactobacillus casei, Arch. Microbiol.,* 154, 249, 1990.
62. **Rammelsberg, M. and Radler, F.,** Antibacterial polypeptide of *Lactobacillus* spp., *J. Appl. Bacteriol.,* 69, 177, 1990.
63. **Deklerk, H. C. and Smit, J. A.,** Properties of *Lactobacillus fermenti* bacteriocin, *J. Gen. Microbiol.,* 48, 309, 1967.
64. **Schillinger, U. and Lucke, F. K.,** Antibacterial activity of *Lactobacillus sake* isolated from meat, *Appl. Environ. Microbiol.,* 55, 1901, 1989.
65. **Mortvedt, C. I. and Nes, I. F.,** Plasmid associated bacteriocin production by *Lactobacillus sake* strain, *J. Gen. Microbiol.,* 136, 1601, 1990.
66. **Mortvedt, C. I., Nissen-Meyer, J., Slettern, K., and Nes, I. F.,** Purification and amino acid sequence of lactocin S, a bacteriocin produced by *Lactobacillus sake* L45, *Appl. Environ. Microbiol.,* 57, 1829, 1991.
67. **Toba, T., Yoshioka, E., and Itoh, T.,** Potential of *Lactobacillus gasseri* isolated from infant faeces to produce bacteriocin, *Lett. Appl. Microbiol.,* 12, 228, 1991.
68. **Anand, S. K., Srinivasan, R. A., and Rao, L. K.,** Antibacterial activity associated with *Bifidobacterium bifidum, Cult. Dairy Prod.,* 20(1), 21, 1985.
69. **Meghrous, J., Euloge, P., Junelles, A. M., Ballongue, J., and Petitdemange, H.,** Screening of *Bifidobacterium* strains for bacteriocin production, *Biotechnol. Lett., 12*, 575, 1990.
70. **Lyon, W. J. and Glatz, B. A.,** Partial purification and characterization of a bacteriocin produced by *Propionibacterium thoenii, Appl. Environ. Microbiol.,* 57, 701, 1991.
71. **Ahn, C. and Stiles, M. E.,** Plasmid-associated bacteriocin production by a strain of *Carnobacterium piscicola* from meat, *Appl. Environ. Microbiol.,* 56, 2503, 1990.
71a. **McKay, A. M.,** Antimicrobial activity of *Enterococcus faecium* against *Listeria* spp., *Lett.Appl. Microbiol., 11, 15*, 1990.
72. **Kramer, J., Keness, J., and Brandis, H.,** Transfer of mini-plasmid determining bacteriocin production and bacteriocin-immunity in *Streptococcus faecium, FEMS Microb. Lett., 20*, 385, 1983.
73. **Hamdan, I. Y. and Mikolajcik, E. M.,** Acidolin: an antibiotic produced by *Lactobacillus acidophilus, J. Antibiot., 27*, 631, 1974.
74. **Shahani, K. M., Vakil, J. R., and Kilara, A.,** Natural antibiotic activity of *Lactobacillus acidophilus* and *bulgaricus.* II. Isolation of acidophilin from *L. acidophilus, Cult. Dairy Prod. J.,* 12(2), 8, 1977.
75. **Vincent, J. G., Veomett, R. C., and Riley, R. F.,** Antibacterial activity associated with *Lactobacillus acidophilus, J. Bacteriol.,* 78, 477, 1959.
76. **Reddy, G. V., Shahani, K. M., Friend, B. A., and Chandan, R. C.,** Natural antibiotic activity of *Lactobacillus acidophilus* and *bulgaricus.* III. production and partial purification of bulgarican from *Lactobacillus bulgaricus, Cult. Dairy Prod. J.,* 19(2), 7, 1984.
77. **Kavasnikov, E. I. and Sodenko, V. I.,** Antibiotic properties of *Lactobacillus brevis, Mikrobiol. Zh. Kyviv,* 29, 146, 1967.
78. **Biswas, S. R., Ray, P., Johnson, M. C., and Ray, B.,** Influence of growth conditions on the production of a bacteriocin, pediocin AcH by *Pediococcus acidilactici* H, *Appl. Environ. Microbiol.,* 57, 1265, 1991.
79. **Ray, B., Nie, H., and Menke, S.,** Mapping of pSMB74, a plasmid encoding bacteriocin pediocin AcH production (Pap^+) phenotype in *Pediococcus acidilactici* H, 91st General Meeting, American Society of Microbiology, Dallas, TX, 031 (Abstr.), 1991.
80. **Steen, M. T., Chung, Y. J., and Hansen, N.,** Characterization of the nisin gene as part of a polycistronic operon in the chromosome of *Lactococcus lactis* ATCC 11454, *Appl. Environ. Microbiol.,* 57, 1181, 1991.

81. **Anon.,** Nisin preparation; affirmation of GRAS status as a direct human food ingredient, *Fed. Register,* 53(56), 11247, 1988.

82. **Delves-Broughton, J.,** Nisin and its uses as a food preservative, *Food Technol.,* 44(11), 100, 1990.

82a. **Benkerroum, N. and Sandine, W. E.,** Inhibitory action of nisin against *Listeria monocytogenes, J. Dairy Sci.,* 71, 3237, 1988.

83. **Rozbeh, M.,** Extension of Shelf-Life in Vacuum-Packaged Fresh Beef, M.S. thesis, University of Wyoming, Laramie, 1990.

84. **Rozbeh, M., Kalchayanand, N., Ray, B., and Field, R. A.,** Shelf-life extension of vacuum-packaged refrigerated beef using starter culture metabolites, Proc. Western Section, American Society of Animal Science, Aug. 6–9, Laramie, WY, 1991.

85. **Holla, S.,** Efficiency of Pediocin AcH on Viability Loss of Pathogenic and Spoilage Bacteria in Food, M.S. thesis, University of Wyoming, Laramie, 1990.

86. **Motlagh, A., Holla, S., Johnson, M. C., Ray, B., and Field, R. A.,** Inhibition of *Listeria* spp. in sterile food systems by pediocin AcH, a bacteriocin from *Pediococcus acidilactici* H, Proc. Western Section, American Society of Animal Science, Aug. 6–9, Laramie, WY, 1991.

87. **Kalchayanand, N.,** Extension of Shelf-Life of Vacuum-Packaged Refrigerated Fresh Beef by Bacteriocins of Lactic Acid Bacteria, Ph.D. thesis, University of Wyoming, Laramie, 1990.

88. **Kalchayanand, N., Ray, B., and Field, R. A.,** Characteristics of *Clostridium laramie,* a psychrotrophic species, associated with spoilage of vacuum-packaged fresh beef, Proc. Western Section, American Society of Animal Science, Aug. 6–9, Laramie, WY, 1991.

89. **Eagan, A. F. and Shay, B. J.,** Significance of lactobacilli and film permeability in the spoilage of vacuum-packaged beef, *J. Food Sci.,* 47, 1119, 1982.

90. **Palumbo, S. A.,** Is refrigeration enough to restrain food-borne pathogens? *J. Food Prot.,* 49, 1003, 1986.

91. **Hurst, A.,** Nisin and other inhibitory substances from lactic acid bacteria, in *Antimicrobials in Foods,* Branen, A .L. and Davidson, P. M., (Eds.), Marcel Dekker, New York, 1983, chap. 11.

92. Mat**tick, A. T. R. and Hirsch, A.,** Powerful inhibitory substance produced by group a streptococci, *Nature (London)*, 154, 551, 1944.

Chapter 9

NISIN OF *LACTOCOCCUS LACTIS* SSP. *LACTIS* AS A FOOD BIOPRESERVATIVE

Bibek Ray

TABLE OF CONTENTS

...nisin has the unique function of being used as a food preservative. There are several reasons for this: it is not used medicinally or in animal food, and, when eaten, it is nontoxic. Since it is a polypeptide, any residue remaining in foods are digested ...It is a mistake to combine together everything that are called antibiotics. Nisin is evidently very different from the antibiotics used in medical and veterinary practices and it has not been used for purposes of growth promotion. Progress in the area of food preservation will be greatly hampered as long as popular confusion remains over all substances called antibiotics being kept out of foods. It might be better to call nisin a biological food preservative and class it together with lactic acid, acetic acid and hydrogen peroxide, the other antibacterial products of the lactic streptococci.[1]

I. INTRODUCTION

Prior to 1951, when nisin was first successfully tested as a food biopreservative to prevent spoilage of Swiss-type cheese by the inhibition of the growth of an anaerobic spore-forming bacterial species,[2] it went through stages of "discovery-forgotten-rediscovery" for over two decades. As early as 1928, Rogers and his fellow researchers[3,4] in the U.S. published reports on metabolites of *Streptococcus lactis* that were inhibitory to other lactic acid bacteria. Then in 1933, Whitehead[5] and his associate in New Zealand observed that during fermentation, some wild strains of streptococci inhibited growth and acid production in milk by lactic cultures during early fermentation in the cheese-making process. They reported that the metabolite of the streptococci that inhibited the growth of the lactic streptococci was polypeptide in nature.[6,7] In 1943, Meanwell in England again recognized that growth and acid production by cheese starter cultures could be adversely affected by other lactic streptococci.[8] Subsequently, other researchers also reported the production of antagonistic metabolites by some lactic streptococci against other lactic starter cultures.[9–12]

Meanwhile, researchers in England in 1947 recognized that this proteinaceous inhibitor was produced by some strains of streptococci from group N, and they coined the name "nisin" to designate the group N inhibitory substance.[13] Several researchers also initially conducted studies on the possible use of nisin as a therapeutic agent to combat bacterial infectious diseases in humans and animals. These and other later studies suggested that nisin, due to its poor solubility at the blood pH level, and due to the loss of activity by the gastric proteolytic enzyme (α-chymotrypsin), could not be used as an effective therapeutic agent by either inoculation or feeding.[14–20] However, initial studies by Hirsch et al.[2] in 1951, and subsequently by many others in the same decade, indicated that nisin could probably be used as a food biopreservative to control growth of Gram-positive spoilage bacteria.[21–26]

During the fifties, a fairly large number of reports were published on the production,[27,28] purification,[29,30] properties,[31–33] assay,[34–36] antibacterial spectrum, and mode of action[37–39] of nisin. It was found to be an acidic peptide (or small protein) with sulfur-containing amino acid and to be effective only against some Gram-positive bacteria. The resistant bacterial strains could produce the enzyme nisinase,[39,40] or the sensitive strain could be trained to develop a resistance against nisin.[41] Since 1960, various aspects of nisin have been studied, covering both basic and applied areas. Some of the most important studies include elucidation of the amino acid sequence of this 34-amino acid peptide by Gross and Morell[42] in 1971, synthesis of nisin molecule,[43] determination of conformation of the molecule,[43a] cloning and sequencing of the gene encoding for nisin-production phenotype[44–47] and the testing of its nontoxicity to animals.[14,48] The results of the applied studies have encouraged its commercial use in several food systems in many countries, including the U.S., and its production and marketing by commercial manufacturers.[49] The results of these studies have been the subject of several reviews.[1,49–53] This chapter is devoted primarily to discussing the effectiveness of nisin as a food biopreservative. However, other relevant information, covering both basic and applied areas, will be included in an effort to make this review current and complete.

II. ANTIBACTERIAL ACTIVITY OF NISIN

A. Measurement of Activity

To measure the antibacterial effectiveness of nisin it is important to know the potency of a preparation. This is also important for standardization of preparations to compare the relative sensitivity of different microorganisms and also to compare different assay methods used in different laboratories.

1. Units Used

Currently, the activity of a nisin preparation is expressed in terms of International Units (IU). Originally, Tamer and Fowler[54] suggested the use of 1 Reading Unit (RU) as the amount of activity present in 1 μg of pure nisin preparation. This was then changed to IU with the conversion of 1 RU = 40 IU, which was accepted by the World Health Organization.[55] The assay method was the end-point dilution of the nisin preparation, and examining for zone(s) of inhibition by an agar diffusion method against the indicator *Micrococcus flavas*.[54] Using *Lactobacillus plantarum* NCDO 955 to compare the potency of preparations of different bacteriocins of lactic acid bacteria, Motlagh et al. observed that 1 IU = 100 AU (activity units).[56,56a] Currently, in different studies, several methods are used to report units of nisin (or commercial preparation, Nisaplin) such as IU/g, mg/kg, ppm, or % nisin in test systems. There is a possibility that in the future all results will use mg/kg of food.[56b] Table 1 presents the relationships between various units now in use.

2. Assay Methods

In 1934, to determine the presence of nisin or nisin-producing bacteria in milk, Cox[57] suggested the use of methylene blue reduction test with cells of starter culture bacteria. Chevalier et al.[57a] suggested the use of agar plate assay method against a strain of *Lactococcus lactis*. Hirsch[34] developed a "lag phase" method that was based on a linear relationship between the length of the lag of a culture of *Streptococcus agalactia* and the nisin concentrations within a range of 5 to 50 IU/ml. Berridge and Barrett[35] later developed a turbidimetric method that was also based on an inverse relationship between the concentrations of nisin and optical.density (OD) at 600 nm of a growing culture of *Streptococcus cremoris*. However, this method could not be used with substances that interfere with OD determination such as many foods do.

Mocquot and Lefebre[36] effectively used a horizontal agar diffusion technique by introducing Tween 80 in the medium to facilitate diffusion of nisin in the agar medium. This was finally improved by Tamer and Fowler[54] when they incorporated 1% Tween 20 in the agar medium seeded with cells of *Micrococcus flavus* and deposited the test material in a well. After overnight incubation at 30°C, the size of the zones of inhibition was measured. The zone diameters and nisin concentrations between 0.5 to 10 IU were linear.

Recently, another method, developed to assay concentrations of another bacteriocin of lactic acid bacteria (pediocin AcH by P*ediococcus acidilactici*),[58] was used to assay nisin concentration in culture broth preparations. This included decimal dilutions of the preparation in sterile deionized water and testing 5 μl from each dilution on an agar-medium plate containing 0.2% Tween 80 and seeded with about 10^6 cells of *Lactobacillus plantarum* NCDO 955. After overnight incubation at 30°C, the highest dilution that gave a zone of inhibition of at least 1 mm was multiplied by 200 to obtain activity units

TABLE 1
Relationship Between Different Units of Nisin Currently Used[56b]

A. Pure nisin: 1 g = 40 × 10^6 IU or 1 × 10^6 RU

1 mg of nisin per kilogram of food	=	40 IU/g of food
	=	1 ppm in food
	=	0.0001%
500 IU nisin per gram of food	=	12.5 mg/kg of food.
	=	12.5 ppm in food
	=	0.00125% in food (or 1.25 mg %)

B. Nisaplin (a commercial preparation): 1 g = 1 × 10^6 IU

1 mg nisaplin per kilogram of food	=	1 IU/g of food
	=	1 ppm of nisaplin or 0.025 ppm of nisin
	=	0.00125% (or 1.25 mg%) of nisaplin or 0.0000312%of nisin (or 0.0312 mg%)
500 IU nisaplin per kiolgram of food	=	500 mg/kg of food
	=	500 ppm nisaplin or 12.5 ppm nisin
	=	0.05% (or 50 mg%) of nisaplin or 0.00125% (or 1.25 mg%) of nisin

C. Nisin preparation in culture broths: 1 IU = 100 activity units (AU)[58,59]

Using *Lactobacillus plantarum* NCDO 955 as an indicator strain on an agar-overlay plate, testing 5 μl portions from decimal dilutions of a heat-treated broth preparation and a commercial preparation of nisin (nisaplin), and looking for highest dilutions that produced a 1 to 2 mm zone of inhibition, indicated that 1 IU = about 100 AU of nisin in the broth preparation (see Chapter 10 for the specific method used to determine AU).

(AU)/ml of the original broth. A comparison between a commercial nisin preparation (Nisaplin) with this broth preparation of nisin indicated that 1 IU ≅ 100 AU.[56,56a,58,59]

Recently an enzyme-linked imino assay (ELISA) method for nisin using polyclonal antiserum has been developed that can detect 1.9 × 10^{-2} IU/ml in samples including foods. The results correlate well with the bioassay method using *M. flavus* as an indicator strain.[60]

B. Antibacterial Spectrum

Nisin is normally ineffective against molds, yeasts, and Gram-negative bacteria, but is effective against a wide range of Gram-positive bacteria. Many of these Gram-positive bacteria are important in foods due to their ability to cause food spoilage and foodborne health hazards. Although most studies have indicated that nisin is ineffective against Gram-negative bacteria, Mattick and Hirsch[13] reported that a few strains of *Nisseria* species are sensitive to it. Currently, several research results have indicated that under certain conditions nisin can be antibacterial against some Gram-negative bacteria.[61,61a,62] These aspects will be discussed in this section.

TABLE 2
Some Gram-Positive Bacterial Species Tested Sensitive to Nisin[1,53,56a,63–65]

A. Non spore formers[a]	B. Spore formers (spores and cells)[a]
Lactococcus lactis subsp. cremoris	*Bacillus coagulans*
Streptococcus agalactia	*Bacillus cereus*
Streptococcus pyogenes	*Bacillus subtilis*
Staphylococcus aureus	*Bacillus stearothermophilus*
Micrococcus luteus	*Bacillus lichiniformis*
Lactobacillus brevis	*Clostridium botulinum*
Lactobacillus bulgaricus	*Clostridium sporogenes*
Lactobacillus casei	*Clostridium butyricum*
Lactobacillus plantarum	*Clostridium bifermentum*
Lactobacillus buchneri	*Clostridium perfringens*
Leuconostoc mesenteroides	*Clostridium pasteurinicum*
Leuconostoc oenos	*Clostridium thermosaccharolyticum*
Pediococcus acidilactici	*Clostridium tyrobutyricum*
Pediococcus damnosus	
Pediococcus pentosaceus	
Listeria monocytogenes	
Listeria ivanovii	

[a] Strain(s) within a species can be resistant. Also, in a sensitive strain some cells can be resistant to bacteriocidal action of nisin.

1. Gram-positive Bacteria and Bacterial Spores

Earlier studies by Mattick and Hirsch[14] indicated that lactic streptococci, pathogenic streptococci from Group B, *Bacillus, Clostridium, and Lactobacillus* species (but not *Staphylococcus* species), were inhibited by a proteinaceous metabolite from a group N *Streptococcus* spp., which was later named nisin. They also observed that all Gram-negative organisms tested were unaffected. Growth of *Streptococcus agalactia* cells was completely inhibited even after diluting 6.4×10^5 times. Subsequent studies showed that *Staphylococcus* spp.[31] and *Mycobacterium tuberculosis,*[13] though found to be 100 times more resistant than streptococci, were sensitive to nisin. Some of these are listed in Table 2.[1,53,63–65] Many of these bacterial species are involved in food spoilage and foodborne diseases. For a sensitive spore former, nisin was found to be inhibitory to both vegetative cells and spores. Among the species and strains in each genus, some could be resistant to nisin (Table 3). The relative frequency of resistance differs among the species. Among 66 lactobacilli, 6 were resistant to nisin and 18 showed slight sensitivity (zone of 0.2 cm), and among *Lactobacillus casei* strains about 50% were resistant to nisin. Also, among the sensitive bacteria, species and strains differ greatly in their relative sensitivity to nisin (Table 4). Thus, among *Staphylococcus aureus,*

TABLE 3
Frequency of Nisin Resistant Species/Strains Among Sensitive Groups

Bacterial genera	Number tested	Number resistant (%)
Lactobacillus spp.	154	23 (14.9)
Leuconostoc spp.	40	3 (7.5)
Pediococcus spp.	67	1 (1.5)
Micrococcus spp.	12	2 (16.7)
Streptococcus spp.	3	0 (0)

Note: Each genus includes many different species and strains. Sensitivity was measured with 100 IU of nisin by well assay; some strains were weakly sensitive and produced less than 2 mm zone (Adapted from Reference 64).

TABLE 4
Variation in Nisin Sensitivity of Vegetative Cells of Bacteria[a]

Bacterial genera	MIC of nisin (IU/ml)[b]
Bacillus spp.	2–4
Clostridium spp.	0.25–80
Sta*phylococcus* spp.[c]	4–128
Streptococcus spp. (Groups A, B, N)	0.25
Cory*nebacterium spp.*	4–120
Mycobacterium spp.	100–500
Pneumococcus spp.	100

[a] Table adapted from References 13 and 31.
[b] Minimum inhibitory concentration (MIC) listed in this table are from studies conducted before the IU was accepted; the results are approximate.
[c] *Staphylococcus aureus, depen*ding on strains, varied in sensitivity to the IU level of nisin (4 to 128 IU/ml).

some strains were sensitive only to 4 IU/ml nisin, while some other strains required as high as 128 IU/ml. *S. aureus* 196, an enterotoxin-producing strain, showed less sensitivity to nisin as 7.2×10^3 IU/ml was required to have a 5-log reduction in cfu (Table 5).[65] *Listeria* species and strains showed great variation in sensitivity to nisin preparations (Table 6).[56] Also, relative sensitivity of a bacterial strain to nisin can be affected by the agar media used in the assay (Table 7).[66] *L. monocytogenes* strain ATCC 7644 required only 7.4×10^2 IU/ml in TSA agar plate to inhibit growth of 10^4 cells, but strain V7 required 1.2×10^5 IU/ml under the same test conditions. In addition, strain 35132 needed only 3.7 IU/ml in MRSA plate, but needed 1.5×10^4 IU/ml in TSA plate to inhibit growth of 10^4 cells. In general, MRSA needed lower

TABLE 5
Influence of Nisin Concentration on Growth and Colony Counts of *Staphylococcus Aureus* 196 (ATCC 13565)[a]

Nisin concentration (IU/ml)	OD[b] $_{686nm}$	cfu/50μl[c]
0	1.7	4×10^8
2.2×10^3	1.7	5×10^7
5.0×10^3	1.3	1.5×10^4
7.2×10^3	0	4×10^3
11.6×10^3	0	1×10^3
13.8×10^3	0	$<1 \times 10^3$

[a] Table adapted from Reference 65.
[b] Cells were inoculated in nisin containing brain heart infusion broth at 37°C for 96 h and OD at 686 nm was taken.
[c] To determine colony forming units (cfu) at the end of 96h incubation, 50 μl of broth from tubes containing different concentrations of nisin were either plated directly or after serial dilution on S110 agar. Cfu were enumerated after 48h incubation of the plates at 37°C.

minimum inhibitory concentration (MIC), which could be due to the presence of Tween 80 in MRS medium.[1] Similarly, clostridial and bacilli spores of different species and strains differ in sensitivity to nisin. Spores of *C. botulinum* type A are more resistant than spores of type B or E, E being the most sensitive[67] (Table 8). Among spores of *Bacillus* species, M spores, i.e, the spores that rupture their envelope by mechanical pressure, were more sensitive to nisin than the L spores (L spores outgrow by a lytic mechanism).[68] These results have indicated that nisin is antibacterial to vegetative cells and spores of many Gram-positive bacteria, although relative sensitivity differ among species and strains.

2. Gram-negative Bacteria

From time to time reports have surfaced showing that nisin can be effective against some Gram-negative bacteria. As early as 1947, Mattick and Hirsch[13] postulated from their studies that the Gram-positive bacteria were sensitive to nisin and that it did not inhibit growth of yeasts, molds, and Gram-negative bacteria, except for a few strains of *Nisseria*. Linnett and Strominger[69] in 1973 observed that, in a cell-free system, 1600 IU/ml of nisin caused 50% inhibition of Peptidoglycon synthesis of *Escherichia coli,* and *B. stearothermophilus.* Ogden et al.[64] in 1988 reported that three *Flavobacterium* strains and one *Hafnia* strain isolated as brewing contaminants were slightly sensitive to nisin. In 1986, Kordel and Sahl[70] reported that *E. coli* became sensitive to nisin when the outer membrane of cells was disrupted. In 1989, Blackburn et al.[61] reported that nisin in 0.1 to 300 IU/ml along with 0.1 to 20 m*M* EDTA prevented growth

TABLE 6
Viability Loss of *Listeria* Spp. by Nisin and Nisaplin

Listeria species/strains	Log_{10} cfu/ml after treating with[a] None(control)	Nisin	Nisaplin
L. monocytogenes			
Scott A	8.2	6.2	4.4
CA	6.3	4.1	3.2
117	7.2	4.9	4.5
Ohio2	8.0	6.4	5.4
V7	8.9	7.7	6.1
$Camp^+/Beta^+$	8.6	6.5	4.8
$Camp^+/Beta^-$	8.6	6.3	5.0
101	8.3	5.9	4.1
103	8.1	6.4	4.3
L. innocua ATCC 35091	8.5	6.2	4.5
L. ivanovii ATCC 19119	8.7	5.5	3.1

[a] Tryptic soy broth (TSB) was inoculated with cells of *Listeria* species containing either a nisin preparation or nisaplin or none (control). Nisin was prepared in our laboratory by growing *L. lactis* subsp. *lactis* ATCC 11454 in a broth. Nisaplin was obtained from Aplin and Barrett Ltd. The concentrations used were 500 IU/ml for nisaplin and about 15 IU/ml for nisin preparation. The broths were kept at 4°C for 1 h and colony forming units were determined by pour plating in tryptic soy agar (TSA). Strains differed greatly in viability loss. The greater reduction of cfu in nisaplin was due to high IU used.

From Motlagh, A. M., Johnson, M. C., and Ray, B., *J. Food Prot.*, 54, 873, 1991. With permission.

of Gram-negative bacteria including *Salmonella* spp., *Escherichia coli, Pseudomonas* spp., *Bacteroides* spp., *Actinobacillus* spp., and *Klebsiella* spp. Sensitivity of *Salmonella* and other enteric pathogens to nisin following EDTA treatment has also been recently reported.[61a] Ray and Kalchayanand[62] also observed the nisin sensitivity of Gram-negative bacteria, including *Salmonella* spp., *Yersinia* spp., *Aeromonas* spp., pathogenic *Escherichia coli,* and *Pseudomonas* spp. following freezing, heating, low pH, and EDTA treatment (Table 9). They explained that the developed sensitivity of Gram-negative bacterial cells to nisin is a consequence of sublethal injury in the outer membrane (in the LPS molecules)[71] of the cells that allows nisin to pass through the outer membrane and come in contact with the cytoplasmic membrane and destabilize its functions. Recently, Gao et al.[72] provided evidence that nisin treatment of liposomes, made from *E. coli* phosphotidylethanolamine, dissipated membrane potential and pH gradient and made it permeable to ions. These results thus indicate that Gram-negative bacteria under certain conditions can be made sensitive to nisin.

TABLE 7
Variation in Sensitivity to Nisin by *Listeria* Species as Influenced by Media of Assay[a]

Listeria species and strains	Inhibitory Zone (mm)[b]	MIC of Nisin (IU/ml) in TSA[c]	MIC of Nisin (IU/ml) in MRSA[c]
L. monocytogenes			
ATCC 7644	18	7.4×10^2	1.9
Scott A	10	1.2×10^4	1.7×10^3
V7	11	1.2×10^5	3.4×10^3
35152	11	1.5×10^4	3.7
L. ivanovii			
C194	11	7.4×10^4	3.7×10^1
KC1714	18	1.5×10^3	3.7
L. saligery			
LA15	12	1.5×10^4	3.7×10^1

[a] Table adapted from Reference 66.
[b] Diameter of zone produced by 70 μl nisin solution equivalent to 2.6×10^3 IU.
[c] Minimum inhibitory concentration (MIC) of nisin required to completely inhibit colony formation by about 10^4 cells in nisin-containing agar plates (Tryptic soy agar or MRS agar).

TABLE 8
Relative Sensitivity of Spore of Sensitive Gram-Positive Bacteria[a]

Bacterial group[b]	MIC (IU/ml) of nisin
Clostridium botulinum	
Type A	5000
Type B	1000 to 1500
Type E	100 to 500
Bacillus subtilis (M)	5
Bacillus coagulans (M)	5
Bacillus licheniforms (M)	10
Bacillus cereus (L)	>100
Bacillus polymixa (L)	>100
Bacillus brevis	>100

[a] Table adapted from References 1 and 67.
[b] Growth inhibition of 10^3 clostridial spores in brain-heart infusion broth containing nisin, after 7 d at 35°C. In bacilli M-spores, the small celled species are more sensitive to nisin than L-spores, the large celled species

TABLE 9
Sensitivity of Sublethally-Stressed Gram-Negative Bacteria to Nisin Preparation

		Log cfu/ml	
Gram-Negative bacteria	Nature of sublethal stress[a]	Without Nisin	With Nisin
E. coli NCSM	Freezing	72×10^3	31×10^2
E. coli 3548-88 (Pathogenic)	Heating	15×10^1	10×10^1
	EDTA	15×10^4	10×10^4
	Na-lactate	11×10^4	4×10^3
S. typhimurium	Freezing	10×10^1	5×10^1
	Heating	35×10^1	10×10^1
	EDTA	12×10^4	11×10^4
	Na-lactate	50×10^3	5×10^3
Y. enterocolitica	Freezing	50×10^1	45×10^1
	Heating	64×10^1	55×10^1
	EDTA	12×10^4	9×10^4
	Na-lactate	85×10^3	17×10^3
P. fluoresens	Freezing	11×10^3	10×10^3
	Heating	30×10^1	20×10^1
	EDTA	35×10^2	15×10^2
	Na-lactate	13×10^3	11×10^2

[a] Overnight culture suspended in peptone at about 10^{3-4}/ml were either frozen (–20°C) overnight and thawed quickly, or heated to 55°C for 10 min and cooled, or treated with 50 m*M* EDTA (pH 6.0 phosphate buffer, 5 m*M*) for 30 min, or treated with 3% Na-lactate (65% solution pH 6.5) for 30 min. The cells were then harvested by centrifugation, resuspended to the same volume with water, and treated with a nisin preparation (pH 5.0, 1600 AU/ml ~ 16 IU/ml), incubated at 4°C for 1 h, and then enumerated for cfu on tryptic soy agar prepoured plates. Each reading is the average of triplicate counts. The presence of nisin during treatments produced more reduction in cfu (Ray, B. and Kalchayanand, N., unpublished data).

C. Mode of Antibacterial Action

The mechanisms by which nisin produces its antibacterial effect at the molecular level against vegetative cells and spores of sensitive Gram-positive bacteria are not clearly understood. Most researchers are in agreement that it is bacteriocidal to the sensitive strains. The available information will be discussed separately for the vegetative cells and for the spores.

1. Vegetative Cells

Nisin behaves as a surface-active cationic detergent, and it is capable of destabilizing the cytoplasmic membrane.[72,73] Its bacteriocidal effect is probably produced by its interference with the membrane functions. However, before it comes in contact with the cytoplasmic membrane, nisin is adsorbed on the cell surface of the sensitive strains but not on resistant strains.[74] The adsorption might be reversible initially for about 10 to 15 min and then it becomes irreversible.[75,76] The adsorption is pH dependent, being maximum at pH 6.5 (100%), which reduces to about 43% at pH 4.5. Cell death following adsorption of nisin is associated with the release of cytoplasmic materials and in some strains cell lysis.[74,77] A recent study by Ruhr and Sahl[78] has confirmed that nisin treatment causes a rapid efflux of amino acids from the cytoplasm of sensitive Gram-positive bacteria. It also decreases the membrane potential of cells and thereby deprives them of the components required for the proton motive force and biosynthesis of cellular materials. At very high concentrations (1500 IU/ml in place of 150 IU/ml) nisin inhibits peptidoglycan synthesis.[79] However, recent results have indicated it to be a secondary effect of nisin.[78] This recent study has also shown that complete membrane depolarization occurs within 1 min following nisin treatment. Anionic detergent can neutralize the bacteriocidal effect of nisin on sensitive cells.[74] Studies in our laboratory have indicated that nisin can be adsorbed by the cells of resistant strains of Gram-positive bacteria, including producer strains that are immune to nisin. Thus, only adsorption of nisin on the cell surface of sensitive Gram-positive bacteria is not associated with the ability of these molecules to destabilize cytoplasmic membrane and cause cell death.[80] It will be necessary to determine the events that occur following adsorption of nisin molecules on cell wall receptors, and its coming in contact with the cytoplasmic membrane, to understand the bacteriocidal modes of action of nisin.

2. Spores

Although the antibacterial effect of nisin on spores of some bacterial species has been demonstrated, it is not clear how nisin causes viability loss of sensitive spores, and if the inhibition is similar on both bacilli and clostridial spores. In general, results of most studies suggest that nisin is sporicidal. However, there are some reports that suggest it is sporostatic. Sublethal heat treatment of sensitive spores by inducing injury is known to increase the antimicrobial effect of nisin.

a. Bacillus Spores

While Campbell and Sniff[81] suggested that nisin was sporostatic and that it prevented the outgrowth of spores of *Bacillus coagulans*, studies by other researchers have shown that spores of sensitive bacilli strains were killed by nisin. Detailed studies by Gould and Hurst,[68] Gould,[82] and Hitchins et al.[83] have strongly indicated that nisin did not inhibit germination swelling, but inhibited the pre-emergence swelling that preceded emergence of cells from the spore envelopes, and that this was followed by the commencement of growth. The small-spored species such as *Bacillus subtilis* seemed to be more sensitive to nisin than the large-spored species like *Bacillus cereus*.[68]

TABLE 10
Inhibition of Growth of 10^5 Spores (or Cells) of Clostridial Species by Nisin[a]

Clostridial species	Treatmen	Growth response after incubation for days:				
		1	2	5	7	14
C. laramie	Control	–	+			
	Nisin	–	–	–	–	–
C. botulinum B (nonproteolytic)	Control	+				
	Nisin	–	–	–	–	–
C. perfringens	Control	+	–	–	–	–
	Nisin	–	–	–	–	–
C. putrefacience	Control	+				
	Nisin	–	–	–	–	–
C. sporogenes	Control	+				
	Nisin	–	+			

Note: Cells for *C. perfringens* and spores for other species were used. About 10^5 cells or spores were inoculated in tubes with 10 ml thioglycolate broth. Treated samples were inoculated with a broth preparation of nisin (prepared from *Lactococcus* strain ATCC 11454). Duplicate tubes for controls and treatment for each strain were used. Nisin was used at 6200 AU/ml which is equivalent to 62 IU/ml. The tubes were incubated at 12°C for C. la*ramie* and 35°C for other species. Growth was observed from visual examination of turbidity. +, growth; –, no growth.

[a] Table adapted from Reference 87.

b. Clostridium Spores

Initial studies by Hirsch and Grinsted[84] indicated that nisin was inhibitory against spores of *Clostridium butyricum, Clostridium sporogenes,* and *Clostridium bifermentans*. However, the degree of inhibition for a given amount of nisin was inversely proportional to the concentrations of spores. The authors also suggested that nisin was sporicidal to *C. butyricum* spores but sporostatic to spores of other clostridia. Sporicidal effect of nisin to *C. butyricum* spores was also confirmed by Ramseier,[74] who reported that spores were killed by nisin after their germination. *Clostridium botulinum* spores were also reported to be sensitive to nisin but required a proportionally higher amount than *C. butyricum* spores.[85] Further, *C. botulinum* spores of type A were more resistant than type E.[67,86,86a] Nisin was also found to inhibit outgrowth of spores of *Clostridium laramie*, a psychrotrophic clostridial species associated with spoilage of refrigerated beef (Table 10).[87]

c. Synergistic Effect with Heat on Spores

Lewis et al.[88] reported that exposing spores of *Clostridium sporogenes* (PA 3679) to nisin reduced the D-value of heat treatment by 55%. Similar reduction in D-value for nisin-treated spores of *C. botulinum, Clostridium thermosaccharolyticum, B. coagulans,* and *Bacillus stearothermophilus* was observed by others.[23,26] It was suggested that reduction in heat resistance was due to the increased adsorption of nisin on spores following heat treatment.[83] Addition of trypsin, which does not inactivate nisin, can reverse the lethal effect of nisin on heat-treated spores. Probably nisin and trypsin have the same receptors on spores, and trypsin, when present, inhibits the binding of nisin.[1,54] The increased sensitivity of heat-treated spores to nisin is currently thought to be due to the sublethal heat injury of spores of bacilli and clostridia, thus resulting in increased sensitivity of the injured spores to nisin.[1,89] However, this hypothesis is difficult to justify if one accepts that nisin kills the cells following spore germnation.[74,82]

3. Membrane Vesicles and Liposomes

Ruhr and Sahl[78] reported that nisin blocked the transport of L-proline in the cytoplasmic membrane vesicles prepared from *Staphylococcus cohonii, B. subtilis,* and *Micrococcus luteus;* also, accumulated amino acids were released from the membrane vesicles following nisin treatment.[78] However, the authors observed that in membrane vesicles from a nonbacterial source, such as asolectin liposomes (from soybean phospholipids), nisin treatment did not cause an efflux of entrapped, labelled amino acids. Although cells of Gram-negative bacteria that were intact were not affected by nisin, cells of *E. coli* became sensitive when the outer membrane was disrupted.[70] Recently, Gao et al.[72] reported that nisin treatments on liposomes, made from *E. coli* phosphotidylethanolamine and from egg phosphotidylcholine, resulted in dissipation of membrane potential and pH gradient, and in proteoliposomes it inhibited oxygen uptake by cytochrome C. They suggested that nisin got incorporated in the membrane, made it permeable to ions, and resulted in the dissipation of membrane potential and pH gradient.

D. Resistance of Gram-Positive Bacteria

Nisin resistance of Gram-positive bacteria can result by three different mechanisms: through developed resistance by growing sensitive strains in a medium containing nisin, by the ability to synthesize nisinase, and due to presence of the gene *nis*^{r} that encodes for nisin resistance phenotype (Nis^r). The relationships between these three types have not been investigated.

1. Developed Resistance

As early as 1950, Hirsch[34] noticed that exposure of *Streptococcus agalactia* cells to nisin resulted in over 99% viability loss. However, the surviving cells were able to grow in a medium containing a 40-fold higher concentration of nisin. Subsequent studies by others also showed that survivors of sensitive strains of Gram-positive bacteria could be "trained" to grow in a medium containing as much as 1000-fold higher concentrations of nisin, and these resistant cells did not produce the enzyme nisinase that could destroy nisin.[39,74] However, when these resistant cells were passaged by growing them in media without nisin, their resistance to nisin was drastically reduced.

2. Resistance Due to Nisinase

Many Gram-positive bacteria, including several lactic acid bacteria, were resistant to nisin due to their ability to synthesize the nisin-hydrolyzing enzyme nisinase.[40,90–95] Nisinase from several *Bacillus* spp. specifically reduced the C-terminal dehydroalanyllysine, of nisin to alanyllysine and this suggested it was a dehydropeptide reductase.

3. Nisin-Resistant Gene (Nis^r)

Several studies have been published indicating the existence of a genetic basis of nisin resistance phenotypes (Nis^r). This trait could be transferred by conjugal mating from a nisin-producing nisin-resistant donor *L. lactis* subsp. *lactis* strain (Nis^+ Nis^r) to other sensitive recipient *L. lactis* subsp. *lactis* strains (Nis^- Nis^s).[96–99] However, these studies failed to show the involvement of a plasmid in this transfer process. Subsequent studies showed that the trait is encoded in a chromosomally linked transposon.[47] The presence of nisin-resistant gene has also been observed in nisin-nonproducing strains of *L. lactis* subsp. *lactis* (Nis^- Nis^r)[100] and *L. lactis* subsp. *lactis* var *diacetylactis*.[101] In these strains, Nis^r gene is encoded in plasmids that could be transferred through conjugal mating. Subsequent studies indicated that the gene encoding for nisin-resistance in the 60 Kb plasmid pNP40 in *L. lactis* subsp. *lactis* var *diacetylactis* DRC3 is located in a 2.6 Kb *EcoR* I-*Xba* I fragment.[102] Current results from their studies[103] have revealed that nisin-resistant gene is located in a 1.2 Kb *Dra* I fragment and encodes for a 35,035 Dal protein. However, the genetic code for nisin-resistant phenotype in this strain is different from that in nisin producing strain of *L. lactis* subsp. *lactis* 11454.[103] Construction of a plasmid vector, pVS40 that carries nisin-resistant gene and confers it to the nisin-resistant phenotype in *L. lactis* subsp. *lactis,* has recently been reported.[104] It will be important to determine the characteristics of the products of nisin-resistant genes from producer and nonproducer strains of *Lactococcus* species, and their differences from nisinase of nonlactococcal strains, to fully understand the nisin-resistant trait of the Gram-positive bacteria.

E. Other Factors Affecting Antimicrobial Function

Several factors besides resistance can reduce the antibacterial effectiveness of nisin against Gram-positive bacteria:

1. Storage Stability

Purified dry nisin powder retains its potency indefinitely under refrigeration storage.[63] However, in food systems, even at refrigeration temperature, the activity is gradually lost. This has been observed in cheese, canned mushrooms, chocolate milk, cooked ham, and other products.[63] In Swiss cheese, initial concentration of 250 IU/g can be reduced to 0 IU/g in 20 d.[63] This aspect will be discussed in more detail in another section dealing with the application of nisin in food preservation.

2. Bacterial Species and Strains

Not all Gram-positive bacteria are sensitive to nisin. Many strains within the same species could be insensitive, and among the sensitive strains the sensitivity can differ greatly. Thus, *Clostridium botulinum* spores were more resistant to nisin than *Clostridium butyricum* spores.[74] Among *Listeria* spp., *L. monocytogenes* strains were more resistant than *L. ivanovii,* and among *L. monocytogenes* strains, strains CA and V7 were more resistant than strain $OHIO_2$ with the same amount of nisin. Even in highly sensitive strains there are some cells that are resistant to nisin.[56]

3. Cell and Nisin Concentrations

As the cell concentration increases for a given concentration of nisin, the amount of cell viability loss decreases. Conversely, as the concentration of nisin increases for a given cell concentration of a sensitive bacterial strain, viability loss increases, until only the resistant fraction remains in the population.[56]

4. Proteolytic Enzymes

Treatment of nisin with α-chymotrypsin, pancreatin, or ficin[105] can destroy its antibacterial property. Sensitivity of nisin to α–chymotrypsin has been reported by others.[1,49] However, trypsin, pepsin, erepsin, elastase, and carboxypeptidase A did not inactivate its action.[63] The stability of nisin against the proteolytic enzymes found in foods of animal, plant, and microbial origins, especially those that are heat stable, should be taken into consideration while using it as a food preservative.

5. pH

Nisin is more soluble and stable at pH below 4 than above it, but at pH above 8 it is destroyed. This property has important implications in the effectiveness of nisin as a food preservative, since most foods that require preservation have a pH above 4. However, even at pH between 5 to 6.5 it has an antibacterial effect, although it is at somewhat reduced levels.[1]

6. Liquid vs. Solid System

Nisin is proportionately more inhibitory to sensitive bacteria in liquid than in solid and semisolid systems. The main reason is probably due to barrier functions of the nonliquid systems. Being hydrophobic, fatty materials could also interfere with uniform dispersion of the nisin molecules. In solid, semisolid, or liquid foods uniform dispersion of the nisin molecules to enable them to come in contact with the bacterial cells should be taken into consideration.

```
 1                  5                   10                  15                  20
NH2-Met-Ser-Thr-Lys-Asp-Phe-Asn-Leu-Asp-Leu-Val-Ser-Val-Ser-Lys-Lys-Asp-Ser-Gly-Ala-Ser-Pro-

23
Arg-
  |
  -Ile-Thr-Ser-Ile-Ser-Leu-Cys-Thr-Pro-Gly-Cys-Lys-Thr-Gly-Ala-Leu-Met-Gly-Cys-Asn-Met-Lys-
    1               5                   10                  15                  20

-Thr-Ala-Thr-Cys-His-Cys-Ser-Ile-His-Val-Ser-Lys-CooH
        25                  30              34
```

FIGURE 1. Amino acid sequence of prepronisin showing 57 amino acids with 23 in leader region and 34 at structural region.

In this respect nisin and other bacteriocins could be different than the water-soluble chemical preservatives.

7. Hydrophilic vs. Hydrophobic System

Uniform dispersion of nisin molecules will be greatly influenced by the lipid content and probably by the nature of distribution of lipids in a food. Being hydrophobic, nisin molecules will be attracted towards the lipid components in a food. In foods with high fat, the nisin molecules may not be uniformly distributed, which in turn can reduce the antibacterial property of nisin.[103a]

III. PROPERTIES OF NISIN

A. Structure and Biological Activity

Nisin is a ribosomally synthesized small protein of 34 amino acids with isoleucine as NH_2 terminal and lysine as carboxy terminal amino acids, and a molecular weight of 3510 Dal.[42,106] It contains four unusual amino acids, dehydroalanine (DHA), dehydrobutyrine (DHB), lanthionine, and ß-methyllanthionine, which form thioether cross linkages to give five rings in the molecule. At the translation level the nisin precursor molecule contains 57 normal amino acids with 23 residues of leader region and 34 residues of structural region.[107] Then, through posttranslation enzymatic modifications, the leader region is removed and serine and threonine in the structural region undergo dehydration at specific positions. This results in the formation of DHA and DHB. Thioether linkages between the dehydrated amino acids and cysteine subsequently produce lanthionine and ß-methyllanthione (Figures 1 and 2 and Table 11).[42,44–46,106,107] The correctness of the amino acid sequence, and the locations of the modified amino acids and thioether linkages, as determined by Gross and Morel[42,108,109] (by chemical degradation method) have recently been confirmed by FAB-MS and FAB-MS/MS[109a] as well as by using 2D-NMR (two-dimensional nuclear magnetic resonance) techniques.[110,111] Also, the NMR data indicated that the molecule has a well-defined three-dimensional structure.[43a,111]

Earlier studies have suggested the possibility that several different types of nisin molecules exist.. Bavin et al.[19] indicated that nisin has two different types of molecules, while Berridge et al.[30] indicated the possibility of four types of

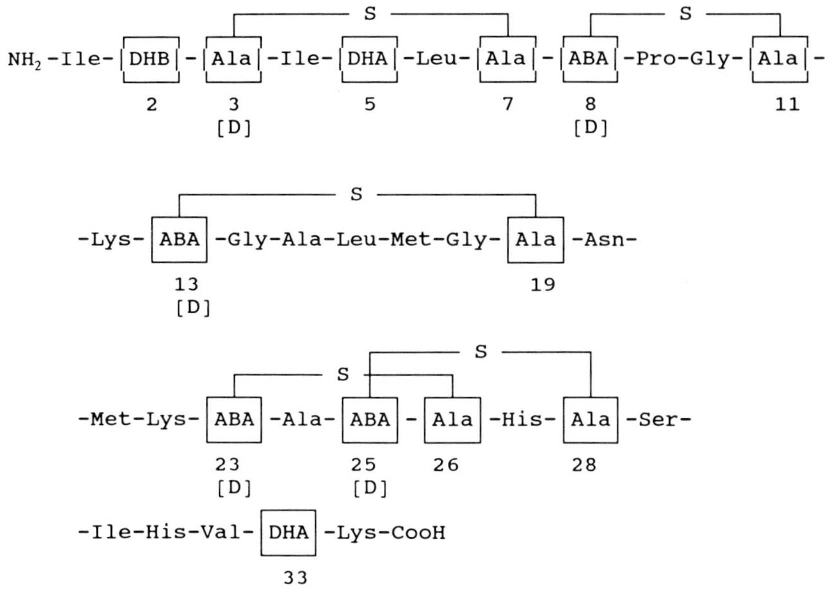

FIGURE 2. Amino acid sequence of nisin showing posttranslation modifications of 13 usual amino acid units to respective unusual amino acids, and formation of thioether linkages in five positions. [D] Stearic configuration.

active molecules, which they designated as A, B, C, and D. They observed that while A and B were biologically highly active, C and D had only one fifth of the activity of A and B. Later, Jarvis and his fellow researchers[94,112] reported a fifth type and designated it as E. Gross and Morell[42] elucidated the amino acid sequence of nisin A to have 34 amino acids and a molecular weight of 3510 Dal. However, previous researchers reported the molecular weight of nisin A to be 7000 and 14,000 Dal.[113,114] The higher molecular weight were the results of formation of dimers and tetramers of the 3500 monomer molecules. The sequences of bases of the gene encoding for the nisin-production phenotype in three *L. lactis* subsp. *lactis* strains by three research groups[44–46] correspond well with the amino acid sequence of nisin A. So it is unlikely that the same nisin-producer strain will produce different types of nisin (types A to E). Mulders et al.[115] recently reported that *L. lactis* subsp. *lactis* NIZO 22186 produces nisin (nisin Z) which differs from A in one amino acid; the amino acid in position 27 is Asn in nisin Z while in nisin A it is His (Figure 1). Studies by Chan et al.[116] have revealed that while freshly prepared nisin has only one type of molecule, "old samples" or fresh nisin treated with hydrochloric acid, and glacial acetic acid produced several different types of molecules. The two predominant types were nisin (1–32) and (des ΔAla5) nisin (1–32). Nisin (1–32) has 32 amino acids without the last two amino acids (DHB 33 and Lys 34) and with an additional amide group in Val 32 (Val–

TABLE 11
Modification of Amino Acids in Nisin Sequence Through Dehydration and Thioether Linkages

Usual amino acids	Position in sequence		Unusual amino acids
Ser; NH_2–CH(–CH_2–OH)–COOH	3, 5, 33	→ H_2O dehydration	DHA[a]; NH_2–C(α)(=CH_2)–COOH (α, β unsaturation)
Thr; NH_2–CH(–CH_2–CH_3)–COOH	2, 8, 13, 23, 25	→ H_2O dehydration	DHB[a]; NH_2–C(α)(=CH–CH_3)–COOH (α, β unsaturation)
Cys; NH_2–CH(–CH_2–SH)–COOH	7	DHA[b] at 3 → thioether linkage	LAN[a] (Ala–S–Ala): NH_2–CH[α](–COOH)–CH_2 [D]–S–CH_2 [L]–CH(NH_2)–COOH

Cys;	11, 19 26, 28	DHB[b] at 8, 13, 23, 25 ——————> thioether linkage	βMLAN[a] (ABA–S–Ala) [α]

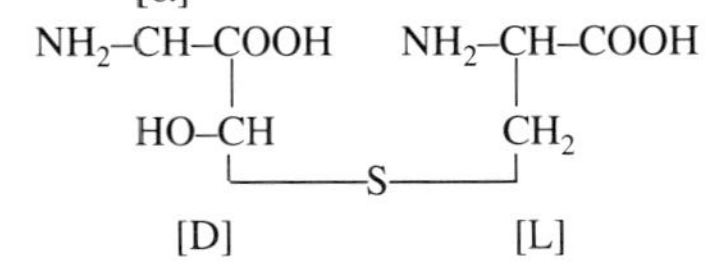

[a] DHA, dehydroalanine; DHB, dehydrobutyrine; LAN, lanthionine; βMLAN, β-methyllanthionine; ABA, aminobutyric acid (CH_3–CH_2–C(NH_2)–COOH); [D] and [L], stereo configuration in relation to α-carbon.

[b] DHA and DHB came from Ala and Thr, respectively.

NH_2). In (des ΔAla5) nisin (1–32) the DHA in position 5 (DHA5 or ΔAla5) has been degraded resulting in N-pyruval substitution on Leu6 (ΔAla5-Leu6- changed to -Pyr5-Leu6-).

A comparison of the antibacterial activity of these species indicated that while nisin (1–32) had the same activity as nisin (1–34), the activity of (des ΔAla5) nisin (1–32) had at least 500-fold less. This finding also suggested that while ΔAla33 (or DHA 33) and Lys 34 at the C-terminal were not required for antibacterial activity of nisin, ΔAla5 (or DHA 5) was important for the antibacterial properties. Earlier studies[30,108,117] that showed reduced activity or loss of activity, following heat treatment of nisin at 110°C in HCl and glacial acetic acid, could be due to the formation of (des ΔAla5) nisin (1–32). The reports that have suggested that autoclaving of nisin in dilute HCl for a short period did not result in significant loss of activity could be due to formation of nisin (1–32) from nisin (1–34). Gross and Morell[42,108,109] suggested that the biological activity of nisin could be due to the DHA (or ΔAla) residues, as they could react with compounds that were metabolically important, such as sulfahydryl-containing enzymes, and produce an antibacterial effect. In addition, α, ß unsaturated amino acids (DHA and DHB) can invoke ion-binding capacity by the polycyclic structure of nisin and play an important role in its biological functions. Recent studies by Liu and Hansen[106] have also supported the idea that dehydro-residues, by reacting with one or more nucleophiles in a sensitive cellular target, can play an important role on the antibacterial property of nisin. It inactivates sulfahydryl residues in germinated bacterial spores.[118] After reacting nisin with nucleophilic compounds to modify the dehydro groups in DHA and DHB residues, the antibacterial property of the molecules is lost. Dehydro residues of nisin, by acting as electrophilic acceptors towards nucleophiles in targets of sensitive cells, produce antibacterial action.

B. Solubility and Stability

1. Solubility

Nisin is insoluble in nonpolar solvents. It is soluble in aqueous media, and the solubility is pH dependent. Hurst reported that at pH 2.5 the solubility of nisin is only 12% and that this is reduced to only 4% at pH 5.0 and almost to zero at neutral and alkaline pH. It has an isoelectric point in the alkaline side.[1] In a technical information sheet of Aplin and Barrett Ltd.,[119] it was reported that the solubility of nisin is 118 mg/ml of 0.02 N HCl (4.7×10^6 IU/ml), 50 mg/ml of distilled water (2×10^6 IU/ml), and 87.5 mg/ml in nonfat milk (3.5×10^6 IU/ml). The common levels of nisin used in food preservation are freely soluble in water and other processing liquids.

2. pH

The solubility of nisin is most at an acidic pH. At pH 2.5 in dilute HCl solutions the activity remains stable for a long time, especially under refrigeration. This could be due to higher solubility at acidic pH. As the pH increases above 7 the molecules undergo irreversible changes with loss of activity; at pH 11 the activity is completely destroyed in 30 min at 63°C.[120]

3. Temperature

Activity of nisin in acidic solution remains stable even after heating at high temperature, such as autoclaving (121°C for 15 min).[1] However, a 40% loss occurs at pH 5.0, and a 90% loss occurs at pH 6.8 after autoclaving. In the presence of food the level of activity loss could be less due to the protective effect of foods to nisin.[49,119] At refrigeration and storage temperature the stability is maintained for a long time.

4. Storage Time

Although the activity of purified nisin at refrigeration or freezing temperature remains stable indefinitely, when added to food its stability starts decreasing with time. In cheese a reduction of 15% occurred in 15 min at 85 to 95°C, and at 30°C a 55 to 70% reduction in 6 months can be expected.[49,119–121] This lead to loss of activity during storage in Swiss cheese, canned mushrooms, chocolate milk, and cooked ham.[49,63,119]

5. Enzymes

Proteolytic enzymes like pancreatin, α-chymotrypsin, and ficin can inactivate nisin.[1,79,105] In general, trypsin, pepsin, and erapsin do not inactivate the antibacterial activity.[63] However, Gross and Morell[42] reported that trypsin could be used to cleave at position(s) between amino acids 1 to 17 in the molecule. Carbohydrate and lipid-hydrolyzing enzymes also do not have any effect on nisin.[42]

C. Toxicological Properties

The results of several studies involving feeding nisin to mice, kittens, rats, and guinea pigs provided evidence to suggest that nisin is not toxic to laboratory animals.[122] Hara et al.[123] conducted an acute toxicity study by feeding 7 g nisin per kilogram of body weight to animals (mice, rats, and kittens), and found that LD_{50} of nisin was similar to that of sodium chloride. Frazer et al.[48] also conducted acute toxicity studies by feeding nisin to rats at a level 1000 times greater than what normally might be consumed, and concluded that orally consumed nisin was nontoxic. In a subchronic toxicity study rats were fed cheese containing as much as 60.2 mg (about 2.4×10^6 IU)/kg body weight per day for 12 weeks[48] or 25 mg nisin per kilogram of body weight per day[124] for 12 weeks. Results of these studies did not indicate any difference in growth, fertility, and pathology between the test and control animals. Similarly, by feeding Wistar female and male rats in F_0 generation 4.9 mg nisin per kilogram of body weight per day for 2 years and subsequently feeding the same amounts of nisin to F_1-offspring, no differences were found between control and treated groups in survival, reproductive, hepatic, renal and gastrointestinal functions, and pathological changes.[48,125] Another study on the effect of feeding nisin to rats on reproductive functions for three generations indicated no adverse effect on survival, growth, reproductive functions, and pathology.[125,126] Oral administration of nisin also failed to sensitize guinea pigs.[124,125] This could be due to degradation of nisin by pancreatin.[127]

D. Immuogenic Properties

Only one study has been reported recently on the immunogenic nature of nisin. Falahee et al.[60] conjugated nisin (1 mg; equivalent to 37 $\times 10^3$ IU) to egg albumen using glutaraldehyde condensation technique, emulsified the nisin-conjugate with Freund's adjuvant, and injected it intramuscularly to Suffolk sheep. Serum obtained 9 months following the injection had an immunospecific antibody.

IV. GENETIC ASPECTS OF NISIN PRODUCTION

A. Location of Nisin Gene

1. Plasmid Linkage

For a long time the exact location of the gene (*nis* A, also designated as *spa* N) encoding nisin production (Nip^+, also designated as Nis^+) phenotype in nisin-producing strains of *L. lactis* subsp. *lactis* has been confusing. Dobrzanski and his associates[128,129] observed that growing some nisin-producing strains under conditions to cure plasmids produced nisin-negative clones and suggested plasmid linkage of the gene for Nip^+ phenotype. LeBlanc et al.[130] reported that the genes for both Nip^+ and Suc^+ (ability to ferment sucrose) phenotypes in strain ATCC 11454 were encoded in a 28 Mdal plasmid. However, Davey and Pearce[131] isolated plasmid-free derivatives of strain H1 that retained the ability to produce nisin, and they suggested that the *nis* A gene was not plasmid-linked. Gasson[96] and his co-researchers reported co-curing both Nip^+ and Suc^+ phenotypes from six strains; but they failed to assign the traits to any plasmid. Later, they found that a gene block controlling the Nip^+, nisin-resistance (Nis^r), and Suc^+ phenotypes could be transferred from eight different donors to a plasmidless recipient strain. They proposed that the genes for the three phenotypes could be located in a plasmid (pSN) of about 30 Mdal. Neve et al.[97] also were able to transfer by conjugation the genetic determinants from *L. lactis* subsp. *lactis* 6F3 to *L. lactis* subsp. *lactis* var *diacetylactis* BU 2-60, but they could neither find a plasmid association with the trait nor cure the donors and transconjugants, and suggested that the trait was encoded in an unstable plasmid. In 1985 Gonzalez and Kunka[98] observed that strain ATCC 11454 when exposed to 35 to 40°C produced variants that were Nip^- and Suc^- and missing a 29 Mdal plasmid. Nip^+ Suc^+ Nis^r traits could be transferred conjugally from donors to both *L. lactis* subsp. *lactis* and *L. lactis* subsp. *lactis* var *diacetylactis* recipient strains. However, no physical evidence of plasmids encoding these three traits could be demonstrated in the transconjugants. The authors suggested that the genes for these traits were either located in a plasmid that has a low copy number, or were unstable to the plasmid isolation techniques, or were in a dynamic state of integration in the recipient chromosome. Steel and McKay[99] could not detect additional plasmids among the transconjugants after the mating of three nisin producing donor strains, ATCC 11454, DL16, and 354/07 and several recipient strains of *L. lactis* subsp. *lactis* and *L. lactis* subsp. *cremoris*. The author suggested that, although the physical evidence of a plasmid could not be demonstrated for the three traits, indirect evidence such as the capability of traits to be transferred to a recipient that is deficient in host mediated homologous recombination (Rec^-), and bilateral plasmid incompatibility of the three traits with plasmids encoding for lactose hydrolyzing (Lac^+) phenotype, were indicative of plasmid linkage of Nip^+ Nis^r Suc^+ phenotypes. Tsai and Sandine[132] from their studies involving curing as well as conjugal mating with nisin-producing donor strain 7962, reported that a 17.5 Mdal plasmid, which could be obscured in the chromosomal fragments, was encoding the Nip^+ Nis^r Suc^+ traits. In 1989, Kaletta and Entian[45] reported that they had cloned and sequenced the *nis* A gene from a very large plasmid in strain 6F3. The plasmid was immobile in 0.7% agarose gels and required endonuclease digestion to obtain distinct bands.

2. Chromosomal Linkage

Davey and Pearce[131] in 1982 were the first to strongly suggest from their studies with nisin-producing plasmidless strain H1 that *nis* A gene could be chromosomally linked. In 1988, Buchman et al.[44] first reported that they had succeeded in cloning and sequencing the gene encoding for the precursor of nisin from the chromosomal DNA in strain ATCC 11454. Dodd et al.[46] confirmed these findings by reporting that in strain FI5876 (originally from NCFB 894) nisin determinant was a component of a large transmissible gene block that also encoded nisin-resistant and sucrose fermenting genes and was located in the chromosome. Donkersloot and Thompson[133] also observed from hybridization studies that in strain K1 Nip^+ and $Suc,^+$ phenotypes are encoded in the chromosome and may be located on a conjugative transposon. Recently, Steen et al.[47] reported that nisin gene in strain ATCC 11454 is part of a polycistronic gene with a size greater than 8.5 Kb; the gene is located as a single copy in the chromosome. The polycistronic gene may contain genes required for the posttranslation processing of prepronisin to nisin.

3. Nisin-Resistant Gene in Nonproducing Strains

Several studies have reported that nisin nonproducing (Nip^-) strains of some lactococci harbor a gene (*nsr*) that confers resistance to nisin (Nis^r). This gene has been demonstrated in a 40 Mdal plasmid in *L. lactis* subsp. *lactis* var *diacetylactis* by McKay and Baldwin.[101] Subsequent studies by this group revealed that *nsr* gene was located in a 2.6 Kb *EcoR* I-*Xba* I fragment of the plasmid.[102] Recently they have cloned and sequenced the *nsr* gene in a 1.3 Kb *Eco*R I-*Nde* I fragment from the original fragment.[103] This fragment could confer nisin resistance to nisin-sensitive lactococcal strains and thus could be used as a selective marker in the construction of food-grade vector and to provide resistance to desirable but originally nisin-sensitive starter culture strains growing along with nisin-producing cultures. Similar nisin-resistant genes have also been reported in plasmids of nisin nonproducing strains of *L. lactis* subsp. *lactis* strain ME2[100] and *L. lactis* subsp. *lactis* var *diacetylactis*

SSD 207.[104] The sequence of these genes could be different from the *nsr* gene present in nisin producing strains of *L. lactis* subsp. *lactis*.[103]

The information on the location and sequence of *nis* A gene and *nsr* gene will be helpful in understanding the mechanism by which the expression of these genes is controlled. In addition, the information on the structure of the different segments of the gene block associated with the production of active nisin will be necessary to understand the role and the mechanisms by which prepronisin is processed. Finally, the information of the base sequence can be used to produce analogues of nisin that have a higher potency and a larger spectrum of activity.

B. Conjugal Transfer

Long before the recognition of the location of genetic determinants of nisin-production trait in the chromosome as a polycistronic gene block of conjugative transposon, several researchers had succeeded in transferring the genetic materials for Nip^+ Nis^r Suc^+ phenotypes from nisin-producing donor strains to nisin-nonproducing and nisin-sensitive recipient strains by conjugal mating (Table 12). Gasson[96] successfully transferred the three traits as a block from six donor strains to the same recipient strains by membrane filter mating. The frequency of transconjugants ranged from 2×10^{-6} to 4.9×10^{-9} per donor cell (colony forming units, cfu). Variation in frequency of transconjugants was also reported by Gonzalez and Kunka[98] using the same donor (derivatives of strain ATCC 11454) but different recipients of *L. lactis* subsp. *lactis*. They were also able to transfer the traits to *L. lactis* subsp. *lactis* var *diacetylactis* and use the transconjugants as donors to transfer the traits to other suitable recipient lactococcal strains. Steel and McKay[99] used solid surface mating to transfer the three traits from three different donors and found that strain DL16 did not produce any transconjugant. Two other donor strains (ATCC 11454 and 354/07), produced transconjugants with recipient strain D6528 but not with LM 2306. However, by treating the donor strains with α-chymotrypsin to destroy any nisin-associated cells both the donor strains and recipient strains produced higher frequency of transconjugants. The transconjugants were also able to transfer the traits to recipient strains of *L. lactis* subsp. *lactis* and *L. lactis* subsp. *cremoris*. The conjugal mating of donor strain 7962 resulted not only in transfer of Nip^+ Nis^r Suc^+ phenotypes to *Leuconostoc dextranicum* 181, but had transconjugants that produced a much higher level of nisin.[132] Broadbent and Kondo[134] compared the ability of donor strains ATCC 11454, DL16, and 7962 to transfer the Nip^+ Nis^r Suc^+ traits to LM2306 by solid-surface mating and found strain ATCC 11454 to be superior. They also developed a direct plate conjugation (DPC) method in which cells of donors and recipients were cross streaked on selective agar media. Strain ATCC 11454 was able to transfer the phenotypes to both *L. lactis* subsp. *lactis* and *L. lactis* subsp. *cremoris* by DPC method at a rate comparable to solid-surface mating.

Development of techniques that will allow the nisin-producing lactococcal strains to transfer the trait to other lactococcal strains as well as other lactic acid bacteria opened up the possibility of developing industrially important starter culture strains that are capable of producing nisin and resistance to nisin. Different lactic acid bacterial strains carrying desirable traits required in fermented foods can then be used to produce nisin during fermentation, and can control the growth of sensitive spoilage and pathogenic bacteria. In addition, strains can be developed that will produce higher levels of nisin in media that are economical.

C. Cloning and Sequencing

1. Sequencing of Region-Producing Prepronisin

Between 1988 and 1989 three groups of researchers succeeded in cloning and sequencing the *nis* A (or *spa* N) gene from three different nisin-producing strains of *L. lactis* subsp. *lactis*. While Buchman et al.[44] and Dodd et al.[46] cloned the genes from the chromosomal DNA of strain ATCC11454 and FI5876 (a transconjugant between strains NCFB894 as a donor and plasmidless MG1614 as a recipient) respectively, Kaletta and Entian[45] reported that they cloned the gene from a very large plasmid that did not migrate during agarose gel electrophoresis in strain 6F3. Although they[45] suggested that the difference could be due either to strain divergence or to differences in DNA purification methods, Steen et al.[47] observed that *nis* A gene in strain ATCC 11454 is located in a large DNA restriction fragment of 1100 Kb, and they ruled out the possibility of plasmid linkage of this gene. Also, a plasmid preparation by $CSCl_2$ density gradient can have chromosomal DNA as contaminants. Whatever may be the differences in strains, DNA isolation methods, or DNA sources, the base sequences of *nis* A genes for nisin precursor (prepronisin) were identical in all three studies (Figure 3). It consists of a 171-base pair (bp) open reading frame (ORF) that encodes a 57-amino acid polypeptide of which 34 amino acids located at the C-terminus of the molecule, correspond precisely to pronisin. Serine, threonine, and cysteine residues are located at the precise positions in this structural gene to generate the dehydro residues and lanthionine and ß-methyllanthionine that are present in the processed molecule (active nisin).[44,46] It is processed from the prepronisin between Arg(–1) and Ile(+1) with a proline residue at –2 (Pro,–2), a structure recognized by proteolytic enzymes.[45] The leader region in the N-terminus, consisting of 23 amino acids may be directly or indirectly involved in the processing of nisin.

2. Processing of Prepronisin

The posttranslation processing of prepronisin to active nisin before it is secreted from the cells involves three types of enzymatic reactions: separation of structural region from the leader region, dehydration of the serine (except Ser29) and threonine residues, and sulfide ring formation with cysteines (a total of five). Ser29 does not undergo dehydration probably because of its close proximity to Cys28, which forms a thioether bridge.[45] One hypothesis is that the enzymatic modifications of the structural molecule occur before the cleavage of the inactive prepronisin, so that the active nisin molecules that may be toxic

TABLE 12
Conjugal Transfer of Nisin Production Phenotype From *L. Lactis* Subsp. *Lactis* Donor Strains to Other Lactococcal Strains

Donor	Recipient[a]	Conjugate method (Donor:Recipient)	Frequency (/donor cfu)	Ref.
SH4353	lactis MG1614	Membrane filter (1:1)	2.0 × 10–6	96
NCDO1404	lactis MG1614	Membrane filter (1:1)	4.9×10^{-9}	96
NCDO0496	lactis MG1614	Membrane filter (1:1)	1.1×10^{-8}	96
6F3	diacetylactis BU2–60	Membrane filter (1:10)	3.5×10^{-7}	97
SLA2.17[b]	lactis SLA2.22	Membrane filter (1:1)	2.7×10^{-4}	98
SLA2.24[b]	lactis SLA1.8	Membrane filter (1:1)	1.0×10^{-6}	98
SLA2.24[b]	diacetylactis SLA3.1	Membrane filter (1:1)	7.0×10^{-7}	98
DL16	lactis D6528	Solid surface (1:2)	0.0×10^{-9}	99
354/07	lactis D6528	Solid surface (1:2)	4.4×10^{-8}	99
354/07	lactis LM2306	Solid surface (1:2)	0.0×10^{-9}	99
ATCC11454	lactis LM2306	Solid surface (1:2)	0.0×10^{-9}	99
ATCC11454	lactis D6528	Solid surface (1:2)	2.2×10^{-8}	99
ATCC11454	lactis LM2306	Solid surface (1:2)[c]	2.2×10^{-6}	99
ATCC11454	lactis D6528	Solid surface (1:2)[c]	1.0×10^{-7}	99
354/07	lactis D6528	Solid surface (1:2)[c]	1.7×10^{-6}	99
JS1[d]	lactis LM2306	Solid surface (1:2)	2.1×10^{-6}	99
JS1[d]	cremoris BC102	Solid surface (1:2)	1.8×10^{-7}	99
7962	dextranium 181	Solid surface (1:4)	4.1×10^{-7}	132
ATCC11454	lactis LM2306	Solid surface (1:2)	1.6×10^{-6}	134
ATCC11454	lactis LM2306	Direct plate	7.0×10^{-6}	134
ATCC11454	lactis LM2306	Direct plate[c]	5.2×10^{-6}	134

DL16	lactis LM2306	Solid surface[e]	0.0×10^{-8}	134
7962	lactis LM2306	Solid surface[e]	0.0×10^{-8}	134
ATCC11454	cremoris SW224	Direct plate	4.0×10^{-7}	134
ATCC11454	cremoris JKEB7	Direct plate	4.5×10^{-7}	134

[a] Recipient strains: lactis, *L. lactis* subsp. *lactis*; diacetylactis, *L. lactis* subsp. *lactis* var *diacetylactis*; cremoris, *L. lactis* subsp. *cremoris*; dextranicum, *Leuconostoc dextrinicum.*

[b] Antibiotic resistant derivatives of *L. lactis* subsp. *lactis* ATCC 11454.

[c] Suspensions of donor cells were treated with α-chymotrypsin (400 μg/ml) to destroy residual nisin.

[d] Transconjugants from mating between strains 11454 × D6528 were used as donors for nisin trait.

[e] No transconjugants were detected by direct plate conjugation.

Initiation codon

ATG	*AGT*	*ACA*	*AAA*	*GAT*	*TTT*	*AAC*	*TTG*	*GAT*	*TTG*	*GTA*	*TCT*	*GTT*	*TCG*	*AAG*
Met	*Ser*	*Thr*	*Lys*	*Asp* 5	*Phe*	*Asn*	*Leu*	*Asp*	*Leu* 10	*Val*	*Ser*	*Val*	*Ser*	*Lys* 15
AAA	*GAT*	*TCA*	*GGT*	*GCA*	*TCA*	*CCA*	*CGC*	*ATT*	*ACA*	*AGT*	*ATT*	*TCG*	*CTA*	
Lys	*Asp*	*Ser*	*Gly*	*Ala* 20	*Ser*	*Pro*	*Arg* 23	Ile 1	Thr	Ser	Ile	Ser 5	Leu	
TGT	ACA	CCC	GGT	TGT	AAA	ACA	GGA	GCT	CTG	ATG	GGT	TGT	AAC	ATG
Cys	Thr	Pro	Gly 10	Cys	Lys	Thr	Gly	Ala 15	Leu	Met	Gly	Cys	Asn 20	Met
AAA	ACA	GCA	ACT	TGT	CAT	TGT	AGT	ATT	CAC	GTA	AGC	AAA	TAA	
Lys	Thr	Ala	Thr 25	Cys	His	Cys	Ser	Ile 30	His	Val	Ser	Lys 34	TER	

Termination codon

FIGURE 3. This figure presents 171 base sequences of 57 amino acids in prepronisin containing 23 amino acids in the N-terminusleader region and 34 C-terminus pronisin (structural)region. Posttranslation modifications involve enzymaticseparation of leader and structural regions, dehydration of serine and threonine, and thioether linkage with cysteine at specific positions in the structural molecules (underlined).[Prepronisin: 57 amino-acid residue; pronisin: 34 amino-acid residue without modifications; nisin (or active nisin): 34 amino-acid residue after modification and contains unusual amino acids.

to producer-cell membrane, do not reach a high concentration but are secreted outside of the cells immediately after cleavage.[45] Also, because N-terminal amino acid sequence of the pronisin (the product following cleavage of leader region) is not conserved, it might not act as a signal sequence for secretion.[45] In contrast, the N-terminal amino acid sequence of the leader region is well conserved, and may act as a signal sequence.[44–46] An analysis of the hydropathic characteristics of the amino acid sequences in the leader and structural regions of prepronisin with other bacteriocins from several other Gram-positive bacteria that undergo posttranslation processing also confirmed that the leader regions, as compared to structural regions, are more conserved. It is speculated that the leader region in the precursor molecule dictates the processing of the structural region to the active state. It could be that all the processing signals are in the leader region of the molecule.[44] However, the leader region of prepronisin does not carry typical signal sequences of procaryotic microorganisms. It might have a different type of signal sequence that directs the prepronisin molecule to a dedicated processing-transport apparatus in the membrane. It is this apparatus in which the prepronisin molecules undergo dehydration of serines and threonines, cross-linking of cysteines, and cleavage of leader peptide to change the molecules to active nisin.[107] Current studies have indicated that in the downstream of the nisin gene there is an open reading frame (ORF) that codes for a membrane-associated protein. This protein could be the apparatus involved in processing and in exporting the nisin molecules.[47]

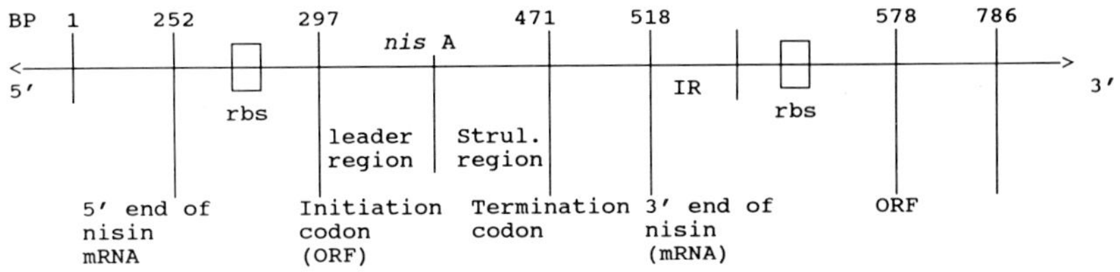

FIGURE 4. Locations of *nis* A gene and flanking regions in the 786 base sequences from 1.7 Kb *Sau*3A fragment from the chromosome of ATCC 11454. Important features: two open reading frames (ORF), two ribosome-binding sites (rbs) and one inverted repeat (IR). No promoter sequence upstream of both ORFs. 3′ end of nisin mRNA is ragged and 3′ end does not have rho-independent terminator. (Adapted from Reference 44.)

3. Characteristics of the Polycistronic Operon

The conjugal transfer of a gene block that encodes for Nip^+ Nis^r Suc^+ phenotypes from a nisin-producing donor to suitable recipient strains leads to the speculation that the genes for these traits could be in an operon system. The operon, in addition to the three genetic determinants, should also encode for regulatory genes and genes encoding for the enzymes involved in posttranslation modification of the prepronisin to active nisin. Finally, the recognition that these genes are chromosomally linked, lead the researchers to speculate that the gene block must contain determinants necessary for transfer of chromosomal DNA from the donors to the recipient cells. Thus, the transferable gene block that houses all these determinants must be relatively large. That the gene block is relatively large was evidenced from the studies of three research groups.[44–47]

Buchman et al.[44] prepared a genomic library with chromosomes from strain ATCC11454. Then, using specific DNA probes, they located the *nis* A gene in a 1.7 Kb *Sau* 3A fragment. Subcloning and partial sequencing of the bases of this fragment revealed that the 5′ end of the nisin-mRNA was not discrete but rather ragged, i.e., terminating at two locations separated by 3 to 5 bases (Figure 4). The ribosomal binding site (rbs) and the initiation codon for the open reading frame (ORF) for *nis* A gene ended after 171 bp with an Ochra-terminating codon. Following this were the 3′ end of nisin mRNA, then an inverted repeat (IR), another rbs, and another ORF. This second ORF did not have a promoter; thus, its expression must be dependent upon the "read-through" *nis* A gene. Also, the *nis* A gene did not have a promoter at the 5′ end; instead about 750 bp upstream there was another ORF with base sequences similar to transpose from the *Escherichia coli* IS2 insertion element. These results strongly suggested that *nis* A gene is part of a polycistronic operon that is transcribed into a long mRNA, only part of which is for prepronisin. Kaletta and Entian[45] cloned a 4.3 Kb *Hind* III fragment from the plasmid DNA of strain 6F3 that contained the *nis* A gene. Partial sequencing of the bases of the

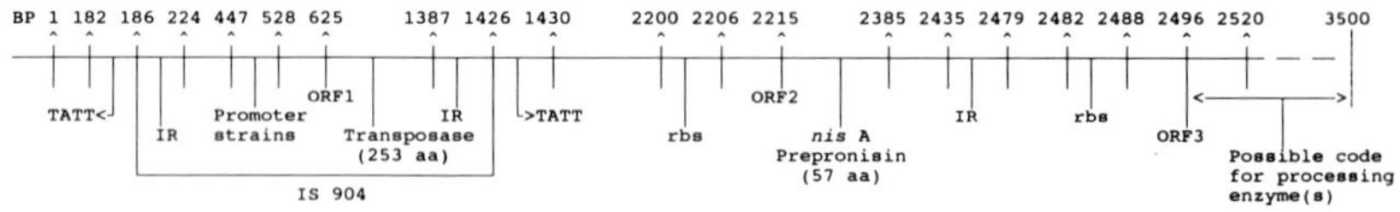

FIGURE 5. Locations of IS904, *nis* A gene and other regions in the 2.5-Kb chromosomal DNA from strain FI 5876. For explanation see text. The distances in base pair (BP) between segments are indicated at the top. (Adapted from Reference 46.)

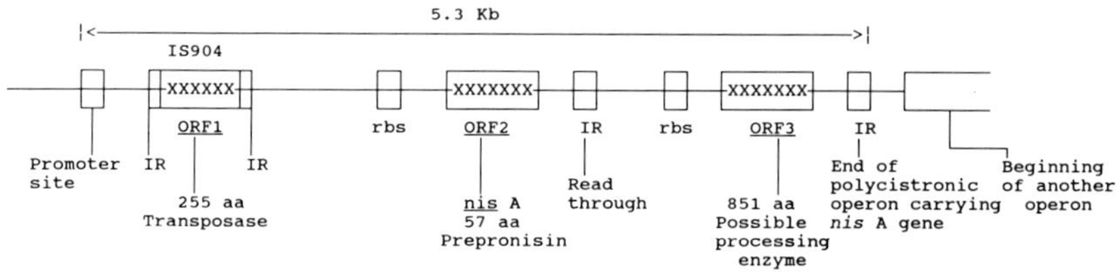

FIGURE 6. A 5.3-Kb segment showing one end (downstream) of a possible 8.5-Kb polycistronic operon carrying genes for transposase, prepronisin, and possible processing genes. The other 3 Kb upstream of the open probably has the open reading frames for *nis*^{r} (nisin resistance), *Suc*$^+$ (sucrose-utilizing) and other necessary genes. (Adapted from information presented in Figures 4 and 5).

fragment located the *nis* A gene that had identical sequences of 171 bases as reported by Buchman et al.[44]

Dodd et al.[46] used a DNA library made from the chromosomes of strain FI5876 and detected the *nis* A gene in a 5.5 Kb *Sal* I fragment. They sequenced a 2.5 Kb segment from this fragment and analyzed its characteristics (Figure 5). It contains three ORF (ORF 1, 2 and 3) of which ORF 2 contains the 171 bp for *nis* A gene; the base sequences were identical to the sequences reported before.[44,45] The flanking regions, both upstream and downstream of *nis* A gene are also identical. ORF 3 is located 108 bp downstream of the termination codon and extends at least 1 Kb. On the upstream of ORF2 is located an insertion sequence element of 1241 bp designated as IS904. It contains inverted repeats (IR) at each end and ORF1 that encodes for a protein of 253 amino acids. Upstream of ORF1 there are two regions that can be possible promoter-consensus sequences of Gram-positive bacteria. Thus this IS904 lies very close to one terminal of the gene block. A similar IS element at the left end of a conjugative transposon encoding genes for Nip$^+$ Nis^r Suc$^+$ phenotypes has been reported in the chromosome of a nisin-producing strain NIZOR5.[135]

Recently, Steen et al.[47] studied the location and organization of nisin locus in the chromosome of strain ATCC11454 by a primer extension study. Their results showed that *nis* A gene is part of a polycistronic operon with a size greater than 8.5 Kb. A promoter of the operon could be present at least 4 Kb upstream from the *nis* A gene (Figure 6). The downstream ORF (ORF3) has a coding region for a protein of 851 amino acids, a mRNA-processing site and rho-independent terminator. This protein could be associated with the cytoplasmic membrane and used in the processing of prepronisin. However, these sequences of bases did not reveal the locations of Nis^r and Suc^+ determinants. It is also not known the functions of the IS904 upstream of *nis* A gene. A recent report suggested that in strain 11454, all the necessary genes are located in the transposon Tn 5307 [*nis* A, IS 904, *Scr* A (Enz II^{Scr} of the Phosphoenolpyruvate-dependent sucrose: phosphotransferase system), *Scr* B (sucrose-6-phosphate hydrolase) and processing gene]. In strain KI-23 these genes along with *Ceo* (N^5-Carboxyethyl-ornithine synthatase) are located in Tn 5306 (a part of an ~ 20 Kb *Eco* R1 fragment).[135a]

D. Improvement of Antimicrobial Spectrum[44,47,107]

Nisin is bacteriocidal to many Gram-positive bacteria and is ineffective against yeasts, molds, viruses, Gram-negative bacteria, and some Gram-positive bacteria. As its antimicrobial property is the function of the amino acid sequences in the pronisin molecule, it is logical to hypothesize that correct modifications of the amino acid sequences can produce nisin analogues with better antimicrobial property. Structural modifications of the molecules of natural products by chemical methods, in general, are difficult, uneconomical, and time consuming, especially if one does not know what modifications will be advantageous. However, the ribosomally synthesized small proteins, such as nisin, are different in this respect. By site-specific mutagenesis of the gene it is possible to produce many analogues of nisin and test their relative antimicrobial properties. By comparing amino acid sequences of other bacteriocins that differ in antimicrobial spectrums with nisin, specific modifications in its structure can be introduced through changes in base sequences in the structural region of *nis* A gene. An understanding of the gene structure and the control mechanism in the expression of the genes will be helpful for logical engineering of the genes.

V. BIOSYNTHESIS OF NISIN

A. Bacterial Strains

Considering that the *nis* A gene, encoding for prepronisin, is chromosomally linked and that a cell contains only one set of *nis* A and other necessary genes in a polycistronic operon, it is difficult to assume that there will be a great variation among strains of *L. lactis* subsp. *lactis* in nisin production per cell. One of the possible reasons for variation in nisin production among strains, is due to variation in total cell numbers or cell mass among strains. In addition, relative stability of the genetic materials and the activity of enzymes involved in posttranslation processing among strains could cause differences in nisin production. Very limited information is available in which different strains have been compared for nisin production per unit cell mass. Hirsch[27] reported that different isolates from a parent strain (354) produced different amounts of nisin in broth and milk cultures. From these he selected a strain (354/07) that produced a higher level of nisin in broth. While strain 354/07 retained its ability to produce high levels of nisin for a long time, in the parent strain

TABLE 13
Nisin Production by *L. Lactis* Subsp. *Lactis* Strains in Four Media

		Strains			
Mediaa	Parametersb	11454	354/07	7962	DL16
TGE+	OD	1.06	0.98	1.24	0.93
	pH	4.37	4.17	4.36	4.19
	AU	12,000	16,000	10,000	2,000
MTGE+	OD	1.12	0.91	1.26	1.24
	pH	5.38	5.65	5.19	5.13
	AU	14,000	18,000	12,000	6,000
MH	OD	1.02	1.04	1.36	1.24
	pH	5.34	5.56	5.10	5.08
	AU	26,000	26,000	14,000	12,000
MM	OD	—	—	—	—
	pH	4.30	4.32	4.34	4.31
	AU	16,000	28,000	20,000	6,000

[a] TGE+: 1.5% trypticase + 2.5% glucose + 1.5% yeast extract + 0.005% Mn^{2+} and Mg^{2+} each.[59]
MTGE+: Modified TGE: TGE + 2.5% Na-acetate + 1.5% Na-citrate + 0.5% K_2HPO_4.
MH: Modified Hirsch: 1% peptone + 2.5% glucose + 1% beef extract + 1% yeast extract + 1.5% Na-acetate + 0.5% K_2HPO_4 + 0.05% cysteine.
MM: 10% nonfat dry milk + 1% whey + 0.2% glucose + 0.005% Mn^{2+} and Mg^{2+} each.
The pH of all media was adjusted to 7.0, autoclaved for 10 min and then supplemented with 0.6 μg/ml pantothenate. Media were distributed in 10 ml portions in triplicate, inoculated with an overnight culture at 1% level, incubated at 30°C for 24 h.

[b] The pH and OD_{600nm} of each tube were read in a pH meter and a Spectronic 20, respectively. To measure activity unit (AU/ml), 1 ml from each of the triplicate tubes was transferred in a sterile tube, heated at 70°C for 10 min, diluted decimally in sterile water from 1:10 to 1:200 dilutions. Five microliters from each dilution was then transferred on the surface of TGE-agar plate overlayed with soft agar containing *Lactobacillus plantarum* NCDO 955 as indicator. The plates were incubated at 30°C overnight and examined for the zone of inhibition. To measure AU/ml, the highest dilution that gave zone of inhibition was multiplied with 200 (100 μl/5 μl). For each strain and each medium AU was determined in duplicate. The results are the average of two readings. Using the same procedure, 100 AU was found to be about 1 IU of nisin. (Ray, B. and Kalchayanand, N., unpublished data.)

(354) production declined rapidly. After frozen storage the strain 354/07 required two or more subcultures before it resumed maximum nisin production. Repeated subculturing would reduce production of nisin. By growing strains in broth or agar media containing nisin, Cheeseman and Berridge[32] were able to increase nisin production of a strain. In our laboratory we compared the relative efficiency of nisin production by four strains of *L. lactis* subsp. *lactis* ATCC 11454, 354/07, DL16, and 7962.[62] The results, presented in Table 13, indicated that in general strain 354/07 produced a higher level of nisin than the other three strains in the media used.

There is one published report indicating that a transconjugant of recipient strain *Leuconostoc dextranicum* 181 produced a thousand times higher level of nisin than the donor strain 7961.[132] However, no information is available since the first report on the stability of this transconjugant in producing such a high level of nisin. Although several researchers have developed nisin-producing transconjugants in other lactic acid bacteria, their nisin-production efficiency or their stability to the trait have not been reported. Gasson[96] has suggested that genetic manipulation could offer the hope for developing strains with higher efficiency of nisin production. One approach will be to increase copy numbers per cell of the gene. However, as the gene *nis* A is part of a large polycistronic transposon in the chromosome, and other genes are necessary for processing inactive prepronisin to active nisin, increasing copy numbers of the gene block can be difficult.

B. Growth Environment

Environment of growth, which includes nutrient availability, temperature and pH of growth, oxygen level, and presence of inhibitory metabolites, influences the production of nisin level greatly. The effect is partially related to their influence on cell multiplication and the cell mass produced per milliter of growth media.

Hirsch[27] designed a basal medium containing peptone-1%, lablemco-1% and glucose-1% (or higher) and supplemented it with different salts and micronutrients to study their influences on nisin production by strain 354/07. In general, 2.5% glucose was more effective in increasing both the growth rate (total cell mass) and nisin production (Table 14). Supplementing the media with Ca-pantothenate (0.6 μg/ml) increased growth and nisin production greatly. Other supplements, such as riboflavin, nicotinic acid, pyridoxin, biotin, xanthin, adenine, and guanine did not have any added influence in the presence of 2.5% glucose and pantothenate together with a buffer salt mixture consisting 1.5% acetate + 1.5% citrate + 0.5% phosphate. While NaCl and $NaNO_3$ reduced nisin production, Na_2SO_3 was effective in increasing it. The beneficial effects of pantothenate and Na_2SO_3 were considered to be due to their influence on coenzyme A, which is necessary for glucose metabolism and supplying "S" to cysteine. Cysteine is present in large numbers in the nisin molecule. The triple buffer salt combination produced higher nisin due to its effectiveness in maintaining the pH between 5.9 and 6.1. Higher nisin production seems to be related with higher cell mass. This was also suggested by Egorov et al.[136] who reported that by optimizing the nutrients for cell growth nisin production could be increased to 2000 IU/ml.

pH has a pronounced effect on nisin production. Maximum nisin is produced when the pH of the broth is maintained above 6.0.[136] Our own experience has indicated that nisin production with strain ATCC 11454 is highest at pH 6.6 to 6.8 and lowest at 4.5 and below. Continuous pH adjustment to maintain one pH is better than intermittent pH adjustment or use of buffer salts.[80] Temperature of incubation also influenced production of nisin. Hirsch[27] reported

TABLE 14
Influence of Media Composition and pH on Nisin Production and Growth of *L. Lactis* Subsp. *Lactis* 354/07[a]

Nutrients added to a basal medium[b]	pH[f]	Nisin (units/ml)	Optical density[g]
Glucose, 1%	Unadjusted	85	0.28
Glucose, 1%	Adjusted	172	0.48
Glucose, 2.5%	Adjusted	470	0.62
Glucose, 2.5% + Pantothenate (0.6 μg/ml)	Adjusted	525	0.71
Glucose, 2.5% + Pantothenate + Other vitamins and N-bases[c]	Adjusted	625	0.72
Glucose 2.5%[d] + Pantothenate + Acetate (1.5%) + Citrate (1.5%) + Phosphate (0.5%)	Adjusted	1963	Not available
+ 1% NaCl[e]	Adjusted	1020	Not available
+ 1% $NaNo_3$[e]	Adjusted	1070	Not available
+ 1% Na_2SO_3[e]	Adjusted	2130	Not available

[a] Table adapted from Reference 27.
[b] Basal medium contained peptone + Lablemco, each 1%.
[c] Vitamins and bases were riboflavin, nicotinic acid, pyridoxin, xanthin, adenine, guanine, and uracil.
[d] Mixture of three buffer salts produced better results than any one or two.
[e] NaCl or $NaNO_3$ or Na_2SO_3 were added to medium "d".
[f] pH was adjusted by adding NaOH solution at selected intervals; unadjusted pH was 4.2 at the end of 18 h at 30°C; adjusted pH was 5.9 to 6.1. In samples with buffered salts, a high pH was maintained by the buffer.
[g] Wave length used was not indicated.

that nisin production was maximum when the culture was incubated between 25 to 30°C as opposed to 37°C. At 37°C, initial growth was rapid but final optical density and nisin production were 0.91 and 386 units per milliter as compared to 1.15 and 542 units per milliter at 26°C, respectively. Aeration also reduced cell growth and nisin production as compared to anaerobic conditions.[27]

C. Synthesis During Cell Growth

Earlier Hurst[137] had previously studied the relationship between cell mass and DNA, RNA, proteins, and nisin concentrations during growth of strain 354/07 in a broth medium at 30°C for 12 h, and reported that synthesis of DNA and RNA attained maximum rates before synthesis of proteins and cell mass began. Nisin synthesis did not begin during early stages of growth (as indicated by increase in protein synthesis and increase in cell mass) but increased sharply after considerable cell growth had occurred (about 50% of the total dry weight at the end of a 12 h period). Thus, nisin could be synthesized as a secondary metabolite. He also reported that initially the cells synthesized nisin that was of larger molecular weight but later (after 7 h) synthesized a type that had lower molecular weight; however, both were active. In light of the present knowledge, the larger molecular species could not be prepronisin but might be the polymeric form (such as a dimer) of the monomer nisin molecules. But why this form was observed during early stages of nisin synthesis was not known. Hurst[137] also reported that as the cells entered the stationary phase, nisin synthesis stopped along with the end of synthesis of DNA, RNA, and proteins. By maintaining the pH constant at 6.8, all the cellular events occurred in the same sequence, but the cells remained longer in the early stationary phase and 30% more nisin was synthesized.[138] In a different type of study, Hurst measured incorporation of radioactive-labelled precursors of mucopeptide, DNA, RNA, and proteins in growing cells in the presence of inhibitors of synthesis of these macromolecules.[139] In the presence of mucopeptide and DNA synthesis inhibitors, synthesis of nisin was not inhibited. Actinomycin D incorporation inhibited RNA synthesis immediately but did not inhibit nisin synthesis until after 60 min. Thus, nisin synthesis is independent of synthesis of new DNA but dependent on mRNA synthesis which generally has a long half-life. This hypothesis may not be correct as recent studies have indicated that mRNA, which carries message of *nis* A gene, has a 7 min half-life.[44,47] Inhibitors of protein synthesis interfered with nisin synthesis, however.[139] The inhibition was immediate and occurred as soon as the inhibitor was added, irrespective of growth phase and nisin production rate. Similar studies by Ingram indicated that inhibitors of protein synthesis inhibited nisin synthesis preferentially.[140] Hurst and Peterson[141] reported that enzymatic processing of nisin-precursor molecules started just before the start of detectable nisin synthesis (about 300 min from the start of incubation) reached the highest rate at about 420 min, and then rapidly declined at 480 min. However, their assumption may not have been correct that enzymes involved in processing of prepronisin to nisin are not present all the time but are synthesized for a brief period towards the end of growth phase.[141] The genes for the processing enzymes and the prepronisin (*nis* A) are all located in a polycistronic operon, and the enzymes in some form should be present when prepronisin is synthesized. Our studies have shown that active nisin is synthesized as a secondary metabolite at a high rate when the cells have reached midexponential phase, and continued to be synthesized during a greater part of the stationary phase when the cells were grown at a constant pH of 6.8 at 30°C for 20 to 24 h.[80]

D. Location in Producer Cells

Prepronisin molecules are processed intracellularly in some specific areas of the cytoplasmic membrane to active molecules, and are secreted outside the cell, probably into the growth medium. However, earlier studies have reported that when the cells are grown at a constant pH of 6.8, more than 80% of the nisin synthesized remained bound to the cells, and at pH below 6.0, 80% remained in culture broth.[138] Subsequent studies revealed that at pH 6.8, cell-bound nisin was sevenfold more per unit dry weight than that at pH 4.2.[142] After breaking the cells and examining different cell fractions, about 60% of nisin was found to be in wall fraction. Total ethanol-soluble fraction (lipid fraction from membrane and other parts and ethanol soluble proteins) contained about 57%, while ethanol fraction from cytoplasmic membrane had only 9% of total nisin.

Nisin is a strong cationic, hydrophobic polypeptide and produces bacteriocidal effects against sensitive cells by destabilizing the structural and functional integrity of cytoplasmic membrane. Initially, however, the molecules bind to the specific receptor on the cell wall. It is now known that nisin binds to the cell wall of nisin-sensitive and nisin-resistant Gram-positive bacteria including the host cells. However, the mechanism by which the cell-wall-bound nisin destabilizes the cytoplasmic membrane of only sensitive strains is not known. Also, the nature of the receptor(s) on the cell wall that binds to nisin and the number of receptor(s) per cell in resistant and sensitive strains are not known. The relative number of such receptors/cell might be dependent on the pH of the environment; at higher pH they are likely to be available in greater numbers. However, this is just speculation that needs to be confirmed by well-designed studies.

E. Purification

Many methods have been used to purify nisin from culture broth. Mattick and Hirsch[13] described a method in which they grew the culture in a broth, adjusted the pH to 1.9, removed the cells, raised the pH of the supernatant fluid to 5.0, and mixed it with 3% chloroform. The chloroform layer was acidified with HCl and distilled off in vacuo to remove $CHCl_3$. The material was filtered at pH 1.5 and precipitated with 2 volumes of ethanol and 1 volume of ether. It was further precipitated with 3 volumes of ethanol and 5 volumes of ether, and then dried. The powder had 1 IU/µg. Several modifications of this method based on differential solubility of nisin in organic solvents and salt solutions were developed by others that yielded nisin in crystalline form and had higher potency.[29] However, the yield was only 10% of the total nisin. Cheeseman and Berridge[32] developed two schemes of purification that gave 50% yield of the total nisin in broth and had very high activity (30 IU/µg of dry powder). This method produced mostly nisin A and small amounts of nisin B. Instead of boiling in 1 M HCl for 10 min, which can destroy some nisin and/or modify its molecular structure, Bailey and Hurst[143] developed a milder treatment procedure that gave a product that had nine times more potency than the previous methods. Briefly, this method consisted of growing the cells at a constant pH of 6.7, harvesting the cells by centrifugation, breaking the cells at natural pH, and extracting broken cells at 2°C with pH 1.8, 0.05 N HCl, and separating on an Amberlite CB-4B anion exchange column at a final pH of 4.0. The filtrate was precipitated with 7 vol acetone at –19°C. The sample was then run through CM-cellulose against pH gradient and recovery of nisin from the characteristic peak. The method yielded more than 71% of the total nisin, with a higher potency.

VI. EFFECTIVENESS OF NISIN IN CONTROLLING SPOILAGE BACTERIA IN FOOD SYSTEMS

Nisin is bacteriocidal to the vegetative cells and outgrowing spores of sensitive strains of Gram-positive bacteria, many of which are associated with food spoilage and foodborne diseases. This and other favorable properties, such as being produced by food-grade bacteria and being a heat-stable protein destroyed by some proteolytic enzymes of the gastrointestinal tract, lead many researchers to study the effectiveness of nisin as a preservative to enhance shelf life and safety of foods. However, while studying the effectiveness of nisin as a food preservative, one should recognize that it is not antimicrobial against all Gram-positive bacteria, and even in a sensitive strain there are cells that are resistant to nisin. It is not effective normally against Gram-negative bacteria, yeasts, molds, and viruses; it is more effective in acid pH, and unstable at higher pH. Also, being a cationic surface-active molecule, its uniform distribution in a system can be influenced by the environment, composition (especially lipid content), and the consistency of a food.

The antibacterial property of nisin against different strains of Gram-positive food-spoilage bacteria including vegetative cells and spores in different food systems, has been studied. These include dairy products, desserts, meat, vegetables, and alcoholic beverages. Studies on each of these food groups will be discussed separately.

A. Dairy Products

1. Hard Cheeses

Hirsch et al.[2] in 1951 first successfully used a nisin-producing strain of *L. lactis* subsp. *lactis* in the preparation of Gruyere cheese to control the blowing defect of this product from the growth of *Clostridium tyrobutyricum* during storage. This cheese has relatively high moisture, low acid (pH 5.5 to 6.0), and low salt, and an internal anaerobic environment which provides an opportunity for the germination and growth of clostridial species. Use of nisin-producing starter cultures helped to reduce the blowing defect dramatically. Products made with a good nisin-producing strain could have 400 to 1000 IU/g.[91] This method was also tested in other similar cheeses with high pH and low salt, such as Dutch and Swiss cheeses. Although the clostridial spoilage was reduced by nisin starter, the growth of other desirable starters necessary for ripening and for producing the normal characteristics of these cheeses (such as eye formation) was inhibited.[91] Use of nisin-resistant starter strains helped overcome this problem.[91] However, the difficulties in maintaining the resistance of a large number of strains to nisin over a long period was complex and the use of nisin-producing strain during processing of cheese did not progress. Current understanding of the genetic determinants for nisin-resistant (Nis^r) phenotype may be helpful to develop nisin-resistant but nisin-nonproducing desirable strains (Nis^r Nip^-) to use along with nisin-producing (Nip^+) strains in fermentation.

2. *Pasteurized, Processed Cheese Products*

The difficulties of having nisin in cheeses that need ripening are not present in processed cheese. Processed cheese products made with cheese and other ingredients are heat treated sufficiently during processing to destroy most vegetable bacteria, but not the spores of aerobic and anaerobic bacteria. The environment of the products, although discourages the growth of aerobic spore formers, encourages the growth of anaerobic clostridial spore formers such as *C. sporogenes, C. butyricum,* and *C. tyrobutyricum* that can be present in concentrations as high as 1000 spores per gram even in a good quality product.[53] Strains of these species are quite sensitive to nisin at the level of 25 to 62.5 IU/g. However, while using nisin in processed cheese products, the following aspects are to be considered in selecting actual nisin concentration: spore type, spore load, and nisin stability. In general, putrefactive clostridia require more nisin than butyric clostridia and nisin level needs to be doubled for every 10-fold increase in the spore load.[53] Also, heat treatment during processing can reduce 15% of the total nisin activity, which in turn can be reduced by 70 to 80% of the original activity level during 8 months' storage at 30°C; at lower storage temperature the retention level is higher. Considering all these factors, the effective range of nisin levels in processed cheese products is 375 to 1250 IU/g (or 375 to 1250 μg/g).[53,119] In processed cheese nisin is used in 28 countries, and it is permitted for use in many other countries including the U.S., U.K., U.S.S.R. (C.I.S.), Australia, New Zealand, and France.[53]

3. *Pasteurized Milk*

In many developing countries, especially those with a warm climate, inadequate refrigeration facilities during transportation and storage of properly pasteurized milk can reduce the shelf- ife considerably. Addition of nisin at 5 to 10 IU/ml can increase the shelf life by 6 d at 15°C, and by 2 d at 20°C.[53]

4. *Canned Evaporated Milk*

Addition of nisin at 80 to 100 IU/ml was found to reduce heat-processing time by 10 min as well as to reduce spoilage of canned evaporated milk from the growth of high-heat-resistant spores during storage.[144–146]

5. *Heat-Sterilized Milk*

Thermophilic bacterial spores can survive commercial sterilization of milk, and can cause spoilage if the storage temperature of milk gets to 40°C or higher even for a brief period. One study showed that addition of nisin at 20 IU/ml completely inhibited spores that survived a 115°C heating for 15 min in milk. Studies in Egypt[147] and India[148] have also shown that shelf life of commercially sterilized milk could be greatly increased by the addition of nisin.

6. High-Heat-Treated-Flavored Milk

Heinemann et al.[149] reported that addition of 2.5 IU/ml nisin to chocolate milk inoculated with *Clostridium* PA 3679 and *Bacillus steathermophilus* spores and heat treated to an F_0 value of 3.2 effectively controlled spoilage. Addition of 80 IU/ml nisin in chocolate milk not only reduced heating time and improved product quality but also reduced spoilage during extended storage at 35°C or 55°C.[149] Similar results were reported by Shehata et al.[147]

7. Yogurt

Gupta and Prasad[150] reported that incorporation of 1000 to 4000 IU/ml (25 to 100 RU/ml) of nisin to yogurt after fermentation considerably reduced the viability and growth of lactic and nonlactic acid bacteria, controlled biochemical reactions of these microorganisms, and increased shelf life during storage at refrigeration temperature as well as at slightly higher temperature (6 to 8°C). They also observed that residual nisin in yogurt decreased rapidly, and after 10 d storage at 6 to 8°C about 80 to 85% of the original activity was lost.[151]

8. Dairy Desserts

These products are not given high heat treatments to maintain their acceptance quality and thus are susceptible to spoilage during storage. The effectiveness of increasing shelf life of different types of dairy desserts by incorporating nisin has been studied. Nisin levels at 1.3 to 3.8 μg/g extended the shelf life of pasteurized cream caramel products by 30 d and chocolate dairy desserts by 31 d, at 12°C, from a shelf life of only 6 d for nonnisin added products.[53,152] Canned milk pudding made with rice, sago, semolina, and tapioca also had extended shelf life and did not require very high heat treatment when nisin at 1.3 to 3.8 μg/g was incorporated in these products.[144] Similarly, incorporation of nisin at 200 IU/g increased the shelf life of kheer, a heat-treated dessert made with milk, sugar, and rice, from 2 d to 10 ds at 37°C.[153]

B. Meat Products

1. Cured Meat Products

Use of nisin to increase shelf life of processed cured meat products produced inconclusive or variable results. Incorporation of nisin in gelatin used in canned ham at 12 μg/g reduced bacterial load substantially.[1] Also, use of nisin at 75 to 100 μg/g together with nitrite (40 ppm) inhibited outgrowth of *C. sporogenes* spores.[154] Houben and Krol[155] successfully inhibited growth of *Enterococcus faecalis* by using 10 μg/g of nisin in ham heated to 68.9°C for 90 min for 180 d at 7°C. However, Gibbs and Hurst[156] observed that 2.5 to 12.5 μg/g nisin did not inhibit growth of lactobacilli in pasteurized bacon during storage at

TABLE 15
Effectivness of Nisin in Reducing Bacterial Population During Storage of Vacuum-Packaged Refrigerated Fresh Beef[a]

Bacterial groups	Treatment	cfu/g at 3°C after weeks			
		1	4	8	12
Aerobic plate counts	Control	4.3	5.9	6.6	7.2
	Nisaplin	0.3	3.3	5.7	6.7
Psychrotrophic counts	Control	3.7	5.8	7.3	7.5
	Nisaplin	0.8	4.0	5.7	6.5
Lactic acid bacteria	Control	3.3	5.5	6.4	7.0
	Nisaplin	0.3	2.8	5.8	6.3

Note: Beef steaks (about 100 g each) from semitendinous muscles were aseptically placed in plastic bags. Nisin (as Nisaplin) in treated samples was added in aqueous solution at 500 IU/g and mixed well. Both control and nisin-treated samples were vacuum packaged and stored at 3°C. At 1, 4, 8, and 12 weeks, 100 ml sterile 0.1% peptone water was added to each sample in the bag, mixed, and the liquid was used for the enumeration by pour plating of aerobic plate counts (in plate count agar, 35°C for 2 d), psychrotrophic counts (10°C for 7 d) and lactic acid bacterial counts (APT agar, 25°C for 2 to 3 d). Each datum is the average of four sets of experiments.

[a] Table adapted from Reference 159.

20°C. In contrast, Calderon et al.[157] observed suppression of lactic acid bacteria in vacuum-packaged bacon by using a combination of nisin (12.5 μg/g) and nitrite (50 ppm) during storage at 5°C with an extended shelf life of 1 week. Also, nisin at 200 IU/g in vacuum-packaged Italian sausages extended the shelf life greatly during 2 months of storage at 2°C. While the products with nisin were all acceptable, controls without nisin developed an acidic smell, bitter taste, and were generally disagreeable.[158]

2. Vacuum-Packaged Fresh Beef

A commercial nisin preparation was incorporated in fresh beef prior to vacuum packaging at 500 IU/g and the meat samples were enumerated for bacterial counts (colony forming units) during storage at 3°C (Table 15). Addition of nisin reduced the initial bacterial population (aerobic plate counts, APC; lactic acid bacteria counts, LAB counts; and psychrotrophic bacteria counts) by 3 to 4 logs. During 12 weeks' storage the untreated controls had higher cfu for all three groups of bacteria than the treated samples.[159] A laboratory preparation of nisin also resulted in reduced growth of these bacterial groups when used at much lower levels.

The discrepancy in results of the effectiveness of nisin as an antibacterial agent could be due to the presence and subsequent growth during storage of bacteria resistant to nisin, unequal distribution of nisin in the solid food systems, poor solubility of nisin at the product pH, and interference of hydrophobic nisin molecules by meat lipids.[49,159] Proper experimental design and techniques will be necessary to resolve the confusion about the effectiveness of nisin against Gram-positive meat-spoilage bacteria associated with processed and fresh meat, especially vacuum-packaged and refrigerated products that are expected to have a long shelf life.

C. Canned Vegetable Products

The effectiveness of nisin in inhibiting growth of surviving thermophilic spores of bacillus and clostridial strains in both high-acid and low-acid canned vegetable products has been studied. Incorporation of 100 to 200 IU/g nisin in high acid (below pH 4.6) products, such as in canned tomato products, greatly controlled growth and spoilage by acid-tolerant *C. pasteurianum* and *Bacillus coagulans*.[160] In low-acid products such as peas, beans, and corn, heat treatment just to kill *C. botulinum* spores will not kill thermophilic spores of *B. steathermophilus* and *C. thermosaccharolyticum* that can subsequently grow and cause product spoilage.[161,162] Incorporation of nisin in the low-acid canned products also greatly reduced spoilage. Addition of nisin also reduced the total heating time required. Similar beneficial effects of nisin were observed for other low-acid canned foods such as potatoes,[163] mushrooms,[52] asparagus,[164] and soups[165].

In canned products, retention of nisin varied greatly with the type of food and storage conditions. Heat processing (115°C for 30 min) of canned peas was reported to retain about 22% of the original activity of nisin; after 30 months of storage in a warehouse the level was reduced to around 8%.[53]

Nisin, however, failed to inhibit spoilage of products with high levels of spore contamination in chow mein, cream style corn, and soups, indicating the importance of using nisin in good quality products. Nisin, thus, cannot be used to mask the use of poor-quality raw materials.

D. Alcoholic Beverages

1. Beer

Prevention of beer spoilage by alcohol-tolerant lactobacilli and pediococci with nisin, either by adding nisin to beer or by treating pitching yeast with nisin to reduce loads of spoilage bacteria, has been extensively studied. In one study, 51 of 56 lactobacilli and 36 out of 37 pediococci, as well as strains of micrococci and streptococci isolated from spoiled beer, were reported to be sensitive to nisin. Nisin did not have any inhibitory effect on brewing yeasts. In a pilot plant trial, addition of nisin (2.5 μg/ml) did not adversely affect the fermentation of yeast; rather, the suppressing of undesirable bacterial growth increased fermentation performance.[166,167]

TABLE 16
Ability of Nisin-Resistant *Leuconostoc Oenos* Strains to Produce Malo-Lactic Fermentation in Wine in the Presence of Nisin

Bacterial strain added	Treatment	*L. oenos* cfu after 3 mos at 18°C	Malic acid after 3 mos (mg/ml)
None	Control	<10	3.5
L. oenos[a] (Nis^s)	Control	7.8×10^7	0.3
L. oenos (Nis^s)	Nisin (100 IU/ml)	<10	3.5
L. oenos (Nis^r)	Control	1.4×10^7	<0.1
L. oenos (Nis^r)	Nisin (100 IU/ml)	6.8×10^6	0.2

[a] *L. oenos* cells were added at 10^{6-7} cfu/ml to sterile Pinot noir wine and stored at 18°C for 3 months before enumerating for cfu of *L. oenos* and determining of malic acid retained. (Adapted from Reference 169.)

2. *Wine*

Lactic acid bacteria (especially *Lactobacillus, Leuconostoc,* and *Pediococcus* species) in acidified grape can reduce the quality of the wine. But addition of 2.5 to 25 μg/ml nisin in the grape-must reduced the growth of some lactobacilli, completely eliminated the other two species, and produced wine with acceptable taste. However, nisin could not be added in wines that depend upon desirable malolactic fermentation.[168] This problem was recently overcome by developing nisin-resistant strains of *Leuconostoc oenos* that can grow and carry on malolactic fermentation in the presence of nisin (Table 16).[169] The addition of nisin to reduce the growth of competitive lactic acid bacteria, and indirectly to favor growth of the fermentating yeast, has been reported to increase alcohol content by 10% in the production of fruit brandies.[170]

VII. EFFECTIVENESS OF NISIN IN CONTROLLING PATHOGENS IN FOOD SYSTEMS

The effectiveness of nisin for controlling growth of Gram-positive foodborne pathogens in food systems has not been studied widely, and information is quite limited. Results of pure culture studies in laboratory media have indicated that nisin cannot produce a 100% kill of a cell suspension of a sensitive Gram-positive pathogenic bacteria. The surviving cells are resistant to nisin, and as nisin does not have bacteriostatic effect, they can grow under suitable conditions. Thus, for a pathogen with zero tolerance, nisin might have very limited importance, especially in foods that are expected to have long shelf life. The available information on the inhibitory effect of nisin against Gram-positive pathogenic bacteria in food systems is presented.

A. Dairy Products

1. Milk[103a]

In a recent study, bacteriocidal effectiveness of nisin against *L. monocytogenes* in milk containing 0% and 12.5% fat was studied. The presence of 10 IU/ml nisin reduced the cfu/ml of strain Scott A by about 5 logs in milk with 0% fat, and by about 0.5 log in milk with 12.5% fat within 2 h. Bioassay indicated that the activity of added nisin was reduced by about 33% in skim milk and 88% in milk with 12.5% fat. This reduction could be attributed to the hydrophobic nature of nisin and its affinity towards lipids. Addition of the nonionic emulsifier Tween 80 counteracted the reduction in nisin activity in high fat milk and increased the viability loss of *L. monocytogenes*.

2. Processed Cheese

Tanaka et al.[171] showed that typical commercial formulations of processed cheese spreads with a moisture level of 51 to 54% will not allow formation of botulinal toxin. However, with higher moisture levels (>55%), or reduced NaCl and phosphate levels, a product might need an additional preservative barrier to ensure safety from the outgrowth of *Clostridium botulinum*. Nisin at a level of 500 IU/g might be effective in providing protection against outgrowth and toxin formation by *C. botulinum* spores. Higher nisin levels will be necessary with higher storage temperature, higher spore levels and longer shelf life of the products. Somers and Taylor[172] also reported that while moisture level, pH, concentrations of NaCl, and disodium phosphate in combination inhibit outgrowth and toxin formation by *C. botulinum* spores, addition of nisin provided an additional antibotulinal means. Addition of nisin also allows some flexibility in the formulation of the products in terms of increasing moisture level and reducing level of phosphate and salt. At a level of 500 to 1000 μg/g nisin either delayed or prevented the growth of, and toxin production by, inoculated spores of *C. botulinum* type A or type B even in cheese spreads with 60% moisture level. But higher levels were required for larger number of spores. Nisin activity level also decreases in processed cheese during storage, especially at higher storage temperature. In determining the level of nisin to be used to ensure safety against botulinal toxin in cheese spread, the above factors should be taken into consideration. Currently the regulatory requirement of nisin levels when added to cheese spread is 250 μg/g.[124]

3. Cheddar Cheese

Addition of nisin at 720 IU/ml to milk inoculated with starter culture and bacteriophage (to cause starter failure) and *Staphylococcus aureus* did not inhibit growth of *S. aureus* and formation of enterotoxin A during Cheddar Cheese processing.[65] Although addition of nisin to brain-heart infusion broth prevented growth of *S. aureus,* addition of whole milk, but not skim milk, to broth (1:10) failed to inhibit the growth. Nisin from an aqueous solution can be removed when mixed with melted butter fat; butterfat can interfere with the antibacterial effectiveness of nisin, probably due to the hydrophobic nature of the nisin molecules.

4. Cottage Cheese

The inhibitory effect of nisin against *Listeria monocytogenes* ATCC 7644 in cottage cheese has been studied by Benkerroum and Sandine.[66] They reported that inoculation of 3.5×10^5 cells of *L. monocytogenes* to either nonsterile or sterile cottage cheese along with 2500 IU/g or 3700 IU/g nisin resulted in more than 5 logs viability loss within 1 d; during 30 d of storage at 4°C no *Listeria* colonies were recovered. However, in this study a strain that is relatively more sensitive to nisin was used, as compared to relatively resistant strains (such as Scott A and V7).

B. Meat Products

Studies involving partial or total replacement of nitrite in processed meat products with nisin to prevent outgrowth and toxin production by *C. botulinum* spores have produced variable results. Rayman et al.[173] reported that up to 550 μg/ml of nisin along with 60 ppm of nitrite failed to prevent outgrowth of type A, and proteolytic and nonproteolytic type B spores of *C. botulinum* in pork slurries at pH 5.8. Although effectiveness of nisin increased at lower pH, outgrowth of *C. botulinum* spores was not inhibited at pH of most cured meat products with a combination of nisin and 60 ppm nitrite. In other studies with vacuum-packaged bacon,[174,175] chicken franks,[175] and other cured products, use of nisin at very high and uneconomic levels inhibited the outgrowth of *C. botulinum* spores. The inability of nisin to control growth of *C. botulinum* in cured meat with low level of nitrite was thought to be due to higher resistance of *C. botulinum* to nisin, binding of nisin to meat, poor solubility of nisin at product pH, uneven distribution of nisin in the product, and probably interference due to high fat content of the products.[176] Caseiro et al.,[158] however, found that growth of *Clostridium perfringens* cells in Wurstle sausage could be inhibited for 38 days at 20°C when nisin at 75 to 100 μg/g was used in combination with 40 ppm of nitrite.

Chung et al.[177] observed that nisin treatment of meat pieces inoculated with *L. monocytogenes* cells caused an immediate loss in cell viability; further incubation did not reduce the cell numbers. Cell viability loss of *Staphylococcus aureus* was also observed. The growth of both pathogens was delayed for at least 1 d at room temperature in the nisin-treated meat samples.

The results of these studies indicate that more well-designed studies need to be done before the effectiveness of nisin against Gram-positive pathogens in fresh meat and meat products can be determined.

C. Fish Products

Fresh fish may contain *C. botulinum* type E spores as contaminants that can grow and produce toxins in fresh or smoked fish that are vacuum-packaged to extend their shelf life. Studies reported by Taylor et al.[178] have indicated that nisin treatment of cod, herring, and smoked mackerel fillets inoculated with *C. botulinum* spores and packaged in 100% CO_2 atmosphere delayed toxin production by 5 d over the control at 10°C, but only by half a day at 26°C. Nisin treatment, however, did not interfere with the growth of nonpathogenic spoilage bacteria. In all three types of control samples at 26°C, and smoked mackerel at 10°C, toxin was formed before the samples were judged to be unacceptable or spoiled.

VIII. LEGAL STATUS OF USE OF NISIN IN FOOD

A. Public Health Consideration

It is now accepted that nisin is safe for human consumption. The nisin-producing *L. lactis* subsp. *lactis* strains occur in milk and in many milk products, and it has been consumed by humans probably for the last 8000 years since the beginning of the consumption of milk that followed the domestication of cattle. Its safety for the health (toxicity) and reproductive functions of animals over three generations have been well documented.[122–126] Other studies have provided information that, although sensitive bacterial strains can be trained to be resistant to nisin by growing in the presence of increasing concentrations of nisin, the property is lost when these strains are subcultured in media without nisin. Also, acquired resistance to nisin does not provide a bacterial strain automatically resistant to therapeutic antibiotics such as penicillin, streptomycin, and tetracycline.[39,91] In addition, when consumed, nisin is degraded by the intestinal proteolytic enzyme (α-chymotrypsin) and does not have any inhibitory effect on the ecology or the microflora of the GI tract.[127] Finally, use of nisin in food will not produce any significant impact in the environment. Considering all these factors, it can be stated with confidence that the use of nisin in food will not cause any public health problems.

B. Legal Status

Nisin was permitted for use as a food preservative in the U.K. in 1959 to prevent blowing of high-moisture, low-acid cheese. Subsequently it has been legalized for use in many other foods. In canned foods its use is permitted only when a product has received a minimal heat treatment to kill *C. botulinum* spores.

In 1969 the Joint Food and Agriculture Organization/World Health Organization (FAO/WHO) Experts Committee on food additives gave international acceptance to nisin.[179] Currently nisin is listed in the FAO/WHO Additives Data System as a fully evaluated preservative. Their recommendation of daily intake is up to 33,000 units per kilogram of body weight, which is about 1 mg of pure nisin (equivalent to 40×10^3 IU). If a food contains 400 IU/g, a person needs to consume 100 g of food for each kilogram body weight to obtain the maximum permitted level of nisin (i.e., a person with 70 kg of body weight would need to consume 7 kg of nisin-containing food every day).

Currently nisin has been permitted for use in different foods in many countries around the world. This information is listed in Table 17. It can be

TABLE 17
Countries Where Specific Approval Exists for the Use of Nisin

Country	Food in which nisin is permitted	Maximum level of addition expressed as mg nisin/kg
Abu Dhabi	Pasteurized milk	No limit
	Flavored milk	
	Long-life milk	
	Processed cheese	
	Cheese	
	Other dairy products	
	Canned vegetables	
Argentina	Processed cheese	12.5
Australia	Cheese	No limit
	Canned tomatoes	
	Canned tomato puree and paste	
	Canned soups	
Bahrain	Pasteurized milk	No limit
	Flavored milk	
	Long-life milk	
	Processed cheese	
	Other dairy products	
	Canned vegetables	
Belgium	Cheese	2.5
Bolivia	Use not prohibited in foods	No limit
Brazil	Pasteurized cheese	12.5
	Processed cheese	
	Requeijao	
Bulgaria	Cheese	5
	Ice for storing fresh fish	
Chile	Food products	100
Colombia	Processed cheese	12.5
Costa Rica	Cheese products	No limit
Cyprus	Cheese	No limit
	Clotted cream	
	Canned vegetables	
Czechoslovakia	Bakery products and fillings	12.5
	Mayonnaise	
	Processed chees	
	Prepared foods	
	Semiprepared foods	
	Canned vegetables	
	Babyfoods — dairy and vegetable	
Denmark	Canned processed cheese	12.5
Dubai	Cheese	No limit
	Pasteurized milk	
	Flavored milk	
	Long-life milk	

TABLE 17 (continued)
Countries Where Specific Approval Exists for the Use of Nisin

Country	Food in which nisin is permitted	Maximum level of addition expressed as mg nisin/kg
	Other dairy products	
	Canned vegetables	
Egypt	Processed cheese	12.5
Eire	Processed cheese	12.5
Finland	Processed cheese	12
France	Processed cheese	No limit
Gibraltar	As in U.K.	No limit
Hong Kong	As in U.K.	No limit
India	Hard cheese	25
	Processed cheese	
Italy	Cheese	12.5
	Canned vegetables	2.5
	Confectionery creams	
Jordan	Processed cheese	12.5
Kuwait	Processed cheese	100
Malaysia	Canned foods and cheese	No limit
Malta	As in U.K.	No limit
Mexico	Food products	No limit
Netherlands	Factory cheese	12.5
	Quarg	
	Certain processed cheese	
	Cheese powder	20
New Zealand	Processed cheese	12.5
Peru	Nisin is a permitted additive	No limit
Philippines	Processed cheese	100
Poland	Ripened natural cheese	100
	Processed cheese	
Portugal	Processed cheese	12.5
Qatar	Milk	No limit
	Milk products	
Saudi Arabia	Some foods and dairy products	No limit
Singapore	Cheese	No limit
	Canned foods	
South Africa	Processed cheese products	12.5
	Some other cheeses	
Spain	Processed cheese	12.5
Sweden	Cheese and margarine	12.5
	Cheese with fresh meat	
Taiwan	Cheese	25
Thailand	Processed cheese	100
Trinidad	Canned foods	No limit
	Cheese	
	Clotted cream	

TABLE 17 (continued)
Countries Where Specific Approval Exists for the Use of Nisin

Country	Food in which nisin is permitted food product	Maximum level of addition expressed as mg nisin/kg
Turkey	Various cheeses	100
United Kingdom	Cheese	No limit
	Canned foods	
	Clotted cream	
USA	Pasteurized cheese spreads and processed cheese spread	6.25
Uruguay	Processed cheese	100
USSR	Dietetic processed cheese	200
	Canned vegetable products	100

seen that there are wide variations in the types of foods and allowable levels of nisin in these foods among various countries.

IX. CONCLUDING REMARKS

There are advantages and disadvantages in using nisin as a food biopreservative. The advantages are its safety for human health and its bacteriocidal property against Gram-positive bacteria, many of which can spoil food and make food unsafe for consumption. As it is produced by food-grade bacteria, its acceptance as a food biopreservative by the regulatory agencies and consumers will be least objectionable. Also, there is a wealth of information available from the research conducted over the last 50 years. Finally, it is commercially available, and the commercial product does not adversely affect the acceptance qualities of a food.

The disadvantages are that it is not effective against Gram-negative spoilage and pathogenic bacteria, yeasts, molds, and viruses, and thus it will have application only in foods where Gram-positive bacteria are of major concern. Even in sensitive Gram-positive strains, the resistant survivors can multiply, as nisin does not have bacteriostatic properties. The food environment such as higher pH, higher lipid content, larger particle size, extended storage time, and ununiform distribution in food can reduce the antibacterial effectiveness of nisin in food systems, especially in solid foods. Finally, the cost of commercial preparation at present may be prohibitive for use in many foods, especially where higher concentrations of nisin need to be used.

Some of these disadvantages can be overcome through the understanding of the genetic control system of nisin production, for example, in developing strains that will produce more nisin, and through the innovation of effective techniques such as uniform distribution of nisin in solid and semi-solid foods. Also, because it is a ribosomally synthesized small protein, protein engineering techniques can be used to produce analogues that will have wider antimicrobial spectrums. In this context it will be interesting to examine if nisin A[1], nisin Z[115] and other currently known nisin variants differ in their bacteriocidal efficiency and antibacterial spectrums. Currently it can be used as an additional barrier against Gram-positive spoilage and pathogenic bacteria along with other methods that include the use of good raw materials, good sanitation practices, effective processing and careful temperature control during postprocessing handling.

REFERENCES

1. **Hurst, A.,** Nisin, *Adv. Appl. Microbiol.*, 27, 85, 1981.
2. **Hirsch, A., Grimsted, E., Chapman, H. R., and Mattick, A. T. R.,** A note on the inhibition of an anaerobic spore former in Swiss-type cheese by a nisin-producing *Streptococcus, J. Dairy Res.*, 18, 205, 1951.
3. **Rogers, L. A. and Whittier, E. O.,** Limiting factors in lactic fermentation, *J. Bacteriol.*, 16, 211, 1928.
4. **Rogers, L. A.,** The inhibiting effect of *Streptococcus lactis* on *Lactobacillus bulgaricus, J. Bacteriol.*, 16, 321, 1928.
5. **Whitehead, H. R. and Riddet, W.,** Slow development of acidity in cheese manufacture, *N. Z. J. Agric.*, 46, 225, 1933.
6. **Whitehead, H. R.,** A substance inhibiting bacterial growth produced by certain strains of lactic streptococci, *Biochem. J.*, 27, 1793, 1933.
7. **Cox, G. A. and Whitehead, H. R.,** Streptococci which produce a substance inhibiting the growth of lactic streptococci, *N. Z. J. Agric.*, 52, 38, 1936.
8. **Meanwell, L. J.,** The influence of raw milk quality on slowness in cheese making, *Proc. Soc. Agric. Bacteriol.*, 19 (Abstr.), 1943; *Nature (London)* 154, 551, 1944.
9. **Hoyle, M. and Nichols, A. A.,** Inhibitor strains of lactic streptococci and their significance in the selection of cultures for starters, *J. Dairy Res.*, 15, 398, 1948.
10. **Baribo, L. E. and Foster, E. M.,** Some characteristics of the growth inhibitor produced by a lactic streptococcus, *J. Dairy Sci.*, 34, 1128, 1951.
11. **Baribo, L. E. and Foster, E. M.,** The production of growth inhibitor by lactic streptococci, *J. Dairy Sci.*, 34, 1136, 1951.
12. **Shattock, P. M. F. and Mattick, A. T. R.,** *J. Hyg.*, 43, 173, 1943; cited in *Nature (London)*, 154, 551, 1944.
13. **Mattick, A. T. R. and Hirsch, A.,** Further observations on an inhibitory substance (nisin) from lactic streptococci, *Lancet*, 2, 5, 1947.
14. **Mattick, A. T. R. and Hirsch, A.,** A powerful inhibitory substance produced by group 'N' streptococci, *Nature (London)*, 154, 551, 1944.
15. **Mattick, A. T. R. and Hirsch, A.,** Sour milk and the tubercle bacillus, *Lancet*, 1, 417, 1946.
16. **Taylor, J. I., Hirsch, A., and Mattick, A. T. R.,** The treatment of bovine streptococcal and staphylococcal mastitis with nisin, *Vet. Rec.*, 61, 197, 1948.
17. **Hirsch, A. and Mattick, A.T.R.,** Some recent applications of nisin, *Lancet*, 2, 190, 1949.
18. **Barber, R. S., Braude, R., and Hirsch, A.,** Growth of pigs given skim milk soured with nisin producing streptococci, *Nature (London)*, 169, 200, 1952.
19. **Bavin, E. M., Beach, A. S., Falconer, R., and Friedman, R.,** Nisin in experimental tuberculosis, *Lancet*, 1, 127, 1952.

20. **Wilkins, K. M. and Board, R. G.,** Natural antimicrobial systems, in *Mechanisms of Action of Food Preservation Procedures,* Gould, G. W., Ed., Elsevier, New York, 1989, 321.
21. **Kooy, J. S. and Pette, J. W.,** The inhibition of butyric acid fermentation in cheese by using antibiotic producing streptococci as starter, *Neth. Milk Dairy J.,* 6, 317, 1952.
22. **McClintock, M., Serres, L., Marzolf, J. J., Hirsch, A., and Mattick, A. T. R.,** Action inhibitrice des streptocques producteurs de nisine sur le developpement des sporules anaerobies dous le fromage de Gruyere fondu, *J. Dairy Res.,* 19, 187, 1952.
23. **O'Brien, R. T., Titus, D. S., Devlin, K. A., Stumbo, C. R., and Lewis, J. C.,** Antibiotics in food preservation. II. Studies on the influence of subtilin and nisin on the thermal resistance of food spoilage bacteria, *Food Technol.,* 10, 352, 1956.
24. **Hawley, H. B.,** Nisin in food technology, *Food. Manuf.,* 32, 370, 1957.
25. **Gillespy, T. G.,** Nisin trials, fruits and vegetables canning, quick-freeze, Research Association Leaflet No. 3, Chipping Campden, England, 1957.
26. **Campbell, L. L., Sniff, E. E., and O'Brien, R. T.,** Subtilin and nisin as additives that lower the heat process requirements of canned foods, *Food Technol.,* 13, 462, 1959.
27. **Hirsch, A.,** Growth and nisin production of a strain of *Streptococcus lactis, J. Gen. Microbiol.,* 5, 208, 1951.
28. **Csiszar, J. and Pulay, G.,** Research on *Streptococcus lactis* producing antibiotics against *Clostridia.* I. The activity spectrum of nisin producing strains and raising the production of antibiotics, *Int. Dairy Congr.,* Proc., 2, 423, 1956.
29. **Berridge, N. J.,** Preparation of the antibiotic nisin, *Biochem. J.,* 45, 486, 1949.
30. **Berridge, N. J., Newton, G. G. F., and Abraham, A. P.,** Purification and nature of the antibiotic nisin, *Biochem. J.,* 52, 529, 1952.
31. **Gowans, J. L., Smith, N., and Florey, H. W.,** Some properties of nisin, *Br. J. Pharmacol. Chemother.,* 7, 438, 1952.
32. **Cheeseman, G. C. and Berridge, N. J.,** Observations on the molecular weight and chemical composition of nisin A, *Biochem. J.,* 71, 185, 1959.
33. **Newton, G. G. F., Abraham, E. P. ,and Berridge, N. J.,** Sulphur-containing amino acids of nisin, *Nature (London),* 171, 606, 1953.
34. **Hirsch, A.,** The assay of antibiotic nisin, *J. Gen. Microbiol.,* 4, 70, 1950.
35. **Berridge, N. J. and Barrett, J.,** A rapid method for the turbidimetric assay of antibiotics, *J. Gen. Microbiol.,* 6, 14, 1952.
36. **Mocquot, C. and Lefebvre, E.,** A simple procedure to detect nisin in cheese, *J. Appl. Bacteriol.,* 19, 322, 1956.
37. **Murray, R. G. E. and Loeb, L. J.,** Antibiotics produced by micrococci and streptococci that show selective inhibition within the genus *Streptococcus, Can. J. Res.,* 28E, 177, 1950.
38. **Eastoe, J. E. and Long, J. E.,** The effect of nisin on the growth of cells and spores of *Clostridium welchii* in gelatine, *J. Appl. Bacteriol.,* 22, 1, 1959.
39. **Carlson, S. and Bauer, H. M.,** A study of problems associated with resistance to nisin, an antibiotic produced by *Streptococcus lactis, Arch. Hyg. Bacteriol.,* 141, 445, 1957.
40. **Galesloot, Th, E.,** Lactic acid bacteria which destroy the antibioticum (nisin) of *S. lactis, Ned. Milk Zdschr,* 10, 143, 1956.
41. **Lipinska, E. and Strazalkowska, M.,** Experiments on training pure cheese cultures to resist antibiotics produced by *Streptococcus lactis, Int. Dairy Congr. Proc.,* 2, 572, 1959.
42. **Gross, E. and Morell, J. L.,** The structure of nisin, *J. Am. Chem. Soc., 93,* 4634, 1971.
43. **Fukase, K., Kitazawa, M., Sano, A., Shimbo, K., Fujita, H., Horimoto, S., Wakamiya, T., and Shiba, T.,** Total synthesis of peptide antibiotic nisin, *Tetrahedron Lett.,* 29, 795, 1988.
43a. **Jung, G.,** Lantibiotics — ribosomally synthesized biologically active polypeptides containing sulfide bridges and α,ß-dihydroamino acids, *Angewandte Chemie,* 30, 1051, 1991.
44. **Buchman, G. W., Banerjee, S., and Hansen, J. N.,** Structure expression and evolution of a gene encoding the precursor of nisin, a small protein antibiotic, *J. Biol. Chem.,* 263, 16260, 1988.
45. **Kaletta, C. and Entian, K.,** Nisin, a peptide antibiotic: cloning and sequencing of the *nis* A gene and posttranslation processing of its peptide product, *J. Bacteriol.,* 171, 1597, 1989.

46. **Dodd, H. M., Horn, N., and Gasson, M. J.,** Analysis of the genetic determinant for production of the peptide antibiotic nisin, *J. Gen. Microbiol.,* 136, 555, 1990.
47. **Steen, M. T., Chung, Y. J., and Hansen, J. N.,** Characterization of nisin gene as part of a polycistronic operon in the chromosome of *Lactococcus lactis* ATCC 11454, *Appl. Environ. Microbiol.,* 57, 1181, 1991.
48. **Frazer, A. C., Sharratt, M., and Hickman, J. R.,** The biological effects of food additives, nisin, *J. Sci. Food Agric., 13,* 32, 1962.
49. **Delves-Broughton, J.,** Nisin and its uses as a food preservative, *Food Technol.,* 44(11), 100, 1990.
50. **Hurst, A.,** Interactions of food starter cultures and foodborne pathogens: the antagonism between *Streptococcus lactis* and spore forming microbes, *J. Milk Food Technol.,* 35, 418, 1972.
51. **Hurst, A.,** Nisin: its preservative effect and function in the growth cycle of the producer organism, in *Streptococci,* Skinner, F. A. and Quesnel, L. B., Eds., Academic Press, New York, 1978, 297.
52. **Heineman, B., Voris, L., and Stumbo, C. R.,** Use of nisin in processing food products, *Food Technol.,* 19(4), 592, 1965.
53. **Fowler, G. G. and Gasson, M. J.,** Antibiotic-nisin, in *Food Preservation,* Russell, N. J. and Gould, G. W., Eds., Van Nostrand Reinhold, New York, 1990, 137.
54. **Tamer, J. and Fowler, G. G.,** Estimation of nisin in foods, *J. Sci. Food Agric.,* 15, 522, 1964.
55. **World Health Organization**, Specifications for identity and purity of some antibiotics, *Food Additive,* 69(34), 53, 1969.
56. **Motlagh, A. M., Johnson, M. C., and Ray, B.,** Viability loss of foodborne pathogens by starter culture metabolites, *J. Food Prot.,* 54, 873, 1991.
56a. **Rozbeh, M., Kalchayanand, N., Ray, B., and Field, R. A.,** Shelf life extension of vacuum-packaged refrigerated beef using starter culture metabolites, Proc., Western Section, American Society of Animal Science, Aug. 6–9, University of Wyoming, Laramie, 50, 1991.
56b. **Anon.,** *The Relationship Between Nisaplin and Pure Nisin,* Technical Information Sheet 3/88, Aplin and Barrett, England.
57. Cox, G. **A.,** A simple method to detect nonacid milk, *N. Z. J. Agric.,* 49, 231, 1934.
57a. **Chevalier, R., Fourmaud, J., Lefebre, E., and Mocquot, C.,** A novel technique for detection of inhibitory and stimulatory streptococci, *Ann. Technol. Agric.,* 2, 117, 1957.
58. **Bhunia, A. K., Johnson, M. C., Ray, B., and Kalchayanand, N.,** Mode of action of pediocin AcH from *Pediococcus acidilactici* H on sensitive bacterial strains, *J. Appl. Bacteriol.,* 70, 25, 1991.
59. **Biswas, S. R., Ray, P., Johnson, M. C., and Ray, B.,** Influence of growth conditions on the production of a bacteriocin, pediocin AcH, by *Pediococcus acidilactici* H, *Appl. Environ. Microbiol.,* 57, 1265, 1991.
60. **Falahee, M. B., Adams, M. R., Dale, J. W., and Morris, B. A.,** An enzyme immuno assay for nisin, *Int. J. Food Sci. Technol., 25,* 590, 1990.
61. **Blackburn, P., Polak, J., Gusik, S. A., and Rubino, S. D.,** Nisin Composition for Use as Enhanced Broad Range Bactericides, *International Patent Application, Patent Application Treaty,* Publication No. WO 89/12399, 1155, AVC Americus, New York, 1989.
61a. **Stevens, K. A., Sheldon, B. W., Klapes, N. R., and Klaenhammer, T.,** Nisin treatment for inactivation of *Salmonella* species and other Gram-negative bacteria, *Appl. Environ. Microbiol.,* 57, 3613, 1991.
62. **Ray, B. and Kalchayanand, N.,** Effectiveness of Nisin to Sublethally Stressed Gram-Negative Bacteria, unpublished data, University of Wyoming, Laramie, 1991.
63. **Hurst, A.,** Nisin and other inhibitory substances from lactic acid bacteria, in *Antimicrobilas in Foods,* Braun, A. L. and Davidson, P. M., Eds., Marcel Dekker, New York, 1983, chap. 11.
64. **Ogden, K., Waites, M. J., and Hammond, J. R. M.,** Nisin and brewing, *J. Inst. Brew., 94, 233, 1988.*
65. **Jones, L. W.,** Effect of butterfat in inhibition of *Staphylococcus aureus* by nisin, *Can. J. Microbiol.,* 20, 1257, 1974.

21

66. **Benkerroum, N. and Sandine, W. E.,** Inhibitory action of nisin against *Listeria monocytogenes, J. Dairy Sci.,* 71, 3237, 1988.
67. **Scott, V. N. and Taylor, S. L.,** Effect of nisin on the outgrowth of *Clostridium botulinum* spores, *J. Food Sci.,* 46, 117, 1981.
68. **Gould, G. W. and Hurst, A.,** Inhibition of *Bacillus* spore development by nisin and subtilin, 8th Int. Congress of Microbiology, A2-11 (Abstr.), 1962.
69. **Linnett, P. E. and Strominger, J. L.,** Additional antibiotic inhibitors of peptidoglycans, *Antimicrob. Agents Chemother.,* 4, 231, 1973.
70. **Kordel, M. and Sahl, H. G.,** Susceptibility of bacterial, eukaryotic and artificial membranes to the disruptive action of cationic peptide pep5 and nisin, *FEMS Microbiol. Lett.,* 34, 139, 1986.
71. **Ray, B.,** Enumeration of injured indicator bacteria from foods, in *Injured Index and Pathogenic Bacteria: Occurrence and Detection in Foods, Water and Feeds,* Ray, B., Ed., CRC Press, Inc., Boca Raton, FL, 1989, chap. 2.
72. **Gao, F. H., Abee, T., and Koonings, W. N.,** Mechanism of action of the peptide antibiotic nisin in liposomes and cytochrome C oxidase containing proteolyposomes, *Appl. Environ. Microbiol.,* 57, 2164, 1991.
73. **Hall, R. H.,** Nisin and food preservation, *Proc. Biochem.,* 1, 461, 1966.
73a. **Henning, S., Metz, R., and Hammes, W. P.**, Studies on the mode of action of nisin, *Int. J. Food Microbiol.,* 3, 121, 1986.
74. **Ramseier, H. R.,** The action of nisin on *Clostridium butyricum, Arch. Microbiol.,* 37, 57, 1960.
75. **Hirsch, A.,** Some polypeptide antibotics, *J. Appl. Bacteriol.,* 17, 108, 1954.
76. **Thorpe, R. H.,** The action of nisin on spoilage bacteria. I. The effect of nisin on the heat resistance of *Bacillus stearothermophilus, J. Appl. Bactetriol.,* 23, 136, 1960.
77. **Hurst, A. and Kruse, H.,** Effect of secondary metabolites on the organisms producing them: effect of nisin on *Streptococcus lactis* and enterotoxin B on *Staphylococcus aureus, Antimicrob. Agents Chemother.,* 1, 277, 1972.
78. **Ruhr, E. and Sahl, H.,** Mode of action of the peptide antibiotic nisin and influence on the membrane potential of whole cells and on cytoplasmic and artificial vesicles, *Antimicrob. Agents Chemother.,* 27, 841, 1985.
79. **Reisinger, P., Seidel, H., Tschesche, H., and Homes, W. P.,** The effect of nisin on murein synthesis, *Arch. Microbiol.,* 127, 187, 1980.
80. **Ray, B.,** Adsorption of Nisin by Cells of Resistant and Sensitive Strains of Gram-Positive Bacteria and Nisin Production, unpublished data, University of Wyoming, Laramie, 1991.
81. **Campbell, L. L. and Sniff, E. E.,** Effect of subtilin and nisin on the spores of *Bacillus coagulans, J. Bacteriol.,* 77, 766, 1959.
82. **Gould, G. W.,** Effect of food preservatives on the growth of bacteria from spores, in *Microbial Inhibitors in Foods,* Molin, N., Ed., Almqvist and Wiksell, Stockholm, 1964, 17.
83. **Hitchins, A. D., Gould, G. W., and Hurst, A.,** The swelling of bacterial spores during germination and outgrowth, *J. Gen. Microbiol.,* 30, 445, 1963.
84. **Hirsch, A. and Grinsted, E.,** Methods for the enumeration of anaerobic spore formers from cheese, with observations on the effect of nisin, *J. Dairy Res.,* 21, 101, 1954.
85. **Denny, C. B., Sharpe, L. E., and Bohrer, C. W.,** Effects of tylosin and nisin on canned food spoilage bacteria, *Appl. Microbiol.,* 9, 108, 1961.
86. **Scott, V. N. and Taylor, S. L.,** Temperature, pH and spore load effects on the ability of nisin to prevent outgrowth of *Clostridium botulinum* spores, *J. Food Sci.,* 46, 121, 1981.
86a. **Scott, V. N. and Taylor, S. L.,** Effect of nisin on the outgrowth of *Clostridium botulinum* spores, *J. Food Sci.,* 46, 117, 1981.
87. **Kalchayanand, N. and Ray, B.,** Inability of *Clostridium laramie* Spores to Grow in Broth Containing Nisin or Pediocin AcH, unpublished data, University of Wyoming, Laramie, 1991.
88. **Lewis, J. C., Michener, H. D, Stumbo, C. R., and Titus, D. S.,** Antibiotics in food processing: additives accelerating death of spores by moist heat, *J. Agric. Food Chem.,* 2, 298, 1954.

89. **Hawley, H. B.,** The uses of antibiotics in canning, in *Antibiotics in Agriculture,* Woodbine, M., Ed., Butterworths, London, 1962, 272.
90. **Kooy, J. S.,** Strains of *Lactobacillus plantarum* which destroys the antibiotic made by *Streptococcus lactis, Ned. Melk Zuiveltijdschr.,* 6, 323, 1952.
91. **Lipinska, E.,** Nisin and its applications, in *Antibiotics and Antibiosis in Agriculture,* Woodbine, M., Ed., Butterworths, London, 1977, 103.
92. **Alifax, R. and Chevalier, R.,** Etude de la nisinase produite par *Streptococcus thermophilus, J. Dairy Res.,* 29, 233, 1962.
93. **Jarvis, B. and Farr, J.,** Partial purification, specificity and mechanism of action of the nisin-inactivating enzyme from *Bacillus cereus, Biochem. Biophys. Acta,* 227, 232, 1971.
94. **Jarvis, B.,** Resistance to nisin and production of nisin-inactivating enzymes by several *Bacillus* species, *J. Gen. Microbiol.,* 47, 33, 1967.
95. **Jarvis, B.,** Enzymatic reduction of the C-terminal dehydroalanyllysine sequence of nisin, *Biochem. J.,* 119, 56, 1970.
96. **Gasson, M. J.,** Transfer of sucrose fermenting ability, nisin resistance and nisin production into *Streptococcus lactis* 712, *FEMS Microbiol. Lett.,* 21, 7, 1984.
97. **Neve, H., Geis, A., and Teuber, M.,** Conjugal transfer and characterization of bacteriocin plasmids in Group N (lactic acid) streptococci, *J. Bacteriol.,* 157, 833, 1984.
98. **Gonzalez, C. F. and Kunka, B. S.,** Transfer of sucrose fermenting ability and nisin production phenotype among lactic streptococci, *Appl. Environ. Microbiol.,* 49, 627, 1985.
99. **Steel, J. L. and Mckay, L. L.,** Partial characterization of the genetic basis for sucrose metabolism and nisin production in *Streptococcus lactis, Appl. Environ. Microbiol.,* 51, 57, 1986.
100. **Klaenhammer, T. R. and Sanozky, R. B.,** Conjugal transfer from *Streptococcus lactis* ME2 of plasmids encoding phage resistance, nisin resistance and lactose-fermentating ability: evidence for a high frequency conjugative plasmid responsible for abortive infection of virulent bacteriophage, *J. Gen. Microbiol.,* 131, 1531, 1984.
101. **McKay, L. L. and Baldwin, K. A.,** Conjugative 40-megadalton plasmid in *Streptococcus lactis* subsp. *diacetylactis* DRC3 is associated with resistance to nisin and bacteriophage, *Appl. Environ. Microbiol.,* 47, 68, 1984.
102. **Forseth, B. R., Herman, R. E., and McKay, L. L.,** Cloning of nisin resistance determinant and replication origin on 7.6 Kb *Eco*R I fragment of pNP40 from *Streptococcus lactis* subsp. *diacetylactis* DRC3, *Appl. Environ. Microbiol.,* 54, 2136, 1988.
103. **Forseth, B. R. and McKay, L. L.,** Molecular characterization of the nisin resistance region of *Lactococcus lactis* subsp. *lactis* biovar *diacetylactis* DRC3, *Appl. Environ. Microbiol.,* 57, 804, 1991.
103a. **Jung, D-S., Bodyfelt, F. W., and Daeschel, M. A.,** Influence of fat and emulsifiers on the efficiency of nisin in inhibiting *Listeria monocytogenes* in fluid milk, *J. Dairy Sci.,* 75, 387, 1992.
104. **von Wright, A., Wessels, S., Tynkkynen, S., and Saarela, M.,** Isolation of a replicating region of a large lactococcal plasmid and use in cloning of a nisin resistance determinant, *Appl. Environ. Microbiol.,* 56, 2029, 1990.
105. **Ray, P.,** Characteristics of bacteriocins from lactic acid bacteria, 40th Int. Science and Engineering Fair, Pittsburgh, Science Service, Washington, D.C., (Abstr.), 1989, 295.
106. **Liu, W. and Hansen, N. J.,** Some chemical and physical properties of nisin, a small-protein antibiotic produced by *Lactococcus lactis, Appl. Environ. Microbiol.,* 56, 2551, 1990.
107. **Hansen, N. J., Banerjee, S., and Bachman, G. W.,** Potential of small ribosomally synthesized bacteriocins in design of new food preservatives, *J. Food Safety,* 10, 119, 1989.
108. **Gross, E. and Morell, J. H.,** The presence of dehydroalanine in the antibiotic nisin and its relationship to activity, *J. Am. Chem. Soc.,* 89, 2791, 1967.
109. **Gross, E. and Morell, J. H.,** Nisin, the assignment of sulfide bridges of ß-methyllanthionine to a novel bicyclic structure of identical ring size, J. *Am. Chem. Soc.,* 92, 2919, 1970.
109a. **Barber, M., Elliot, G. J., Bordoli, R. S., Green, B. N., and Bycroft, B. W.,** Confirmation of the structure of nisin and its major degradation product by FAB-MS and FAB-MS/MS, *Experimentia,* 44, 266, 1988.

110. **Chan, W. C., Lian, L. Y., Bycroff, B. W., and Roberts, G. C. K.,** Confirmation of the structure of nisin by complete H-N.M.R. resonance assignment in aqueous and dimethyl sulfoxide solution, *J. Chem. Soc. Perkin Trans.*, I, 2359, 1989.
111. **Slijper, M., Hilbers, C. W., Konings, R. N. H., and van de Ven, F. J. M.,** NMR studies of lantibiotics: assignment of the H-N.M.R. spectrum of nisin and identification of interresidual contacts, *FEBS Lett.*, 252, 22, 1989.
112. **Jarvis, B., Jeffcoat, J., and Cheeseman, G. C.,** Molecular weight distribution of nisin, *Biochem. Biophys. Acta, 1*68, 153, 1968.
113. **Cheeseman, G. C. and Berridge, N. J.,** An improved method of preparing nisin, *Biochem. J.*, 65, 603, 1957.
114. **Ingram, L., Tombs, M. P., and Hurst, A.,** Mobility-molecular relationships of small proteins and peptides in acrylamide-gel electrophoresis, *Anal. Biochem.*, 20, 24, 1967.
115. **Mulders, J. W. M., Boerrigter, I. J., Rollema, H. S., Siezen, R. J., and deVos, W. M.,** Identification and characterization of the lantibiotic nisin Z, a natural nisin variant, *Eur. J. Biochem.*, 201, 581, 1991.
116. **Chan, W. C., Bycroft, B. W., Lian, L.Y. and Roberts, C. K.,** Isolation and characterization of two degradation products derived from the peptide antibiotic nisin, *FEBS Lett.*, 252, 29, 1989.
117. **Gross, E. and Morrell, J. L.,** The number and nature of α, ♣–unsaturated amino acids in nisin, FE*BS Lett.*, 2, 61, 1968.
118. **Morris, S. L., Walsh, R. C., and Hansen, N. J.,** Identification and characterization of some bacterial membrane sulfahydryl groups which are targets of bacteriostatic and antibiotic action, *J. Biol. Chem.*, 259, 13590, 1984.
119. **Anon.,** Nisaplin in the *Preservation of Processed Cheese and Processed Cheese Products*, Technical Information, Ref. 10/88, Aplin and Barrett, England, 1988.
120. **Anon.,** *The Food Preservative Nisaplin,* Technical Information, Ref. 1/88, Aplin and Barrett, England, 1988.
121. **Tamer, J.,** The inhibitory action of nisin on *B. stearothermophilus*, in *Microbial Inhibitors in Foods,* Molin, N., Ed., Almquist and Wiksell, Stockholm, 25, 1964.
122. **Anon.,** The Safety of Ni*sin As a Food Additive,* Technical Information Sheet no. 2/88, Aplin and Barrett, England, 1988.
123. **Hara, S., Yakazu, K., Nakakawaii, K., Takenchi, T., Kobayashi, T., Sata, M., Imat, Z., and Shibuya, T.,** An investigation of toxicity of nisin, *Tokyo Med. Univ. J., 20,* 175, 1962.
124. **Food and Drug Administration,** Nisin preparation: affirmation of GRAS status as a direct human food ingredient, *Fed. Regist.*, 53(66), 11247, 1988.
125. **Pesquera, T. I.,** Nisin — its use: estimation and toxicity in sterilized milk, *Revista Espanola de Lecheria,* 59, 16, 1966.
126. **Huntington Research Center**, *Effect of Nisaplin on Reproductive Functions of Multiple Generations in Rats,* unpublished reports, 1984, cited in ref. 124.
127. **Heinemann, B. and Williams, R.,** Inactivation of nisin by pancreatine, *J. Dairy Sci.*, 49, 312, 1966.
128. **Kozak, W., Rajcher-Trzl, M., and Dobrzanski, W. T.,** The effect of proflavin, ethidium bromide and elevated temperature on the appearance of nisin-negative clones in nisin-producing strain of *Streptococcus lactis, J. Gen. Microbiol.*, 83, 295, 1974.
129. **Fuchs, P. G., Zajde, J., and Dobrzanski, W. T.,** Possible plasmid nature of the determinant for production of the antibiotic nisin in some strains of *Streptococcus lactis, J. Gen. Microbiol.*, 88, 189, 1975.
130. **LaBlanc, D. J., Crow, V. L., and Lee, L. N.,** Plasmid mediated carbohydrate catabolic enzymes among strains of *Streptococcus lactis,* in *Plasmids and Transposons: Environmental Effects and Maintenance Mechanisms,* Stuttard, C. and Rozee, K. R., Eds., Academic Press, New York, 1980, 31.
131. **Davey, G. P. and Pearce, L. E.,** Production of diplococcin by *Streptococcus cremoris* and its transfer to nonproducing group streptococci, in *Microbiology–1982,* Schlessinger, D., Ed., American Society of Microbiology, Washington, D.C., 1982, 221.

132. **Tsai, H. and Sandine, W. E.,** Conjugal transfer of nisin plasmid genes from *Streptococcus lactis* 7962 to *Leuconostoc dextranium* 181, *Appl. Environ. Microbiol.,* 53, 352, 1987.
133. **Donkersloot, J. A. and Thompson, J.,** Simultaneous loss of N^5-(carboxyethyl) ornithine synthase, nisin production and sucrose-fermenting ability, *J. Bacteriol.,* 172, 4122, 1990.
134. **Broadbent, J. R. and Kondo, J. K.,** Genetic construction of nisin-producing *Lactococcus lactis* subsp. *cremoris* and analysis of a rapid method for conjugation, *Appl. Environ. Microbiol.,* 57, 517, 1991.
135. **Rauch, P. J. G., Beerthuyzen, and deVos, W. M.,** Nucleotide sequence of IS904 from *Lactococcus lactis* subsp. *lactis* strain NIZO R5, *Nucleic Acidss, Res.,* 18, 4253, 1990.
135a. **Thompson, J., Nguyen, N. Y., Sackett, D. L., and Donkersloot, J. A.,** Transposon encoded sucrose metabolism in *Lactococcus lactis, J. Biol. Chem.,* 266, 14573, 1991.
136. **Egorov, N. S., Baranova, I. P., and Kozlova, Y. I.,** Optimization of nutrient medium composition for the production of the antibiotic nisin, *Mikrobiologia,* 40, 860, 1971.
137. **Hurst, A.,** Biosynthesis of the antibiotic nisin and other basic peptides by *Streptococcus lactis* grown in batch culture, *J. Gen. Microbiol.,* 45, 503, 1966.
138. **Hurst, A.,** Function of nisin and nisin-like basic proteins in the growth cycle of *Streptococcus lactis, Nature (London),* 214, 1232, 1967.
139. **Hurst, A.,** Biosynthesis of the antibiotic nisin by whole *Streptococcus lactis* organisms, *J. Gen. Microbiol.,* 44, 209, 1966.
140. **Ingram, L. A.,** Ribosomal mechanism for synthesis of peptides related to nisin, *Biochem. Biophys. Acta,* 224, 263, 1970.
141. **Hurst, A. and Peterson, G. M.,** Observations on the conversion of an inactive precursor protein to the antibiotic nisin, Can. J. Microbiol., 17, 1379, 1971.
142. **White, R. J. and Hurst, A.,** The location of nisin in the producer organism *Streptococcus lactis, J. Gen. Microbiol.,* 53, 171, 1968.
143. **Bailey, F. J. and Hurst, A.,** Preparation of a highly active form of nisin from *Streptococcus lactis, Can. J. Microbiol., 17,* 61, 1971.
144. **Anon.,** Preservation of *Milk and Milk Products With Nisaplin,* Technical Information Sheet 11/88, Aplin and Barrett, England, 1988.
145. **Radaeva, I. A., Efimova, V., and Anotskya, A.,** Effect of nisin on microorganisms causing spoilage of evaporated milk, *Molochnaya Promyshlennost Part I* cited in Technical Information Sheet 11/88, Aplin and Barrett, England, 1966, 13.
146. **Gregory, M. E., Henry, K., and Kon, S. K.,** Nutritive properties of freshly prepared and stored eveporated milks manufactured by a normal commercial procedure or by reduced thermal process in the presence of nisin, *J. Dairy Res.,* 31, 113, 1964.
147. **Shehata, A. E., Khalafalla, S. M., Magdoub, M. N. I., and Hofi, A. A.,** The use of nisin in the production of sterilized milk drinks, *Egypt J. Dairy Sci.,* 4, 37, 1976.
148. **Wajid, H. R. A. and Kalra, M. S.,** Nisin as an aid for extending the shelf life of sterilized milk, *J. Food Sci. Technol. (Mysore),* 13(1), 6, 1976.
149. **Heinemann, B., Stumbo, C. R., and Scurlock, A.,** Use of nisaplin in preparing beverage quality sterile chocolate flavored milk, *J. Dairy Sci.,* 47, 8, 1964.
150. **Gupta, R. K. and Prasad, D. N.,** Incorporation of nisin in stirred yogurt, I. Effect on lactic and nonlactic organisms during storage, *Cult. Dairy Prod., 23(*3), 17, 1988.
151. **Gupta, R. K. and Prasad, D. N.,** Incorporation of nisin in stirred yogurt. III. Quantitative estimation of residual nisin, *Cult. Dairy Prod.,* 24(2), 11, 1989.
152. **Anon.,** *Extending Shelf Life of Chilled Dairy Desserts With Nisaplin,* Technical Information Sheet 12/88, Aplin and Barrett, England, 1988.
153. Sukum**ar, D., Thompkinson, D. K., Gahlot, D. P., and Mathur, D. N.,** Studies on a method of preparation and preservation of kheer, *Indian J. Dairy Sci.,* 26, 316, 1976.
154. **Rayman, M. K., Aris, B., and Hurst, A.,** Nisin: a possible alternative or adjunct to nitrite in the preservation of meats, *Appl. Environ. Microbiol.,* 41, 375, 1981.
155. **Houben, J. H. and Krol, B.,** *Meat Sci.,* 14, 205, 1985 (cited in Fowler, G.G. and Gasson, M.J., Antibiotic-nisin, in *Food Preservation,* Russell, N. J. and Gould, G. W., Eds., Van Nostrand Reinhold, AVI Publishing, New York, 1990, 137.

156. **Gibbs, B. M. and Hurst, A.,** Limitations of nisin in the food industry, in *Microbial Inhibitors in Foods,* Molin, N., Ed., Almqvist Wiksell, Stockholm, 1964, 151.
157. **Calderon, C., Collins-Thompson, D. L., and Usborne, W. R.,** Shelf life studies of vacuum-packaged bacon treated with nisin, *J. Food Prot.,* 48, 330, 1985.
158. **Caserio, G., Stecchini, M., Pastore, M., and Gennani, M.,** The individual and combined effects of nisin and nitrite on the spore germination of *Clostridium perfringens* in meat mixtures subjected to fermentation, *Ind. Alimentari, (Pinerolo, Italy),* 18, 894, 1979.
159. **Rozbeh, M.,** Extension of Shelf-Life in Vacuum-Packaged Fresh Beef, M.S. thesis, University of Wyoming, Laramie, 1990.
160. **Vas, K.,** Use of nisin in the thermal preservation of tomato products, *Fruehtstaft-Ind. Confructa, 8(2), 73,* 1963.
161. **Eyles, M. J. and Richardson, K. C.,** Thermophilic bacteria and food spoilage, *CSIRO Food Res. Q.,* 48, 19, 1988.
162. **Vas, K., Kiss, I., and Kiss, N.,** Use of nisin for shortening the heat treatment in the sterilization of green peas, *Z. Lebensmit. Unterschung U. Forschung, 133(3), 141, 1967.*
163. **Maslennikova, N. M. and Loshina, P. B.,** The use of nisin in canned potatoes, *Konserv. I Ovosteshus,* 1(1,) 12, 1968; cited in *J. Food Prot.,* 44, 100, 1990.
164. **Anon.,** *The Use of Nisaplin in the Preservation of Canned Foods,* Technical Information Sheet 13/88, Aplin and Barrett, England, 1988.
165. Bards**ley, A.,** Antibiotics in food canning, *Food Technol. (Australia),* 14, 532, 1962.
166. **Ogden, K.,** Nisin: a bacteriocin with a potential use in brewing, 92, 379, 1986.
167. **Ogden, K. and Waites, M. J.,** The action of nisin on beer spoilage lactic acid bacteria, *J. Inst. Brew,* 92, 463, 1986.
168. **Radler, F.,** Possible use of nisin in wine making. II. Experiment to control lactic acid bacteria in the production of wine, *Am. J. Enol. Viticul.,* 41(1), 7, 1990.
169. **Daeschel, M., Jung, D. S., and Watson, B. T.,** Controlling wine malolactic fermentation with nisin and nisin-resistant strains of *Leuconostoc oenos, Appl. Environ. Microbiol.,* 57, 601, 1991.
170. **Henning, S., Metz, R., and Hammes, W. P.,** New aspects for the application of nisin to food based on its mode of action, *Intl. J. Food Microbiol.,* 3, 135, 1986.
171. **Tanaka, N., Traisman, E., Plantinga, P., Finn, L., Flom, W., Meske, L., and Guggisberg, J.,** Evaluation of factors involved in antibotulinal properties of pasteurized processed cheese spreads, *J. Food Prot.,* 49, 526, 1986.
172. **Somers, E. B. and Taylor, S. L.,** Antibotulinal effectiveness of nisin in pasteurized processed cheese spreads, *J. Food Prot.,* 50, 842, 1987.
173. **Rayman, K., Malik, N. and Hurst, A.,** Failure of nisin to inhibit outgrowth of *Clostridium botulinum* in a model cured meat system, *Appl. Environ. Microbiol.,* 46, 1450, 1983.
174. **Taylor, S. and Somers, E.,** Evaluation of antibotulinal effectiveness of nisin in bacon, *J. Food Prot.,* 48, 949, 1985.
175. **Taylor, S., Somers, E. and Krueger, L.,** Antibotulinal effectiveness of nisin-nitrite combinations in culture medium and chicken frankfurter emulsions, *J. Food Prot.,* 48, 234, 1985.
176. **Bell, R. G. and DeLacy, K. M.,** The efficiency of nisin, sorbic acid and monolaurin as preservatives in pasteurized cured meat products, *Food Microbiol.,* 4, 277, 1987.
177. **Chung, K. T., Dickson, J. S., and Crouse, J. D.,** Effects of nisin on growth of bacteria attached to meat, *Appl. Environ. Microbiol.,* 55, 1329, 1989.
178. **Taylor, L., Cann, D. D., and Welch, B. J.,** Antibotulin properties of nisin in fresh fish packaged in an atmosphere of carbondioxide, J. *Food Prot.,* 53, 953, 1990.
179. **World Health Organization,** Specification for the identity and purity of food additives and their toxocological evaluation: some antibiotics, 12th Report of the Joint FAO/WHO Expert Committee on Food Additives, WHO Technical Report, Series 430, World Health Organization, Geneva, 1969.

Chapter 10

PEDIOCIN(S) OF *PEDIOCOCCUS ACIDILACTICI* AS A FOOD BIOPRESERVATIVE

Bibek Ray

TABLE OF CONTENTS

Several *Pediococcus acidilactici*...strains used in meat fermentation were found to have antimicrobial activity against different food pathogens and spoilage bacteria. The antimicrobial substances are produced extracellularly, are proteinaceous in nature, heat stable and susceptible to proteolytic enzymes.[1]

I. INTRODUCTION

Pediococcus acidilactici and *Pediococcus pentosaceus*, previously designated as *Pediococcus cerevisiae*, are phylogenetically closely related in the genus *Pediococcus*. The genus also includes several other species that are phylogenetically heterogeneous.[2,3] Both strains have been associated with the fermentation of meat and vegetables[4,5] and with the development of desirable flavor in Cheddar cheese.[6,7] Recent studies indicated that the two species could be useful as a secondary flora in accelerating the ripening of Cheddar cheese.[8,9]

P. acidilactici and *P. pentosaceus* species differ with each other only in a few characteristics. Homology study for a 1340-nucleotide region of 16S rRNA between the reference strains of the two species revealed 95% similarities.[2] However, DNA/DNA homology between the two species is only 20 to 30%.[3,10] *P. pentosaceus* possesses a pseudocatalase that can give a false-positive result when colonies are grown in a low glucose medium and tested with H_2O_2. Some *P. acidilactici* strains may possess a heme-requiring catalase.[3,11] The optimum temperature of growth of *P. pentosaceus* is 28 to 32°C, while for *P. acidilactici* it is 40°C; *P. acidilactici* can grow at 50°C; *P. pentosaceus* cannot. Also, *P. pentosaceus* ferments maltose but *P. acidilactici* cannot.[3] Finally, *P. acidilactici* cells are slightly more heat resistant (needing 70°C for 10 min or more) than *P. pentosaceus* (65°C for 8 min).[3]

The antagonistic effect of two strains of *P. pentosaceus* (FBB 61 and L 7230, designated as *P. cerevisiae)* against other Gram-positive bacteria, but not against Gram-negative bacteria, was reported in the sixties.[12,13] Subsequent studies with strain FBB 61 revealed that the inhibitory substance is bacteriocidal to sensitive cells, nondialyzable, stable at high temperatures and freezing, and is inactivated by pronase. Thus, the inhibitory effect was considered to be due to a bacteriocin-like compound.[14] Other studies indicated that the bacteriocin designated as pediocin A was inhibitory to many lactic acid bacteria, *Clostridium* spp. and *Staphylococcus aureus*, and that the trait is encoded in a 13.6 MDal plasmid.[15] Another *P. pentosaceus* strain FBB 63 and its several derivatives were also reported to produce a bacteriocin-like compound, and the phenotype was encoded in a 10.5 MDal plasmid.[16] However, no further studies were reported on the molecular weight of these bacteriocins, the effect of environmental parameters on their production, the genetic control systems, their effectiveness in food systems, and other information that would be necessary for the use of these bacteriocins from *P. pentosaceus* as food preservatives.

Bacteriocins produced by several strains of *P. acidilactici* were first reported by Bhunia et al. in 1987.[1] Subsequently, they designated this plasmid-encoded inhibitory protein of about 2700 Dal, pediocin AcH.[17–19] In 1987, Gonzalez and Kunka also reported the production of a 16,000-Dal molecular weight bacteriocin by *P. acidilactici* PAC.1[20] and designated it pediocin PA-1. As opposed to the pediocins of *P. pentosaceus* strains, pediocins of *P. acidilactici* strains, especially Pediocin AcH, and to some extent pediocin PA-1, have been studied in detail for important characteristics and effectiveness in food systems since they were first reported in 1987.[1,20] As their effectiveness in food systems is known, this chapter will include a detailed discussion on the properties of pediocins of *P. acidilactici* strains. Pediocins of *P. pentosaceus* strains are discussed in the next chapter, as are the bacteriocins of several other lactic acid bacteria for which more information beyond just identifying a strain as a bacteriocin producer is available, but whose effectiveness in food systems have not been tested.

II. ANTIMICROBIAL ACTIVITY OF PEDIOCIN OF *P. ACIDILACTICI*

A. Assay of Activity and Units

1. Isolation of Producer Strains

To identify a producer strain, Bhunia et al.[1] initially grew a *P. acidilactici* strain in MRS broth at 37°C for 20 to 24 h, serially diluted it, and pour plated 12 plates to have 20 to 30 colony forming units (cfu) per plate with about 6 to 8 ml of MRS agar (MRS broth and 1.5% agar). After solidification each plate was overlaid with 5 ml of MRS agar. This was done so as to have a relatively thin layer of agar in each plate and to prevent formation of surface colonies. The plates were incubated for 20 to 24 h at 37°C. Twelve nonbacteriocin-producing strains of lactic acid bacteria from genera *Lactococcus, Pediococcus, Leuconostoc,* and *Lactobacillus* that were capable of growing at pH 4.5 or below were selected as indicators. They were grown in MRS broth overnight at their optimum growth temperatures; 10 μl of each was added to melted MRS-soft agar (MRS broth and 0.8% agar) at 45°C and poured over the surface of a plate containing *P. acidilactici* colonies. The plates were incubated overnight at the optimum growth temperatures of each indicator and then examined for zones of inhibition around the colony forming units. A strain producing a zone against one or more indicators was then grown in MRS broth, and the supernatant was further examined to verify that the growth inhibition of an indicator was not due to acid or H_2O_2. Later, for isolation from natural sources, the following four strains were used simultaneously as indicators: *P. acidilactici* L, *Leuconostoc mesenteroides* Ly; *Lactobacillus plantarum* WSO-39, and *L. plantarum* NCDO 955.

Gonzalez and Kunka[20] assayed production of *P. acidilactici* PAC 1.0 by spotting cells on MRS agar, incubating the plates at 35°C for 18 h, exposing the plates to chloroform vapor for 30 min, and overlaying the surface with 0.75% soft agar seeded with *P. pentosaceus* PEP1.0 cells. Following incubation at 32°C for 18 h, the presence of clear zone around the colonies was considered to be due to bacteriocin production by the strain. Hoover et al.[21,21a] used *L. mesenteroides* 8293 as an indicator against the *P. acidilactici* strains by agar plate diffusion[22] using APT agar to examine the presence of zone of inhibition.

2. *Measurement of Activity Units (AU)*

Gonzalez and Kunka[20] grew *P. acidilactici* PAC 1.0 in an MRS broth statically at 35°C for 18 h, adjusted the pH of the broth to 6.0, removed the cells by centrifugation at 16,300 × g for 15 min at 5°C, filtered the supernatant through a 0.45-μm membrane filter and used the filtrate to measure pediocin activity. They diluted the filtrate through twofold dilution series and transferred 5 μl from each dilution onto MRS plates overlaid with soft agar seeded with *P. pentosaceus* FBB 63 (previously known as *P. cerevisiae* FBB 63[16]). These plates were then incubated at 32°C and examined for the zone of growth inhibition. They defined the arbitrary unit (AU) as a 5 μl of the highest dilution of the filtrate yielding a definite zone of growth inhibition on the indicator lawn, and expressed the titer of the filtrate as the reciprocal of the highest dilution showing inhibition.[20,23]

Bhunia et al.[18] initially used a disc assay method to titer pediocin AcH using *L. plantarum* WSO 39 as the indicator. They seeded about 10^6 cells of the indicator in 5 ml casein-glucose (CG) soft agar[17] and poured it over a CG agar plate. After 30 min sterile paper discs were placed on the agar surface. Cell-free supernatant of the broth preparations that were to be used for measuring the activity titer were serially diluted twofold, and 40 μl from each dilution was transferred to each disc. After incubation at 30°C for 18–24 h the plates were examined. The reciprocal of the highest dilution that produced a 2 mm zone of growth inhibition was reported to be the antimicrobial titer. Later, they used *L. plantarum* NCDO 955 as an indicator strain as it was more sensitive to pediocin AcH than *L. plantarum* WSO-39. They used the procedure of Pucci et al.[23] to express the titer as arbitrary units (AU)/ml from the highest dilution from a serial twofold dilution that gave an inhibitory zone against 5 μl of the diluted pediocin.[24] Recently, Biswas et al.[25] and Motlagh et al.[26,27] expressed the titer of a preparation of pediocin AcH in activity units (AU). They prepared serial dilutions of a sample containing pediocin AcH in sterile deionized water and examined 5 μl from each dilution to determine the highest dilution that gave a clear zone of 2 mm on the lawn of the indicator *L. plantarum* NCDO 955. To prepare the lawn they inoculated about 10^5 to 10^6 from an overnight culture (5 μl) in 5 ml melted TGE-soft agar (45°C) and overlaid on the surface of a TGE-agar plate with a well-dried surface. They refrigerated the plate for a minimum of 30 min for the indicator cells to go to the lag-phase, spotted 5 μl just underneath the soft agar layer, and incubated the plates at 30°C for 18 to 20 h. To determine AU/ml of the preparation the highest dilution that gave the clear zone was multiplied by 200 (AU/ml = dilution factor × 1000

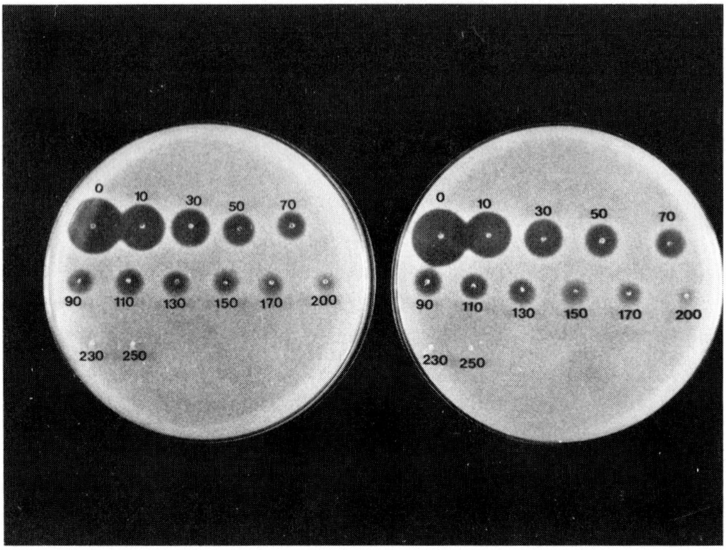

FIGURE 1. Estimation of activity unit (AU) of a pediocin AcH preparation. The heat-treated broth was serially diluted with sterile water between 1:0 and 1:250, and 5 µl from the dilutions were spotted on the surface of duplicate agar plates with an overlay seeded with about $10^{5\text{-}6}$ per plate of indicator cells of *L. planatarum* NCDO 955. After incubation overnight, the plates were examined for zone of growth inhibition around the spots. The highest dilution that gave a distinct zone (about 2 mm diameter) was multiplied by 200 (1000 ml/5 ml) to obtain AU/ml of the preparation. This preparation produced zones at highest dilution of 1:200; the titer of the preparation was 40,000 AU/ml using the same procedure 1 IU of nisin ≃ 100 AU of pediocin AcH. (Kalchayanand, N., unpublished data).

µl/5 µl). This method was found to give very consistent results (Figure 1). In different studies to measure the bacteriocidal effect of pediocin AcH against sensitive bacteria, these researchers have used the method just described as a standard (AU/ml against *L. plantarum* NCDO 955 in TGE agar).[28–31] Other researchers have reported the use of different strains of *Listeria monocytogenes* as indicator bacteria in measuring the arbitrary unit of a pediocin preparation from *P. acidilactici* strains.[32,33]

To compare the antimicrobial properties of pediocin preparations from *P. acidilactici* strains in studies conducted in different laboratories, it will be important to use one method of assay. This means measuring the titer of a preparation by the same procedure using the same indicator strain. Alternately, a preparation of a new bacteriocin can be compared for potency with a standard, such as pure nisin.

B. Antimicrobial Spectrums

Partially purified pediocins from culture broths have been used to determine their ability to inhibit growth of different Gram-positive bacteria. Very limited studies have been conducted with bacterial spores and Gram-negative bacteria. There is some confusion now as to whether the pediocins of *P. acidilactici* strains reported by different research groups are the same or different. Our current studies have indicated that the pediocin produced by the strains H, E, F, and M is the same, and it has been designated pediocin AcH. The similarities or differences of pediocin AcH produced by strains H, E, F, and M with pediocin PA-1 produced by *P. acidilactici* strain PAC1.0, and other pediocins produced by other *P. acidilactici* strains, have not yet been compared. Due to this the studies conducted on antibacterial spectrums of pediocins from different strains will be presented separately.

1. Gram-Positive Bacterial Cells and Spores

Pediocin AcH from *P. acidilactici* H, E, F, and M either in culture broth or in partially purified form has been reported to be inhibitory against vegetative cells of Gram-positive bacteria from many genera (Table 1). These include nonpathogenic species and strains of *Lactobacillus, Leuconostoc, Enterococcus, Pediococcus, Propionibacterium, Bacillus, Brochothrix, Clostridium laramie,* and *Listeria* spp. other than *L. monocytogenes.* Many of these nonpathogenic strains were associated with spoilage of meat and dairy products. Among the pathogens, species and strains that were found to be sensitive included *L. monocytogenes, Bacillus cereus, Clostridium perfringens,* and *Staphylococcus aureus.*[18,19,24,26,27,33–36] In most sensitive species there were some resistant strains. Pediocin PA-1 from *P. acidilactici* PAC 1.0 was also found to be inhibitory against *Pediococcus, Lactobacillus, Leuconostoc,* and *Listeria* species and strains (Table 1).[23,32,37] Pediocin from *P. acidilactici* PO_2 has also effectively inhibited the growth of *Pediococcus, Enterococcus, Listeria monocytogenes, Bacillus cereus,* and *Staphylococcus aureus.*[21] *P. acidilactici* JD1-23 also inhibited growth of *L. monocytogenes.*[38,39]

The limited studies conducted only with pediocin AcH have revealed that it inhibits germination and outgrowth of spores of *Clostridium laramie,* a strain of *Clostridium botulinum* E, a strain of *Bacillus steathermophilus,* and some *Bacillus* species obtained from spoiled heat-treated milk[28,36,39a] However, pediocin AcH was reported to be ineffective against some other *Clostridium botulinum* spores.[40]

The results of the limited studies included in this section suggest that pediocin(s) from *P. acidilactici* strains is(are) antibacterial against Gram-positive bacteria from many genera. Systematic studies against other Gram-positive bacteria as well as spores of interest will be helpful to determine their antibacterial spectrums and their possible effectiveness as food biopreservatives.

2. Gram-Negative Bacteria

Earlier studies have suggested that pediocin AcH preparations from *P. acidilactici* H was antibacterial to *Pseudomonas putida*[18] and *Aeromonas hydrophila*[18,35] strains as determined by the zone of growth inhibition. However, a later study reported that pediocin AcH molecules did not adsorb the cells of those species.[24] Thus, it appears that the inhibition as reported for these strains of Gram-negative bacteria might not be due to the direct effect of pediocin AcH.

TABLE 1
Species/Strains of Bacterial Cells and Spores Tested Sensitive by Pediocin of *P. Acidilactici* Strains

Bacterial Species	PAS[a] tested	Pediocin designation	Ref.
	Gram-positive cells[b]		
Lactobacillus plantarum	H,E,F,M	Pediocin AcH[c]	18,19,24,34
Lactobacillus leichmanni	H	Pediocin AcH	18
Lactobacillus viridescens	H,E,F,M	Pediocin AcH	18,19,34
Lactobacillus strains[d]	H	Pediocin AcH	18
Leuconostoc mesenteroides	H,E,F,M	Pediocin AcH	18,19,24,34
Leuconostoc strains[f]	H	Pediocin AcH	18
Pediococcus acidilactici (nonproducer)	H,E,F,M	Pediocin AcH	18,35,UP[e]
Enterococcus faecalis	H	Pediocin AcH	24
Propionibacterium spp. (dairy)	H	Pediocin AcH	UP[d]
Brochothrix thermosphacta	H	Pediocin AcH	18
Clostridium laramie	H	Pediocin AcH	28,36
Listeria monocytogenes	H,E,F,M	Pediocin AcH	18,19,26, 27,33–35
Listeria ivanovii	H	Pediocin AcH	26,27
Listeria innocua	H	Pediocin AcH	26,27
Bacillus cereus	H	Pediocin AcH	18
Clostridium perfringens	H	Pediocin AcH	18,19,24,34
Staphylococcus aureus	H,E,F,M	Pediocin AcH	18,19,24,34,35
Pediococcus pentosaceus	PAC 1.0	Pediocin PA-1	20
Pediococcus acidilactici (nonproducer)	PAC 1.0	Pediocin PA-1	20
Lactobacillus plantarum	PAC 1.0	Pediocin PA-1	20
Lactobacillus casei	PAC 1.0	Pediocin PA-1	20
Lactobacillus bifermentum	PAC 1.0	Pediocin PA-1	20
Leuconostoc mesenteroides	PAC 1.0	Pediocin PA-1	20
Listeria monocytogenes	PAC 1.0	Pediocin PA-1	23,32,37
Listeria ivanovii	PAC 1.0	Pediocin PA-1	32
Listeria innocua	PAC 1.0	Pediocin PA-1	37
Pediococcus pentosaceus	PO_2	None	21
Pediococcus acidilactici (nonproducer)	PO_2	None	21
Enterococcus faecalis	PO_2	None	21
Leuconostoc mesenteroides	PO_2	None	21
Listeria monocytogenes	PO_2	None	21
Bacillus cereus	PO_2	None	21
Staphylococcus aureus	PO_2	None	21
Listeria monocytogenes	JD1-23	None	38,39

TABLE 1 (continued)
Species/Strains of Bacterial Cells and Spores Tested Sensitive by Pediocin of *P. Acidilactici* Strains

Bacterial Species	PAS[a] tested	Pediocin designation	Ref.
	Spores[f]		
Clostridium laramie	H	Pediocin AcH	28,36
Clostridium botulinum E	H	Pediocin AcH	UP
Bacillus steathermophilus	H	Pediocin AcH	UP
Bacillus spp.[d]	H	Pediocin AcH	UP
	Gram-negative cells[g]		
Pseudomonas putida	H	Pediocin AcH	UP
Pseudomonas fluorescens	H	Pediocin AcII	UP
Escherichia coli	H	Pediocin AcH	UP
Escherichia coli (pathogenic)	H	Pediocin AcH	UP
Salmonella anatum	H	Pediocin AcH	UP
Aeromonas hydrophila	H	Pediocin AcH	UP
Yersinia enterocolitica	H	Pediocin AcH	UP

[a] PAS tested: *P. acidilactici* strain(s) producing pediocin were tested.
[b] For most species more than one strain was tested. Among the species there were strains that were resistant. Also, each sensitive strain contained some cells that were resistant to pediocin.
[c] Current studies in our laboratory have suggested that all four strains produce the same pediocin, so the designation "pediocin AcH" has been used for all.
[d] Strains isolated from spoiled foods of meat and dairy origins.
[e] UP: unpublished data from the author's laboratory.
[f] Some *C. botulinum* spores were not sensitive to pediocin AcH[40]
[g] Gram-negative cells were sublethally injured to determine their developed sensitivity to pediocin AcH (Ray, B. and Kalchayanand, N., unpublished data).

However, Gram-negative bacteria including *Pseudomonas, Escherichia* (including pathogenic strain), *Salmonella, Yersinia,* and *Aeromonas* species/strains can be made susceptible to pediocin AcH when they are sublethally stressed. This aspect will be discussed later in this chapter.

C. Mode of Antibacterial Action

Pediocin of *P. acidilactici,* like nisin and other bacteriocins of lactic acid bacteria, is bacteriocidal to sensitive Gram-positive bacteria as evidenced by rapid reduction in cfu in pediocin-treated samples.[18,24] The mechanisms by which pediocin AcH from *P. acidilactici* H produces bacteriocidal effect has been studied and will be included in this discussion.

1. Gram-Positive Bacterial Cells

Pediocin AcH is a hydrophobic protein and its antibacterial property is produced through the destabilization of cytoplasmic membrane. However, before pediocin AcH molecules come in contact with the bacterial cell membrane they are adsorbed by specific receptor(s) on the cell surface.

a. Adsorption of Pediocin AcH on Cell Surface

Bhunia et al.[24] mixed known AU of pediocin AcH with approximately the same cell concentrations of different bacterial species, incubated and removed the cells by centrifugation, and measured the cell-free supernatant for the remaining AU of pediocin AcH. Their report showed that molecules were adsorbed by cells of Gram-positive bacteria that were sensitive as well as resistant to pediocin AcH. Even the strains producing pediocin AcH (resistant to pediocin AcH) adsorbed pediocin AcH as did the resistant variants of sensitive strains.[24,41] Adsorption of pediocin PA-1 and other bacteriocins to both sensitive and resistant Gram-positive bacterial strains has also been observed.[23,42,43] The efficiency of adsorption of pediocin AcH among the sensitive and resistant Gram-positive bacterial cells was very high (Table 2). In general, the Gram-negative bacterial cells did not adsorb pediocin AcH.

The adsorption of pediocin AcH to sensitive cells is very rapid. It is optimum at pH 6.0 and occurs at the same level at O°C and 25°C, as well as by the heat-killed cells and in the presence of many chemicals (Table 3). Adsorption of pediocin AcH to sensitive cells was adversely affected in the presence of several other chemicals, most important of which were chloride and phosphate salts. This effect is concentration dependent, and for a given concentration of chloride or phosphate the effect can be neutralized by increasing the level of pediocin AcH (Table 4).[24,41]

The inability of pediocin AcH molecules to adsorb or bind to the cell surface receptors in the presence of chloride or phosphate ions is probably due to the blocking of the binding site(s) in the pediocin AcH molecules, or duc to blocking of the receptor(s) on the bacterial cell surface.

b. Damage of Cytoplasmic Membrane by Pediocin AcH

Following the adsorption or binding of pediocin AcH molecules to specific receptors on the cell surface of sensitive cells, the molecules rapidly come in contact with the cytoplasmic membrane and destabilize the membrane functions probably by their surface active property. How the pediocin AcH comes in contact with the cytoplasmic membrane following binding to the cell wall receptor only in sensitive cells of Gram-positive bacteria, is not clearly known. A hypothesis will be given later. The membrane destabilization of pediocin AcH was demonstrated from the loss of K^+ into the environment by the cells. In sensitive *L. plantarum* cells with intracellular ß-galactosidase, addition of ONPG (an analogue of lactose) in the environment following pediocin AcH treatment will cause rapid efflux of ONP (orthonitrophenol). This is indicated

TABLE 2
Adsorption of Pediocin AcH by Sensitive and Resistant Bacterial Strains

Sensitive bacteria	Adsorption[a] %	Resistant bacteria	Adsorption %
Lactobacillus plantarum NCDO955	100	*Lactobacillus plantarum* NCDO955r[b]	100
Pediococcus acidilactici L	100	*Pediococcus acidophilus* H[c]	100
Leuconostoc mesenteroides Ly	100	*Salmonella anatum*	0
Enterococcus faecalis	100	*Escherichia coli*	0
Clostridium perfringens	100	*Yersinia enterocolitica*	25
Staphylococcus aureus	66	*Aeromonas hydrophila*	0
		Pseudomonas fluorescens	0

[a] Adsorption was measured by adding known amounts (AU/ml) of pediocin AcH to a cell suspension, removing the cells by centrifugation after 30 min incubation at 37°C, and measuring the supernatant for residual activity. Absence of activity was considered to be due to 100% adsorption by cells using the amount of pediocin AcH added as 100%.

[b] *L. plantarum* NCDO 955r, a resistant variant isolated by growing the wild sensitive strain in the presence of pediocin AcH.

From Bhunia, A. K., Johnson, M. C., Ray, B., and Kalchayand, N., *J. Appl. Bacteriol.*, 70, 25, 1991. With permission.

by the yellow color of the environment.[24] Rapid entrance of ONPG in pediocin AcH-treated cells in water suspension and in the absence of an energy source, and release of ONP from the cell interior to the outside environment, were regarded to be due to membrane destabilization.

In some sensitive strains, pediocin AcH treatment not only caused the viability loss of cells due to membrane destabilization but also resulted in cell lysis in specific strains of *Leuconostoc, Listeria monocytogenes,* and other species (Figures 2–4)[24,26,41]. The cell lysis is neither a function of higher sensitivity of a strain to pediocin AcH nor due to the lytic effect of pediocin AcH. The strains that were lysed after pediocin AcH treatment probably have autolytic systems. Following viability loss by pediocin AcH action the autolytic system in these strains was activated, which in turn caused the lysis. In sensitive strains without an autolytic system the membrane damage by pediocin AcH is extensive enough to affect its functions and viability loss. Pucci et al.[23] also reported cell lysis of *L. monocytogenes* strain and a *P. pentosaceus* strain following treatment with pediocin PA-1.

TABLE 3
Influence of Treatment of Cells and Cell Wall Preparations of *L. Plantarum* NCDO955 With Different Chemical Agents on Subsequent Adsorption to Pediocin AcH

	Treatments	Adsorption (%)[a]
Cell:		
	None	100
	NaCl (200 mM)	0
	KCl (300 mM)	25
	NH_4Cl (300 mM)	25
	$MgCl_2$ (300 mM)	25
	Na-acetate (0.5%)100	
	NH_4 citrate (0.1%)	100
	EDTA (0.01%)	100
	Temperature (4 or 93°C)	100
	SDS (1%)	100
	Triton X-100 (2%)	100
	Ethanol (80%)	100
	Enzymes[b]	100
Cell wall:		
	None	100
	Ethanol (80%)	100
	Hexane (80%)	100
	Acetone (80%)	100
	SDS (1%)	25
	Methanol + Chloroform + TCA (40+40+20), 4°C, 2 h	100
	Methanol + Chloroform + TCA (40+40+20), 93°C, 5 min	0

[a] Cell and cell-wall suspensions were treated with different chemicals and tested for adsorption to a known AU of pediocin AcH.

From Bhunia, A. K., Johnson, M. C., Ray, B., and Kalchayand, N., *J. Appl. Bacteriol.*, 70, 25, 1991. With permission.

c. Possible Mechanism of Membrane Destabilization of Sensitive Cells by Pediocin AcH

The sequence of events that results in viability loss of sensitive Gram-positive bacterial cells following pediocin AcH treatment are (1) adsorption or binding of pediocin AcH molecules on specific receptor(s) on the cell surface, (2) entrance of the pediocin AcH molecule(s) through the wall, (3) contact of pediocin AcH molecules with the membrane, (4) destabilization of the functional integrity as evidenced by rapid influx and efflux of small molecules, and (5) in some strains, loss of structural integrity of membrane as evidenced by cell lysis.

Three aspects need proper clarification to understand the bacteriocidal

TABLE 4
Effect of Pediocin AcH Concentrationon the Death of *L. Plantarum* NCDO955 Cells in the Presence of NaCl or NaH_2PO_4

	100 m*M* NaCl[a]		100 m*M* NaH_2PO_4[a]	
Pediocin AcH (AU/ml)	Colony count/ml	Cell death(%)	Colony count/ml	Cell death(%)
0	1.4 × 107	0	1.4×10^7	0
2560	1.3×10^7	4.4	1.3×10^7	8.1
5120	9.6×10^5	93.0	5.6×10^6	58.8
10,240	5.0×10^5	96.4	1.4×10^6	90.0

[a] About 1.4×10^7 cfu/ml of *L. plantarum* NCDO955 either in 100 mM NaCl or in 100 m*M* NaH_2PO_4 (pH 6.0) were treated with 0 to 10,240 AU/ml of pediocin AcH, incubated at 37°C for 1 h and enumerated for cfu. Cell death was determined from the differences in cfu between the pediocin AcH-untreated control (0%) and the treated samples. In the absence of NaCl or NaH_2PO_4, 2560 AU/ml pediocin AcH produced >99.99% cell death of 1.4×10^7 cfu/ml as compared to 4.4% and 8.1% in the presence of 100 m*M* NaCl and 100 m*M* NaH_2PO_4, respectively. Also, adsorption of 2560 AU/ml to cells was reduced from 100% to <1% in the presence of NaCl and NaH_2PO_4. By increasing the pediocin AcH concentrations, its adsorption to cells and cell death in the presence of NaCl and NaH_2PO_4 was achieved.

From Bhunia, A. K., Johnson, M. C., Ray, B., and Kalchayand, N., *J. Appl. Bacteriol.*, 70, 25, 1991. With permission.

property of pediocin AcH: (1) why pediocin AcH is not adsorbed in Gram-negative bacteria but adsorbed on both sensitive and resistant cells of Gram-positive bacteria; (2) why among Gram-positive bacteria some are sensitive, while others are resistant even though both types adsorb pediocin AcH; and (3) why even in the sensitive strains, some cells are resistant.

The inability of Gram-negative bacteria to adsorb pediocin AcH has been at least partially determined by Bhunia et al.[24,41] They observed that when a cell wall preparation from a pediocin AcH-sensitive *L. plantarum* strain was extracted with a mixture of methanol, chloroform, and TCA (40:40:20) at 93°C for 5 min, the remaining wall materials lost their ability to adsorb pediocin AcH (Table 3). Since the hot extraction procedure removed the lipotichoic acid (LTA) from the cell wall,[43a] it was suspected that LTA molecules could have the receptor site of pediocin AcH. To verify this they did adsorption studies of pediocin AcH by sensitive cells in the presence of purified LTA from two Gram-positive bacteria. LTA inhibited adsorption of pediocin AcH to the cells. If the assumption that LTA has the receptors for pediocin AcH is correct, it explains why all Gram-positive bacteria with LTA on the cell wall do adsorb, while Gram-negative bacteria without LTA on the cell wall (outer membrane) do not adsorb pediocin AcH.

The possible mechanism(s), or predisposing factor(s), that makes some Gram-positive strains (or cells) sensitive and other Gram-positive strains (or

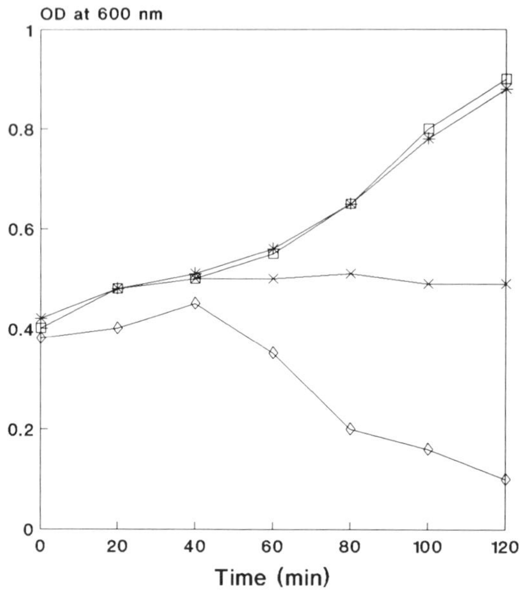

FIGURE 2. Lysis of growing *Leuconostoc mesenteroides* Ly, but not *L. plantarum* NCDO 955 by pediocin AcH: (◇), treated and (□), untreated *Leuconostoc;* (×), treated and (*), untreated *Lactobacillus*. Untreated cells grew. Treated *Lactobacillus* stopped growing, treated *Leuconostoc* lysed. (From Bhunia, A. K., Johnson, M. C., Ray, B., and Kachayanand, N., *J. Appl. Bacteriol.*, 70, 25, 1991. With permission.)

cells) resistant to pediocin AcH, can be explained as follows: pediocin AcH molecules bind to the specific receptor site in LTA on the cell surface of all Gram-positive bacteria. In the sensitive strains (or sensitive cells in a strain) the association between pediocin AcH and LTA brings about a change in the structural configuration of the wall. This change allows other pediocin AcH molecules (probably not bound to LTA) to pass through the wall and come in contact with the cytoplasmic membrane resulting in its destabilization (Figure 5). This hypothesis is extremely speculative. However, it is known that some Gram-positive bacteria, such as *Lactobacillus acidophilus* that are normally resistant to bile salts and other surface active agents, become susceptible and are killed by these agents when the cells are sublethally injured by freezing. Freezing causes alteration in the cell wall architecture in such a way that its barrier functions against many chemicals are lost and the injured cells become vulnerable to many agents to which they are normally resistant.[44,44a,44b] Also, it is known that among *L. acidophilus* and other bacteria, strains vary greatly in their resistance to bile salts and many other chemicals, entry of which into the cell is controlled by the cell wall.[44a,44b] These differences in susceptibility among strains are considered to be due to some structural differences in their wall.[44,45] This aspect will be further discussed while discussing the susceptibility of sublethally stressed Gram-negative bacteria to pediocin AcH.

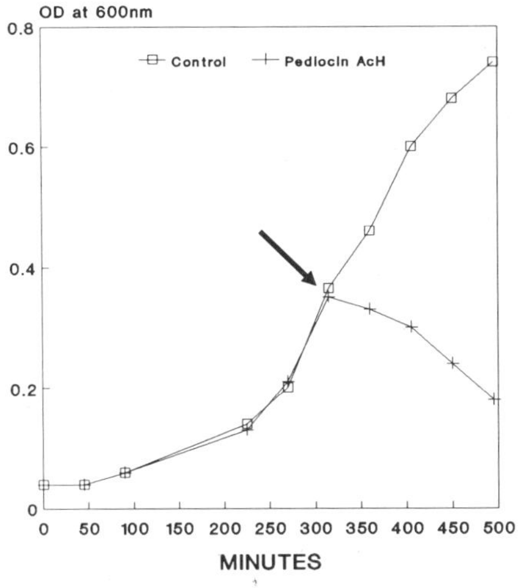

A

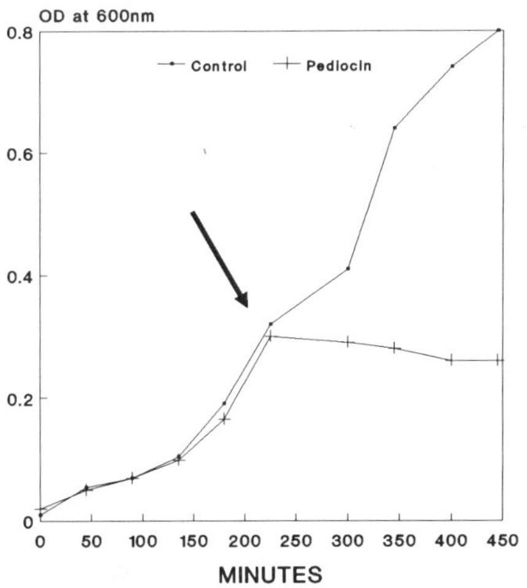

B

FIGURE 3. Lysis of *L. monocytogenes* strain 117 (A) and growth inhibition without lysis of *L. monocytogenes* strain ATCC 35091 (B) by pediocin AcH. Untreated cultures grew.[31]

2. Spores

Only one study has been conducted to determine the possible mechanism by which pediocin AcH produces inhibitory effect against spores of sensitive bacterial strains. Kalchayanand[28] used two methods to study the inhibition by pediocin AcH on germination and growth of spores of *Clostridium laramie,* a psychrotrophic clostridial species associated with spoilage of vacuum-packaged meat.[28,36,46] In the first method he inoculated 10^5 spores per milliter in 10 ml

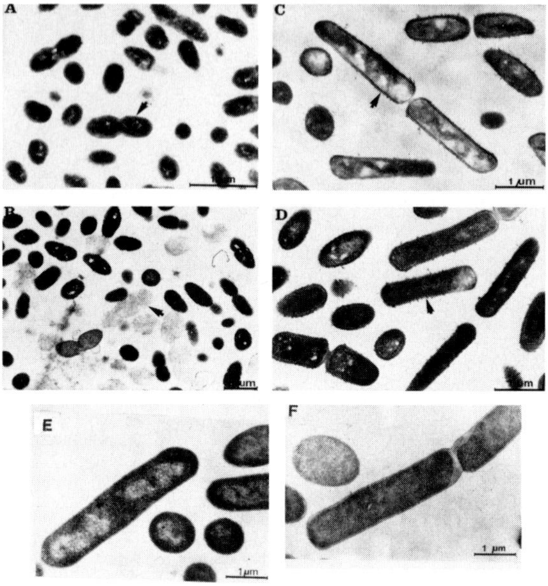

FIGURE 4. Transmission electron micrographs of cells of *Leuconostoc mesenteroides* Ly (A and B), *Lactobacillus plantarum* NCDO 955 (C and D), and *Clostridium laramie* (E and F). Pediocin-treated samples showed lysed ghost cells (B) and darker cytoplasm (D and F). (Taken from Bhunia, A. K., Johnson, M. C., Ray, B., and Kachayanand, N., *J. Appl. Bacteriol.*, 70, 25, 1991; Kalchayanad, N., Extension of Shelf-Life of Vacuum-Packaged Refrigerated Fresh Beef by Bacteriocins of Lactic Acid Bacteria, Ph.D. thesis, University of Wyoming, Laramie, 1990. With pemission.)

fluid thioglycollate broth containing 0 (control) or 10,000 AU/ml pediocin AcH; he then incubated the tubes at 10°C up to 14 d, and examined visually for growth and gas formation. All control tubes showed growth within 2 to 3 d, while none of pediocin AcH-treated cells showed growth in 14 d (Table 5). In the second method, 10^5 spores/ml in duplicate sets were exposed to 10,000 AU/ml pediocin AcH for 1 h at 10°C. The samples were centrifuged, the spores were washed and resuspended in the same volume of thioglycollate broth. One set of tubes were inoculated again with pediocin AcH (10,000 AU/ml), while another set was used as a control without pediocin AcH. The tubes were incubated at 10°C and examined for growth as before. The control tubes showed growth within 2 to 3 d, while the tubes with pediocin AcH did not show growth up to 14 d. Kalchayanand[28] suggested that pediocin AcH does not kill spores, but probably kills the cells following germination and outgrowth of spores by the same mechanism it kills vegetative cells of sensitive Gram-positive bacteria.

3. Gram-Negative Bacterial Cells

Normally, Gram-negative bacteria are resistant to the bacteriocidal action of pediocin AcH. Limited studies in our laboratory have indicated that if the

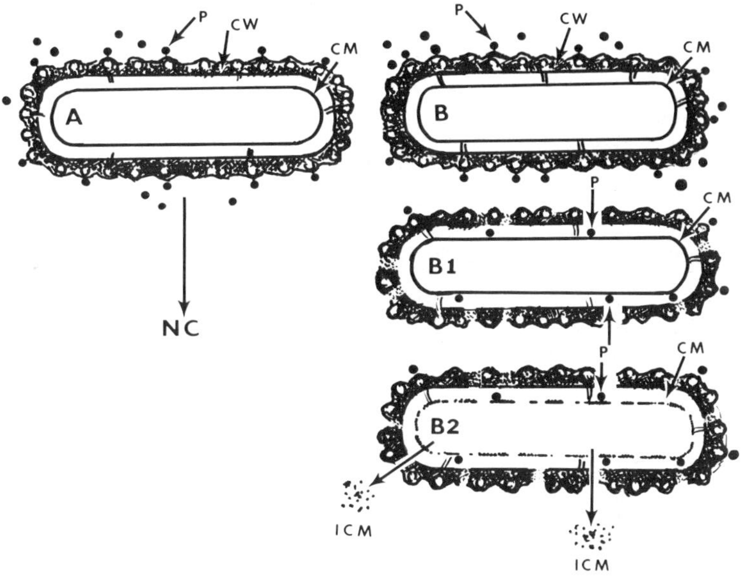

FIGURE 5. A hypothetical model of the adsorption of pediocin AcH on the cell-surface receptors of Gram-positive bacteria and destabilization on cytoplasmic membrane in sensitive, but not in resistant, strains. (A) adsorption of pediocin AcH (P) on cell wall (CW) receptors on resistant cells with no change in wall conformation. (B) adsorption on CW receptors in sensitive cell; (B1) changes in wall conformation resulting in passage of P through the wall; (B2) destabilization of the cytoplasmic membrane (CM) causing loss of intracellular materials (ICM) (drawn by A. Bhunia).

TABLE 5
Effect of Pediocin AcH Treatment on *Clostridium Laramie* Spores Following Treatment With Pediocin AcH During 14 Days Incubation at 12°C

	Growth of spores (10^5/ml)[28]
Untreated control	+
Pediocin AcH treated	–
Pediocin AcH treated and washed	+
Pediocin AcH treated, washed, retreated with pediocin AcH	–

Gram-negative bacterial cells are injured by sublethal stresses, they become sensitive to pediocin AcH. Results of developed sensitivity of several freeze-stressed Gram-negative pathogenic, index and spoilage bacteria to pediocin AcH are presented in Table 6. The unfrozen cells were resistant to pediocin AcH. However, when the cells were frozen along with pediocin AcH, more

TABLE 6
Developed Sensitivity of Gram-Negative Bacteria to Pediocin AcH Following Sublethal Injury as Indicated by the Viability Loss

Species[a]	cfu/ml following treatment: Unfrozen	Frozen (control)	Frozen with pediocin AcH (% reduction)[b]
Aeromonas hydrophila	16×10^4	8×10^4	9×10^3 (89)
Yersinia enterocolitica	38×10^3	21×10^3	5×10^3 (78)
Salmonella typhimurium	8×10^4	6×10^3	28×10^2 (53)
Escherichia coli NCSM	14×10^4	6×10^4	19×10^3 (68)
Escherichia coli JM 108	31×10^3	10×10^1	1×10^1 (90)
Pseudomonas putida	39×10^4	18×10^4	19×10^3 (89)
Pseudomonas fluorescens	11×10^4	7×10^4	38×10^3 (46)

[a] Cells were grown in tryptic soy broth overnight at 30°C, centrifuged, washed, and resuspended in water to contain about 10^{4-5} cells per milliter. Cell suspensions in 0.9 ml portions were added either with 0.1 ml deionized water (control) or 0.1 ml of a pediocin AcH preparation (Treated; 5000 AU/ml). One set of control and treated samples was frozen at –40°C for 2 h, thawed, and enumerated for cfu on tryptic soy agar plates. Data presented represents the average of three readings. Pediocin AcH was prepared by precipitating a culture broth supernatant of *P. acidilactici* H with 70% ammonium sulfate using the method of Bhunia et al.[18] Unfrozen cells did not show reduction in cfu following treatment with pediocin AcH.

[b] Percent reduction in cfu in pediocin AcH-treated samples in relation to the frozen control (100%). The differences in cfu between unfrozen and frozen controls were viability loss due to freezing. Gram-negative cells that survived freezing also became sensitive to bile salts, deoxycholate, and SDS, to which the unfrozen cells are resistant.

cells were killed as compared to cells frozen without pediocin AcH. Cells injured by other sublethal treatments, such as exposure to EDTA, lactic acid, or low heat, also became sensitive to pediocin AcH.

Developed sensitivity of sublethally stressed Gram-negative bacterial cells to hydrophobic chemicals, such as bile salts, deoxycholate, and sodium dodecyl sulfate, were found to be due to changes in the three-dimensional conformation of the lipolysaccharide (LPS) molecules in the outer membrane (OM). Normally the LPS molecules act as a barrier and exclude the hydrophobic molecules (and many other types of chemicals) to pass through the OM. Following sublethal stresses the LPS molecules lose their barrier function thus allowing the hydrophobic molecules to pass through the OM and come in contact with the cytoplasmic membrane (inner membrane). This causes viability loss of the injured cells.[45] The developed sensitivity of the injured Gram-negative cells to pediocin AcH, which is also a hydrophobic protein, and their action on the membrane, may be the result of a similar mechanism. These aspects are currently being studied in our laboratory.

D. Factors Affecting Antibacterial Funcion

The bacteriocidal efficiency of pediocins from *P. acidilactici* strains varies greatly under different conditions; some are due to the physical and chemical properties of the molecules, while some are due to environmental factors. The influence of the environmental factors on the antimicrobial activity of pediocins from *P. acidilactici* will be discussed in this section.

1. Storage Stability[28–31]

Pediocin AcH preparations produced by growing the producer strain in a suitable broth for 20 h at 37°C, heating the broth at 70 to 95 C for 15 min, and storing it either as liquid, concentrate (5 to 10 times) or in freeze-dried form, retained the AU/ml without any loss at –20°C for 6 months or more. At 4°C the stability of the dried and concentrated products remained unchanged for 12 weeks or longer. However, trials with liquid preparations (not concentrated) at 4°C produced variable results. In some trials there was no loss in activity for 12 weeks, while in other trials as high as 50% loss has been observed. The possible rcasons for this variation are now being examined. At room temperature the liquid preparation was found to remain stable up to about 6 weeks and then showed gradual reduction; by 12 weeks more than 50% of the activity was lost. Unheated broth with the cells lose activity rapidly after 24 h; however, cell-free supernatant retains the activity like that of heated broth with cells.

Partially purified pediocin AcH prepared by precipitation with NH_4 sulfate or acetone lose activity rather rapidly either during storage at room temperature or at refrigeration temperature.[41] The dried preparations retained stability more under vacuum storage than in the presence of air.

The storage stability of pediocins from other *P. acidilactici* strains has not been reported.

2. Indicator Bacterial Species and Strains

Gram-positive bacteria showed great variation in sensitivity to the bacteriocidal effect of pediocins from *P. acidilactici* strains. The studies of Gonzalez and Kunka,[20] Bhunia et al.,[18,24] Hoover et al.[21] and Lewis et al.[35] have indicated that different strains of the same species differ greatly from being extremely sensitive, to being less sensitive, to being completely resistant to pediocins. Even among the highly sensitive strains, some cells in the population can be resistant to pediocin AcH. Bhunia et al.[24] isolated a variant, *L. plantarum* NCDO 955 r, from the extremely sensitive parental strain NCDO 955 by growing it in broth containing pediocin AcH. Similar resistant variants have been isolated from sensitive strains of *Leuconostoc, L. monocytogenes, Enterococcus,* and *Lactobacillus* strains. The great variation in sensitivity of *Listeria* species and *L. monocytogenes* strains to pediocin AcH is presented in Table 7.[26,31] The sensitivity to 1500 AU/ml of pediocin AcH in these species

TABLE 7
Relative Sensitivity of *Listeria* Species and Strains to Pediocin AcH as Measured by Reduction in CFU

Listeria species/strains	Control (Log_{10} cfu/ml[a])	Pediocin AcH (Log_{10} cfu/ml[a])	Reduction in log_{10}
L. ivanovii	8.7	3.1	5.6
L. innocua	8.5	4.9	3.6
L. monocytogenes			
117	7.2	3.1	4.1
$Ohio_2$	8.0	4.0	4.0
101	8.3	5.0	3.3
$Camp^+/Beta^-$	8.6	5.3	3.3
$Camp^+/Beta^+$	8.6	5.3	3.3
103	8.1	5.8	2.3
Scott A	8.2	5.9	2.3
CA	6.3	4.1	2.2
V7	8.9	7.4	1.5

[a] *Listeria* cell suspensions in tryptic soy broth were treated with pediocin AcH (1500 AU/ml) and enumerated after 1 h at 4°C. Untreated cell suspensions were also enumerated as controls.

From Motlagh, A. M., Johnson, M. C., and Ray, B., *J. Food Prot.*, 54, 873, 1991. With permission.

and strains ranged from as low as 1.5 log to as high as 5.6 log under similar study conditions. The differences in sensitivity of *L. monocytogenes* strains to pediocin AcH is also reflected by their minimum inhibitory concentrations (MIC). With the same population level the MIC of pediocin AcH for strain $Ohio_2$ was only 150 AU/ml, but for strain Scott A it was 600 AU/ml.[26] Differences in MIC have also been observed among sensitive species and strains against pediocin PA-1.[23,32] Other studies have shown that pediocin may not be effective against some strains of *Clostridium botulinum*.[40]

3. Cell Concentrations of Sensitive Strains and Pediocin Concentration

It is not known how many molecules of pediocin AcH are necessary to kill a single bacterial cell of a sensitive strain. Differences in MIC of pediocins among strains tend to suggest that the number of molecules of pediocin necessary to kill a sensitive cell differ between strains. Also, the bacteriocidal efficiency of a given concentration of pediocin differs with the differences in cell concentrations of a bacterial strain. Conversely, with a given cell concentration the bacteriocidal efficiency will differ with different amounts of pediocin. These relationships have been studied with strains of *Lactobacillus*, *Leuconostoc*, and *L. monocytogenes* against pediocin AcH.[27,28,30,31] The

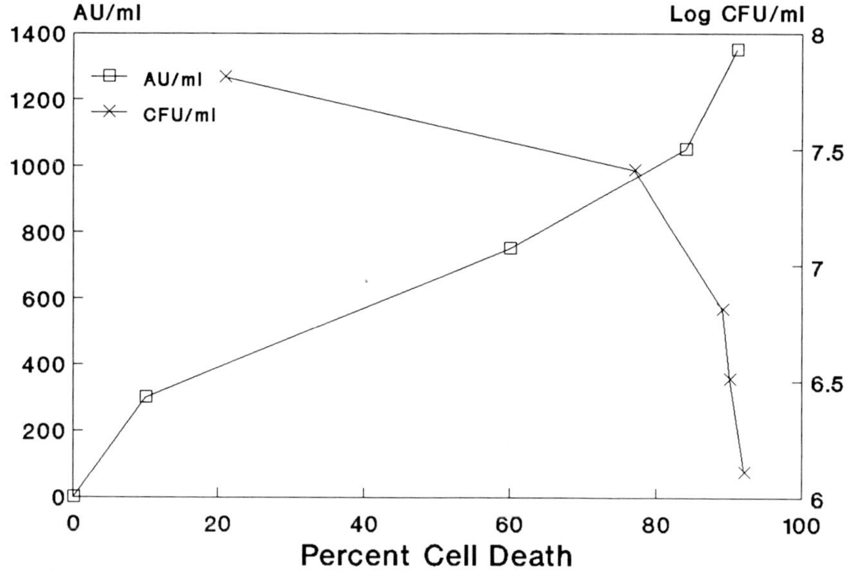

FIGURE 6. The influence of cell concentrations and pediocin AcH concentrations on viability loss of *L. monocytogenes* Scott A. To determine the influence of pediocin AcH concentrations 1.4×10^7 cfu/ml were exposed to 0 to 1350 AU/ml pediocin AcH (□), and to determine the influence of cell concentrations 1.3×10^6 to 6.5×10^7 cfu/ml cell suspensions were exposed to 1350 AU/ml pediocin AcH (×). Cfu were enumerated after 1 h at 4°C in tryptic soy agar. (Adapted from References 30 adn 31.)

bacteriocidal efficiency of pediocin AcH as influenced by cell concentrations of *L. monocytogenes* is presented in Figure 6. It can be seen that viability loss of a given cell concentration of *L. monocytogenes* increases, as the AU/ml of pediocin AcH increases and for a given concentration of pediocin AcH, viability loss reduces as the cell concentration increases.

4. Food Types

The bacteriocidal property of a pediocin preparation varies greatly with the characteristics, composition, and storage environment of foods. These include consistency of a food, amount of lipids, concentrations of NaCl and phosphate, oxygen availability, pH, and storage temperature. This will be discussed further in another section of this chapter dealing with the effectiveness of pediocins in controlling spoilage and pathogenic bacteria in food systems.

III. PROPERTIES OF PEDIOCIN OF *P. ACIDILACTICI*

A. Molecular Weight and Amino Acid Composition

Gonzalez and Kunka[20] used an ascending Sephacryl S-200 column and 50 m*M* Tris maleate buffer at pH 6.5 to elute pediocin PA-1. By comparing the elution volumes of pediocin PA-1 fraction with that of standard proteins, they

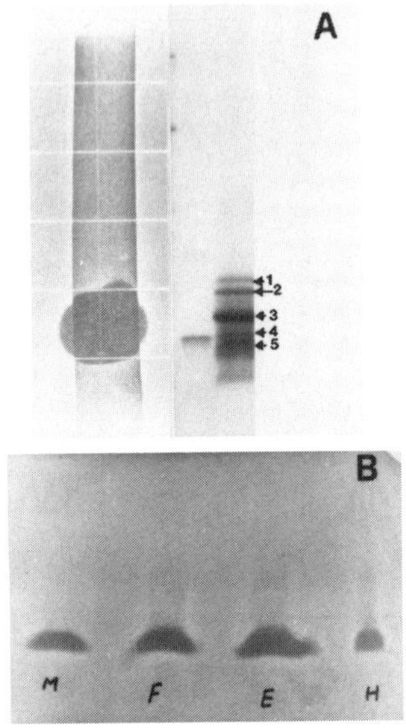

FIGURE 7. Direct detection of the band of pediocin AcH on sodium dodecyl sulfate polyacrylamide gel by staining and by growth inhibition of *L. plantarum* NCDO 955. (A) Strain *P. acidilactici* H with molecular weight standard (1) 16,950; (2) 14,400; (3) 8160; (4) 6210; and (5) 2510 Dal. (B) Strains P. *acidilactici* H, E, F, M. (Bhunia, A. and Ray, P., unpublished data).

suggested the molecular weight of PA-1 to be 16,500 Dal. Bhunia et al.[17] developed a procedure to identify directly the band formed by pediocin AcH on a SDS-PAGE gel by the ability of the band to form a clear zone of growth inhibition of an indicator bacterial strain. They ran a crude preparation of pediocin AcH on a SDS-PAGE gel in duplicate. After electrophoresis the gel was cut into two halves; one half containing molecular weight-standards was stained with Coommassie blue while the other half was first washed in deionized water for 8 to 12 h to remove SDS, then placed on a prepoured agar-medium plate and overlaid with soft agar medium seeded with cells of an indicator bacteria. The plate was incubated and examined for zone of growth inhibition. By comparing the position of the zone with the stained half of the gel the position of pediocin AcH band was recognized and its molecular weight was calculated from the standards (Fig.7). By this procedure pediocin AcH produced by *P. acidilactici* strains H, E, F, and M was found to be about 2700 MDal.[17,34,36] Recently, the molecular weight of pediocin AcH, as determined by the amino acid composition and sequence, was reported to be 4,628 Dal and an estimated pI, 9.6.[47]

Bhunia[41] reported the purification of pediocin AcH by FPLC on a reverse phase column using 10 to 35% gradients of acetonitrile with 0.1% trifluoroacetic acid. Biologically active pediocin AcH fraction was hydrolyzed with 6N HCl, and the amino acid composition was analyzed by HPLC. Along with normal amino acids the acid hydrolyzed material produced three unknown amino acids. These could be the amino acids resulting from post-translation modification of serine, threonine, cysteine and alanine as has been reported for nisin.[48,48a] The absence of pediocin AcH activity in the broth in which *P. acidilactici* H actively grew above pH 5.0[25], lead us to suspect that a low pH dependent (below pH 5.0) posttranslation modification might be involved in changing the inactive prepediocin AcH to active pediocin AcH. These aspects are currently being investigated.

A recent report has indicated that pediocin AcH is a basic polypeptide (pI 9.6) consisting of 44 amino acids with two disulfide bonds. The molecule is translated as a prepediocin consisting of 62 amino acids. Thus, during posttranslation processing the 18 amino acids (leader region) at the N-terminal end is removed to produce prepediocin. This is then processed to form two disulfide bonds to produce the active 44 amino acid pediocin AcH.[47,49] The 44 amino acid sequence of the prepediocin AcH molecule is:

NH_2-Lys-Tyr-Tyr-Gly-Asn-Gly-Val-Thr-Cys-Gly-Lys-
1 5 10
His-Ser-Cys-Ser-Val-Asp-Trp-Gly-Lys-Ala-Thr-Thr-
15 20
Cys-Ile-Ile-Asn-Asn-Gly-Ala-Met-Ala-Trp-Ala-Thr-
25 30 35
Gly-Gly-His-Gln-Gly-Asn-His-Lys-Cys-COOH
40 44

During posttranslation modification, enzymatic cleavage occurs between -Gly-Lys- at position between 18 and 19. The active molecule of pediocin AcH has 25% nonpolar, 56.8% polar, 2.3% acidic and 15.9% basic amino acids.[47] Recently, two reports have indicated that active pediocin PA-1 has also 44 amino acids which is processed from a 62 amino acid inactive molecule. The active molecule has also two thioether rings.[49a,49b] The limited information has indicated that the amino acid sequence of both pediocin AcH and pediocin PA-1, at least for the first 24 amino acids, is the same.

B. Solubility and Stability

1. Solubility

Partially purified pediocin AcH prepared from a broth culture by NH_4-sulfate precipitation and extensive dialysis is acidic, indicating that the molecules have a net positive charge (cationic). In dilute solution it remains stable, but in concentrated form the molecules aggregate and precipitate out from solution. This effect is more evident at refrigeration temperature than at room temperature. The aggregates retain their antibacterial activity.[41]

2. pH

The antibacterial activity (AU/ml) of a preparation of pediocin AcH is higher at low pH, but lower at pH 6.0 and above. It is not clear if the higher activity at lower pH (pH 5.0 and below) is due to greater solubility of pediocin

AcH at low pH, or higher susceptibility of bacterial strains to pediocin AcH at lower pH.

The retention of antibacterial activity of pediocin PA-1 was studied by exposing a partially purified preparation in buffer at pH ranging from 2 to 11 for 18 h. Subsequent testing revealed that the activity was most stable at pH 4 to 7, partially lost at pHs 2, 3, 9, and 10, and mostly lost at pH 11.[20] Similar studies for pediocin AcH also indicated that the activity was retained after exposure to pHs ranging from 2.5 to 9 at 25°C for 24 h or at 93°C for 15 min, but was lost at pH 10. At pH 12 even a 2 h exposure at 25°C resulted in the complete loss of activity.[18] At this pH, the molecules probably undergo irreversible changes either in their configuration or in the amino acid sequence that is necessary for their antibacterial activity.

3. *Temperature*

Although many reports have indicated that the antibacterial activity of bacteriocins from lactic acid bacteria are heat stable, high heat treatment for a long time can reduce their activity. This has been reported for nisin.[50] The loss of activity of pediocin PA-1 following heating was: 68% at 80°C for 60 min, or 100°C for 10 min, or 121°C for 15 min.[20] Pediocin AcH preparations also retained activity even after heating at 121 C for 15 min, but had a loss of about 30 to 40% of original activity.[18,28] No loss was observed after heating at 70 to 80°C for 15 min. High heat treatment might cause some changes in the molecular structure and reduce the potency of pediocins.

The influence of storage temperatures (25 to –20°C) on the retention of pediocin AcH preparatioins has been discussed before.

4. *Enzymes*

The activity of pediocin PA-1 was not adversely affected by lipase, phospholipase C, lysozyme, DNase or RNase, but was completely lost following incubation for 1 h with protease, papain, and α-chymotrypsin.[20] Similar studies for 1 h with pediocin AcH from different *P. acidilactici* strains have indicated that the antibacterial activity of the molecules is destroyed following treatment with trypsin, protease IV, protease XIV, protease XXIV, protease K, papain, α-chymotrypsin, and ficin, but not with RNase, lysozyme, lipase, catalase, and DNase.[18,19,34] Thus, almost all the proteolytic enzymes tested inactivated the antibacterial activity of pediocins of *P. acidilactici* strains.

5. *Organic and Inorganic Chemicals*

Pediocin AcH retained its antimicrobial activity following treatment with: formaldehyde, chloroform, acetone, isopropanol, ethyl alcohol, hexane, acetonitrile, sodium-dodecyl sulfate, sodium chloride, sodium phosphate, and ammonium sulfate.[18]

C. Toxicological and Antigenic Properties

No toxicological study report involving feeding of pediocins of *P. acidilactici* strains to laboratory animals is available at present. However, the pediocin-producing strains of *P. acidilactici* either have been isolated from fermented meat products[18,19] or are used for commercial production of fermented meat products.[20,21,34] The fermented food products produced from the use of these strains will contain pediocins. No adverse effect from the consumption of these products has been recorded. *P. acidilactici* H strain that produces pediocin AcH was used in a study to produce fermented lamb sausages and the products were consumed to evaluate acceptance quality by a group of taste panelists (12 to 14) ten times or more over a period of 6 months. No adverse effect to the taste panelists was reported.[51,51a]

In another study, laboratory animals, consisting of mice and rabbits, were given subcutaneous, intravenous, and interperitonial injections to produce antibodies.[52] Localized or general adverse reactions in these animals were not observed.

To study the antigenic property, partially purified pediocin AcH was used either as an oil adjuvant-based antigen or by conjugating with bovine serum albumin to immunize mice. For immunization of rabbits, pediocin AcH was mixed with incomplete Freund's adjuvant and used. In the ELISA test, serum from immunized mice and rabbits gave positive titer against partially purified pediocin AcH. However, the results of immunoblotting showed that there was no specific antibody against the pediocin AcH in either mice or rabbit serum (Figure 8). The antibodies produced, as evidenced by the positive ELISA test, were against two relatively large proteins that were present in the partially purified pediocin AcH. The results thus indicate that pediocin AcH was not immunogenic to mice and rabbits by the methods used in this study.

Recently, in another study, three 6-week-old BALB/C mice were immunized with broken heat-killed cells of *P. acidilactici* H (10^{10} cfu/ml) for 2 months and 5 d (1 injection per week; a total of 10 injections with 250 μl per injection). Both the broken heat-killed cells and the extract from the cells had pediocin AcH, as examined by determining SDS-PAGE gel electrophoresis and zone of growth inhibition of *L. plantarum* NCDO 955. The injection did not produce any undesirable effect in the mice. The mice were sacrificed at the end of the immunization period to produce hybridoma cells. None of the hybridoma produced antibodies against pediocin AcH as determined by Western blotting.[49] These data are in agreement with our previous studies.[52]

In addition, Bhunia[49] also tested the effect of pediocin AcH treatment on the survival of murine hybridoma cells. Murine hybridoma cells (10^5 cells per milliter) were treated with 50 μl pediocin AcH crude preparation and incubated under CO_2 at 37°C for 48 h. No viability loss was observed when both the control and the treated cells were examined under a phase contrast microscope.

IV. GENETIC ASPECT OF PEDIOCIN PRODUCTION IN *P. ACIDILACTICI*

A. Location of Genes

1. Plasmid Linkage of Pediocin Production Trait (Bac$^+$, Pap$^+$)

Studies conducted by three groups of researchers have convincingly proved that pediocin-production phenotype in all *P. acidilactici* strains so far tested

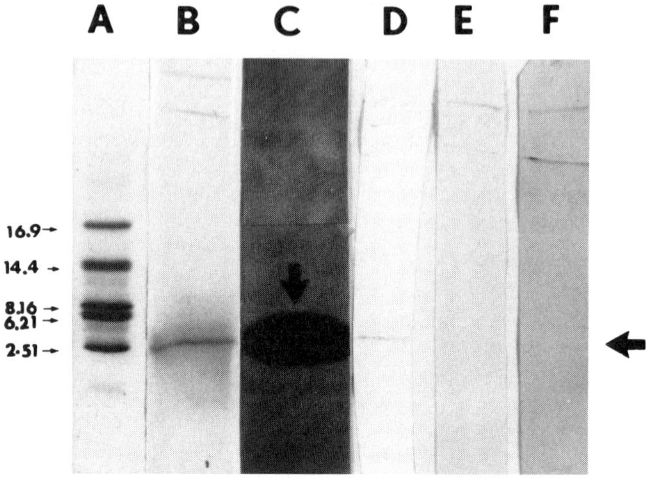

FIGURE 8. Sodium-dodecyl sulfate polyacrylamide gel electrophoresis and immunoblotting of pediocin AcH. Lane A, molecular weight standards (16950, 14400, 8160, 6210, and 2510 Dal); lane B, pediocin AcH stained with Coommassie blue; lane C, zone of inhibition of *L. plantarum* NCDO 955; lane D, amino black staining of nitroculture blot of pediocin AcH; lanes E and F, nitrocellulose blot reacted with mouse serum and rabbit serum respectively, showed no antibody reaction against pediocin AcH; see arrow. (From Bhunia, A. K., Johnson, M. C., Ray, B., and Belden, E. L., *J. Appl. Bacteriol.*, 69, 211, 1990. With permission).

is encoded in plasmids.[19-21,53,54] However, the size of the plasmids showed great variations in different studies ranging from 5.5 MDal to 8.0 MDal or about 8.25 Kb to 12.0 Kb (Table 8).[19–21,53–59] This has created some confusion. It is not clear whether the differences in size are because of the techniques used in different laboratories, or due to other reasons. Surprisingly enough, all three groups used plasmids from *E. coli* V517 as molecular weight standards. This strain has eight plasmids of molecular weight 1.4, 1.8, 2.0, 2.6, 3.4, 3.7, 4.8, and 35.8 MDal. The molecular weight of the plasmids encoding for pediocin are located between 4.8 and 35.8 MDal. In estimating plasmid size using kilobase ladder we recently observed that in strain H, E, F, and M the plasmid sizes were all about 8.9 Kb.[57,58] However, using molecular weight standards from *E. coli,* our results varied from 11.1 to 12.0 Kb.[19,34] The strain PAC 1.0 showed two different plasmid sizes, 4.7 and 6.2 MDal, by the same group in two different studies.[20,54] Also, studies with PAC 1.0 (originally strain NRRL B-8627) indicated a plasmid size of 6.2 MDal and 5.5 MDal by two research groups.[20,21] Likewise, if the PC strain that was used by Hoover et al.[21] and Kim et al.[34] was the same, both should have 8.9 Kb plasmid encoding for the pediocin production trait.[57,58] This confusion can be resolved by determining the plasmid size, their restriction maps, and finally the base sequence of the plasmids from the strains. Some relevant information for the

TABLE 8
Strain Designation, Plasmid Size, and Sources of Pediocin Producing *P. Acidilactici* Strains

P. acidilactici strains	Plasmid designation	Pediocin designation	mol wt of plasmid MDal (Kb)	Source and/or other designation of strain	Ref.
PAC 1.0	pSRQ11	Pediocin PA-1	4.7 (7.05) 6.2 (9.30)	Microlife, FL NRRL-B-18050 NRRL-B-5627	20, 53, 54
PO_2	None	None	5.5 (8.25)	Diversitech, FL, J.Bacus	21
PC	None	None	5.5 (8.25)	Diversitech, FL, J.Bacus	21
B5627			5.5 (8.25)	A. Lyons, Peoria, IL	21
MC-03 (*P. pentosaceus*)	None	None	5.5 (8.25)	Pepperoni isolate	21
H	pSMB74	Pediocin AcH	7.4 (11.10)	Fermented meat	18, 19
F	None	None	7.8 (11.70)	Fermented meat	19
E	None	None	8.0 (12.00)	Fermented meat	19
M	None	None	7.4 (11.10)	Marschel Division Miles Lab (designated as PC)	34, 56
H,E,F,M	pSMB74	Pediocin AcH	8.9 Kb	University of Wyoming (Food Microbiol. Lab.)	57
JBL 1095 H	pSMB74	Pediocin AcH	7.4 (11.10)	University of Wyoming (Food Research Inst.)	33
JD1-23	None	None	Not tested	Commercial culture suppliers	38, 39

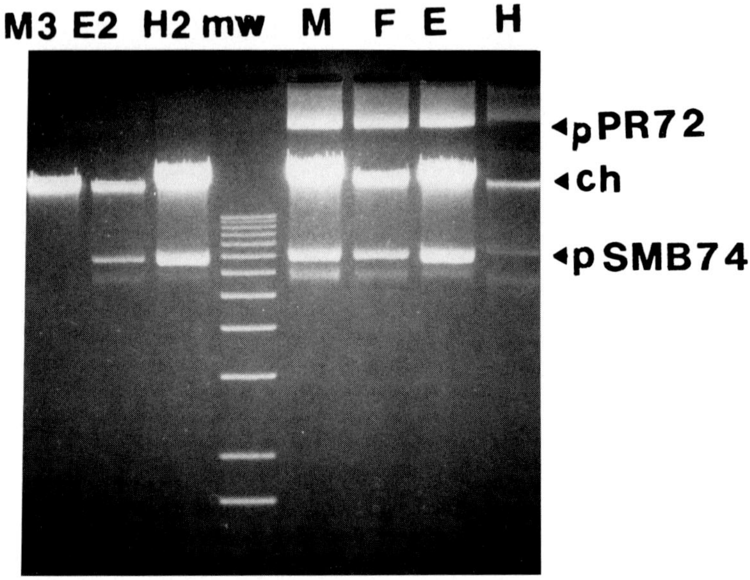

FIGURE 9. Plasmid profiles of pediocin AcH-producing *P. acidilactici* strains and pediocin AcH-nonproducing cured variant. Strains H, E, F, M are Pap$^+$ (pediocin AcH production, pSMB74; 8.9 Kb) and Suc$^+$ (sucrose hydrolysis, pPR72; 53Kb). Strains H2 and E2 are Pap$^+$ Suc$^-$ and M3 is Pap$^-$ Suc$^-$ variants of respective wild strains.

pediocin-producing strains, their origins, and the plasmids encoding pediocin-production trait are listed in Table 8, which can be useful in reducing the confusion.

2. *Stability of the Pediocin-Producing Plasmid*

To assign the pediocin-production phenotype (Bac$^+$[19,20] or Pap$^+$[58]) to specific plasmids, standard curing experiments were conducted by all three groups.[19–21] These included either growing the cultures in a broth at elevated temperature (48 to 50°C)[19,20] or in the presence of one of the following chemicals: acriflavine,[19–21] acridine orange,[21] ethidium bromide,[18] or novobiocin[19]. Following pour plating, colonies were examined for loss of pediocin-production against an indicator lawn by the method described earlier. Cells from the colonies without a zone of growth inhibition were then used along with the parental strain to compare plasmid profiles and to assign the pediocin production trait to a specific plasmid (Figure 9). In our studies with strains H, E, F, and M curing treatment generated variants with the loss of pediocin production trait (Bac$^-$ or Pap$^-$) at relatively high frequencies ($\leq 10^{-2}$). There were also some spontaneously generated Bac$^-$ (Pap$^-$) variants from each strain. The frequency of spontaneous loss of the plasmid increased when a broth culture was maintained for a long time.[31] Spontaneous loss of this plasmid was also observed during storage by refrigeration, freezing, and freeze-drying of a producing strain.[31] It appears that the plasmid(s) encoding pediocin production phenotype may not be very stable.

3. Plasmid Linkage of the Gene Encoding for Resistance or Immunity to Pediocin (Bac^r or Pap^r)

Gonzalez and Kunka[20] reported that the loss of pediocin plasmid (pSRQ11) in strain PAC 1.0 (Bac$^+$) though it resulted in the loss of pediocin production (Bac$^-$), the Bac$^-$ derivatives were resistant to pediocin. Thus they concluded that gene encoding for pediocin resistance (Bac^r) was not present in pSRQ11. Ray et al.[19] observed however, that some Bac$^-$ derivative of strain H became sensitive to pediocin AcH (Bac^s), while other Bac$^-$ isolates from the same strain, and the Bac$^-$ variants from strain E and F, remained resistant to pediocin AcH. Kim[34] also observed that among Bac$^-$ variants derived by plasmid curing of strain M some were resistant to pediocin AcH (Bac$^-$ Bac^r), while some were sensitive to it (Bac$^-$ Bac^s). Ray et al.[59] in another study observed that transfer of pSMB74 (the plasmid-encoding pediocin AcH-production phenotype) to a Bac$^-$ Bac^s recipient strain of *P. acidilactici* produced Bac$^+$ Bac^r strain that harbored pSMB74.

Recently, controlled studies by Motlagh[31] showed that cells of some Bac$^-$ variants of *P. acidilactici* H did not produce a zone of inhibition against a particular indicator strain but harbored pSMB74 in very low copy numbers and were resistant to pediocin (Bac$^-$ Bac^r). Only when a Bac$^-$ variant had lost all the copies of pSMB74, did it become sensitive to pediocin (Bac$^-$ Bac^s). This aspect is now being studied with other strains.

It has been previously indicated that many resistant Gram-positive bacterial cells adsorbed pediocin AcH. A model for this resistance has been proposed, but the genetic, physiological, and biochemical basis of this resistance is not known. It could be the immunity protein in producing strain that helps to maintain the cell wall conforms in such a way that pediocin AcH cannot enter inside to destabilize the membrane.

4. Other Traits in Plasmid Encoding for Pediocin Production

Hoover et al.[21] indicated that the pediocin plasmid in strains PO_2, PC and B5627 did not encode carbohydrate fermentation genes. However, they did not mention the carbohydrates that were tested. Ray et al.[19] reported that pSMB74 did not encode for the fermentation of sucrose, fructose, xylose, cellobiose, trehalose, and mannose, nor for resistance to antibiotics vancomycin, polymyxin, streptomycin, kanamycin, cycloserine, amphotericin, nalidixic acid, colistin, and griseoflavin. Curing of cells of pSMB74 did not result in either loss of fermentation of the carbohydrates or loss in resistance to the antibiotics listed.

However, as pediocin PA-1 molecules are translated to prepediocin and undergo enzymatic posttranslation processing involving the shortening of molecules from 62 to 44 amino acids, and with the formation of sulfide rings, the gene(s) for the enzyme(s) that are required for posttranslation processing are likely to be encoded in the same plasmid. Current studies in our laboratory have indicated the enzymatic processing of the inactive precursor of pediocin AcH into active pediocin AcH is enzymatic in nature and pSMB74 encode for the trait.[59a] Also, the transportation of the processed molecule from the cytoplasm to the exterior of the cells could be mediated by a protein, and the gene for this could also be encoded in the same plasmid. In some *L. lactis* subsp. *lactis* strain-gene(s) all these functions for nisin production are thought to be encoded in the same polycistronic transposon.[48a]

B. Transfer of Plasmid Encoding Pediocin Production Trait

Only one study has been reported on the conjugal transfer of plasmid pSMB74.[59] In this study, *P. acidilactici* H, which is sensitive to genetamycin (Gm, 300 μg/ml) and neomycin (Nm, 1.5 mg/ml) with phenotype Bac^+ Bac^r Gm^s Nm^s, was used as a donor, and a plasmidless strain *P. acidilactici* LB42-315 with phenotype Bac^- Bac^s Gm^r Nm^r was used as a recipient. Solid-surface mating was used on agar plates prepared with components of MRS broth, but containing only 0.25% glucose to prevent pediocin production by *P. acidilactici* H that could otherwise kill *P. acidilactici* LB 42-315 during mating. Possible transconjugants were selected on MRS agar plate containing crude pediocin AcH (10 mg/ml), Gm (300 μg/ml), and Nm (1.5 mg/ml). The colonies (Bac^r Gm^r Nm^r) were purified and examined for pediocin AcH production (Bac^+) against a lawn of a pediocin AcH-sensitive strain *L. plantarum* WSO-39. Most of the isolates were Bac^- Bac^r Gm^r Nm^r, but a few colonies (frequency about 3.8×10^{-8} per input donor cell) were Bac^+ Bac^r Gm^r Nm^r. Examination of the plasmid profile indicated only one plasmid was the same as pSMB74, as opposed to two or three plasmids of the donors. These were considered to be transconjugants (strain LB 42-923). It was suspected that a large plasmid in the donor, present in very low copy numbers may be involved in the cotransfer of pSMB74, which is only about 8.9 Kb and probably does not carry genes necessary for its own transfer by conjugation.

Recently Kim et al.[56] reported that their efforts to introduce purified plasmid pSMB74 into plasmidless pediocin AcH-sensitive *P. acidilactici* recipients by electroporation were not successful. Although by this technique they were able to introduce four different plasmids ranging in size from 4.4 to 53.7 Kb at frequencies ranging from 0.8 to 92.5/μg DNA, electrotransformants for pSMB74 were not detected. They suggested that the plasmid (pSMB74) was probably transformed but the transformants could not be identified among the recipients by the selective techniques they used.

C. Mapping of Plasmid Encoding Pediocin Production

1. RE Map of pSMB74

Recently Ray et al.[57,58] reported the mapping of plasmid pSMB74 in an effort to locate the gene(s) associated with pediocin AcH production. They obtained purified pSMB74 plasmid from the transconjugant *P. acidilactici* LB42-923 by CsCl density gradient, and used it for digestion with several restriction endonuclease (RE). It was observed that pSMB74 did not have cleavage sites for *Ava* I, *BamH* I, *Pst* I, and *Xma* I, but did have unique sites

TABLE 9
Molecular Weights of the Fragments Generated After Digestion of pSMB74 with Several R.E.

R.E. Used	Fragments	Mol wt of fragments (Kb)
Bgl I	1	9.0
Sal I	1	9.0
*Eco*R V	2	7.1, 1.8
Pvu II	2	5, 3.9
Hind III	3	3.5, 2.7, 2.7
Cla I	4	6.0, 1.1, 1.1, 0.7
*Eco*R I	1	9.0
Bgl I - *Pvu* II	2	5.0, 3.9
Bgl I - *Sal* I	2	4.9, 4.0
Bgl I - *Eco*R V	3	4.2, 2.9, 1.8
Bgl I - *Hind* III	4	3.5, 2.6, 2.3, 0.4
Bgl I - *Cla* I	4	5.8, 0.95, 0.95, 0.7
Bgl I - *Eco*R I	2	1.95, 7.05
Pvu II - *Eco*R I	3	2.0, 3.0, 3.9
Hind III - *Sal* I	4	2.9, 2.8, 2.3, 1.8
Sal I - *Eco*R I	2	3.4, 5.5
*Eco*R V - *Hind* III	5	2.8, 2.6, 1.8, 1.1, 0.6
*Eco*R V - *Eco*R I	3	1.8, 2.3, 4.8
Hind III - *Pvu* II	5	3.8, 2.3, 1.9, 1.7, 0.5
Hind III - *Eco*R I	4	1.2, 1.5, 2.7, 3.5
Cla I - *Eco*R V	6	3.1, 2.7, 1.1, 0.8, 0.65, 0.45
Cla I - *Hind* III	7	2.9, 2.65, 1.0,0.8,0.6,0.5,0.4
Cla I - *Pvu* II	6	5.5, 1.1, 1.1, 0.8, 0.3, 0.1
Cla I - *Sal* I	5	5.9, 1.1, 0.8, 0.75, 0.35
Cla I - *Eco*R I	5	0.75, 1.1, 1.1, 1.95, 4.0
*Eco*R V - *Sal* I	3	7.1, 1.05, 0.75
Sal I - *Hind* III	4	2.8, 2.8, 2.1, 1.2
Sal I - *Pvu* II	3	3.5, 5.0, 0.4

Note: Unique sites for *Bgl* I and Sal I. 2 to 4 cutting sites with *EcoR* V, *Pvu* II, *Hind* III, and *Cla* I. Vectors with unique sites for any of these R.E. can be used for cloning and locating gene for pediocin AcH production (Ray, B., unpublished data).

for *Bgl* I, *Sal* I, *Eco*R I, and *Sac* I. Several other enzymes have two to seven cleavage sites. Using the molecular weight of the fragments generated by the single and double digestion with the restriction enzymes, a map of pSMB74 was constructed. The molecular weight of the plasmid was calculated to be 8.9 Kb. The results of RE digestion and the partial map generated from these data are presented in Table 9 and Figure 10 respectively.

2. Cloning of RE Fragments of pSMB74[31,57,58]

Complete digestion of pSMB74 with *Hind* III produced three fragments, with molecular weight of one being 3.5 Kb and the other two being 2.7 Kb. These were designated as pRBI, pRBII, and pRBIII, respectively. Among the

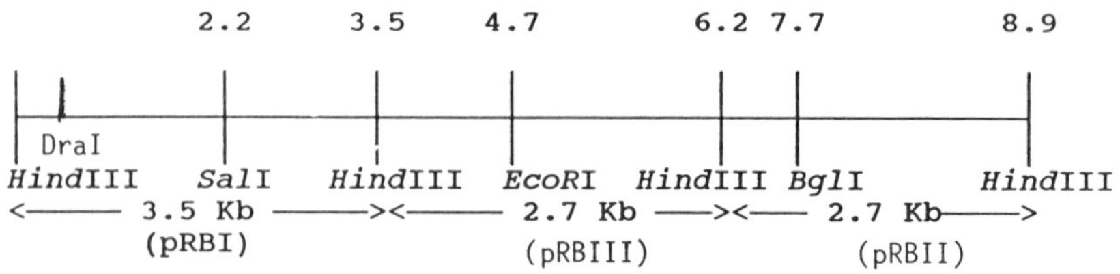

FIGURE 10. Linear map of pSMB74 showing three fragments generated by Hind III (pRBI, 3.5 Kb; pRB2, 2.7 Kb, and pRBIII, 2.7 Kb). The two 2.7-Kb fragments can be identified from the differences in unique sites for *Eco*R I in pRBIII and *Bgl* I in pRBII. The *pap* gene is located in pRBI in a 0.3 Kb fragment between *Dra*I and *Hind* III, Contains 186 bp with start codon towards *Dra*I.

two 2.7 Kb fragments pRBII has *Bgl* I site while pRBIII has *EcoR* I site. Incomplete digestion of pSMB74 with *Hind* III produced four more fragments from the combination of the three original fragments. They were designated as pRBIV (pRBI + pRBII), pRBV (pRBI + pRBIII), pRBVI (pRBII + pRBIII), and pRBVII (pRBI + pRBII + pRBIII). Each of the fragments were cloned in vector pUC 119 which has Amp^r gene, a multiple cloning site that includes a *Hind* III site, α-complement of Lac Z′ and origin from phage M13. The cloned vectors were transformed in *E. coli* XL-1 blue. The transformants carrying fragments pRB1 through pRB7 in pUC 119 were identified and grown in a broth containing 2% glucose. The culture broth and the colonies of the seven types of transformants were examined for pediocin AcH activity against a lawn of sensitive *L. plantarum* NCDO 955, but no zone of inhibition was observed. It was suspected that as the pH of the broths following growth of *E. coli* did not go below 5.0, it is probable that active pediocin AcH was not present in the broth (see section on pediocin AcH production). Alternatively, the gene(s) were not expressed in *E. coli* host, or none of the fragments had intact promoters for pediocin-production gene (*Pap*) and genes for posttranslation modification. Attempts were also made to introduce pUC 119 carrying fragments pRB1 to pRB7 in *P. acidilactici* LB42 and in *P. acidilactici* F3 (a plasmid-cured variant of *P. acidilactici* F) by electroporation. However, no Amp^r colonies were recovered, though when *E. coli* XL 1 blue was used as a control recipient, Amp^r transformants were obtained. This could be due to the inability of pUC 119 carrying pSMB74 fragments to express the Pap^+ trait in *P. acidilactici* recipients.

Subsequent attempts to sequence all three *Hind* III fragments from both ends in pRBI, pRBII and pRBIII by the primer extension method and comparing the sequence with the amino acid sequence obtained from amino acid analysis enabled us to locate the *Pap* gene in a 0.3 Kb fragment between *Dra*I and *Hind* III in pRBI. The gene with the start condon towards *Dra*I had 186 bp and codes for 62 amino acids of inactive prepediocin AcH.[47]

Successful cloning of the gene associated with pediocin PA-1 production in plasmid pSRQ11 has been reported recently.[49a,49b] The limited information indicates that a 5.6-Kb fragment from the plasmid carries the gene(s) for the production of pediocin PA-1. The fragment carries three open reading frames (ORF) and ORF1 encodes for the prepediocin PA-1 containing 62 amino acids.

TABLE 10
Probes Used to Determine Base Sequence of the Three *Hind* III Fragment of pSMB74 Cloned in pUC 119

A. Original probes for all three fragments (17 mer)
   Forward: 5′ G T T T T C C C A G T C A C G A C
   Reverse: 5′ C A G G A A A C A G C T A T G A C
B. New probes for the three fragments (16 mer)
   1. pRBI (3.7 Kb)
      Forward: 5′ G A A T G T T T G C C A C A A G
      Reverse: 5′ A C G C T A A A T C T T T T T A
   2. pRBIII (2.7 Kb)
      Forward: 5′ C A A C T T A T C T G G T G G G
      Reverse: 5′ A T T C T C T G C A C A A A G C
   3. pRBII (2.7 Kb)
      Forward: 5′ A C C A G C A A A C G C C T G A
      Reverse: 5′ C T C A A A A A A A C C A A C T

3. Sequencing of Plasmid pSMB74

Currently, we are sequencing the whole plasmid. This information will also be helpful in determining the location of immunity gene(s) and processing gene(s) as they are present in pSMB74, and also in understanding the control systems in pediocin AcH production.

Initially we used a forward and a reverse primer each of 17 base for all three fragments (pRBI, pRBII, and pRBIII; Table 10). This enabled us to sequence 250 nucleotides from each side in each fragment. Subsequently we used new probes for each fragment.

D. Improvement of the Antimicrobial Spectrum of Pediocin by Protein Engineering

Some Gram-positive bacterial cells and spores are not sensitive to pediocin(s) of *P. acidilactici* strains. As the bacteriocidal action is dependent upon the molecular structure of a pediocin molecule, specific modifications of the structure may produce analogues that are more effective as antimicrobial agents. These modifications can be carried out at the molecular level through the use of protein engineering. Pediocin(s) are not only relatively small, but they are ribosomally synthesize proteins. The genes encoding the proteins are in a small plasmid which can be present in some strains in high copy numbers. Thus, producing site specific mutation to change the sequence of the molecule can be done relatively easily to produce many different types of analogues and to test the analogues in a relatively short time. It may be that some of these will be effective against Gram-negative bacteria and fungi.

V. BIOSYNTHESIS OF PEDIOCIN

Pediocins are ribosomally synthesized proteins. The molecules are translated from the mRNA probably in inactive form (proprepediocin) and then they undergo posttranslation modification to active form (first to inactive prepediocin, then to active pediocin). The influences of different parameters on the production of pediocin by *P. acidilactici* strains have been reported only for pediocin AcH. The results of these studies will be included in this section. Appropriate references to other published works will be included when available.

A. *P. Acidilactici* Strains

Studies in our laboratory have indicated that pediocin AcH production varied greatly among strains. Among the four strains H, E, F, and M, strain H produced more pediocin under identical conditions. Even when cells from colonies of the four strains producing large zones of growth inhibition against the same sensitive indicator bacteria were used over a period of time, strain H produced more pediocin AcH (AU/ml).

We observed that at least two factors are responsible for these differences. First, strain H grows faster and produces more cells before the pH drops below 4.5. At a low pH necessary for the production of active pediocin AcH more cells were able to produce more pediocin AcH during the 18- to 24 h incubation period. Second, the plasmid pSMB74 is present in fairly high copy numbers per cell. However, the copy number can decrease easily under different conditions, including low pH. Strain H seems to maintain the high copy numbers better than the other three strains. Other factors may also be involved, especially the genetic control mechanisms in pediocin AcH production. These aspects are currently being studied.

B. Growth Environment

1. Media Composition

Many different broth and agar media have been used by several research groups to study production of active pediocin by *P. acidilactici* strains. These include MRS broth,[19,20,23,32,54] CG broth,[17,18,24,52] APT broth[21] and TGE broth.[25,27,33] However, only one study has been reported that examined the influence of media composition on the production of pediocin AcH by *P. acidilactici* H.[25] MRS broth and several modifications of MRS broth were compared with TGE broth for the production of pediocin AcH; TGE broth was found to be better (Table 11). The superiority of TGE broth was due to its lower buffering capacity. This aspect will be discussed in the section under pH influence. The concentrations of the individual ingredients used in the TGE broth greatly influenced the level of pediocin AcH production. The production increased as the concentrations of Trypticase, glucose, yeast extracts, and Tween-80 increased. Mn^{2+} increased production up to a certain level (Figure 11, Table 12). For economic reasons we used the composition listed in Table 11. Among the several carbohydrates metabolized by *P. acidilactici* H, glucose resulted in highest pediocin AcH production, whereas in the case of trehelose, arabinose and raffinose, no production was detected (Figure 12). A comparison

TABLE 11
Growth, pH, and Pediocin AcH Production by *PediococcusAcidilactici* H at 37°C in Different Broths

Growth media	OD[a] (600nm)	pH	% AU[b]
Lactobacilli MRS broth	4.1	4.3	85
Modified MRS brothc	3.9	4.5	77
TGE brothd	3.6	3.7	100

[a] OD was determined by diluting the broth tenfold in the same medium.
[b] AU: Activity units per milliter were highest in TGE broth (40 × 1000/ml) and was used as 100%. Addition of acetate, citrate, and phosphate to TGE broth at the levels used in other two broths reduced AU of pediocin AcH. *L. plantarum* NCDO 955 was used as indicator.
[c] Among the several modifications, the one in the table has: Trypticase, 1%; glucose, 1%; yeast extract, 1%; Tween 80, 0.1%; ammonium citrate, 0.2%; Na-acetate, 0.5%; $MgSo_4$, 0.005%; $MnSO_4$, 0.005%; K_2HPO_4, 0.2%; pH, 6.8.
[d] TGE broth: Trypticase or tryptone, 1%; glucose, 1%; yeast extract, 1%; Tween 80, 0.2%; $MnSO_4$, and $MgSO_4$, each 0.005%; pH, 6.5.

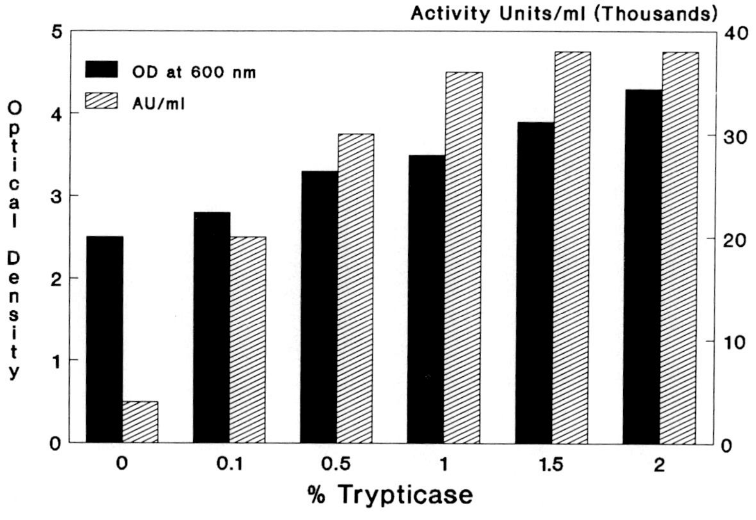

FIGURE 11. The effect of different concentrations of (A) Trypticase (B) yeast extract (C) Tween 80, and (D) manganese in TGE broth on growth (OD_{600nm}) and pediocin activity (AU/ml). In all concentrations of the ingredients above 0%, pH of TGE broth was reduced to 3.8 to 3.6. (From Biswas, S. R., Ray, P., Johnson, M. C., and Ray, B., *Appl. Environ. Microbiol.*, 57, 1265, 1991. With permission.)

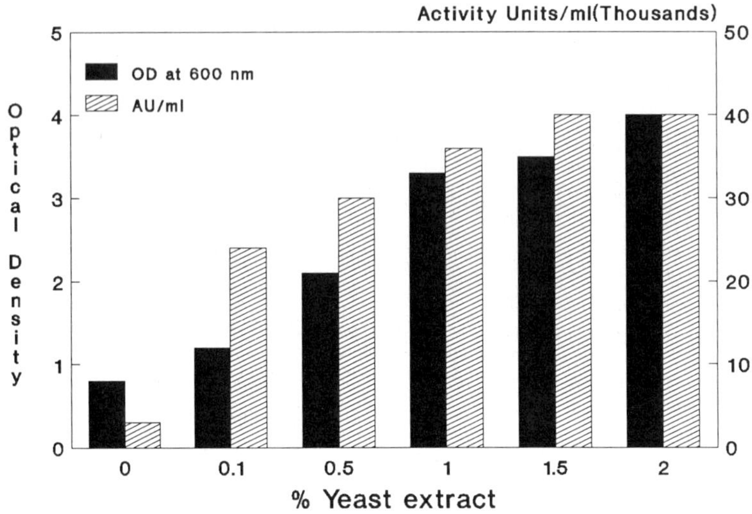

FIGURE 11B.

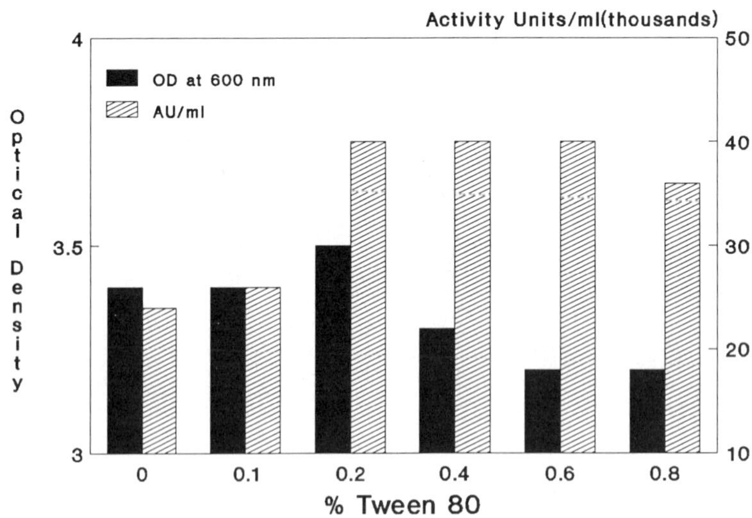

FIGURE 11C.

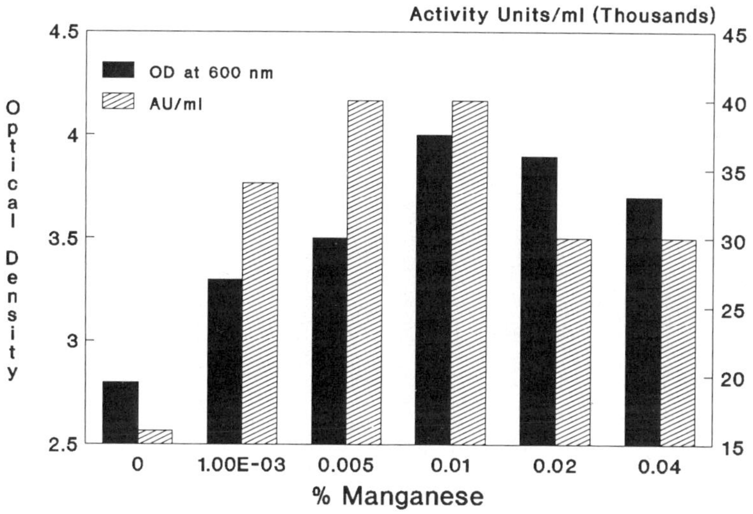

FIGURE 11D.

TABLE 12
Effects of Glucose Concentration on Growth, pH, and Pediocin AcH Production by *P. Acidilactici* H at 37°C for 16 h

Concentration m*M* (%)	OD (600nm)	Final pH	AU[a] (x 10^4)/ml
Control (0)	0.8	5.9	0.0
20 (0.36)	3.0	4.5	1.0
40 (0.72)	3.0	3.8	3.0
60 (1.08)	3.5	3.7	3.6
80 (1.44)	3.9	3.7	3.6
100 (1.80)	4.0	3.7	3.8
120 (2.16)	4.3	3.7	3.8

a AU: Activity units.

From Biswas, S. R., Ray, P., Johnson, M. C., and Ray, B., *Appl. Environ. Microbiol.*, 57, 1265, 1991. With permission.

of growth rate (OD_{600nm}), rate of acid production (pH; primary metabolites), and rate of pediocin AcH production during a 24 h incubation period in TGE broth suggested that considerable amounts of pediocin AcH are produced during the long stationary phase; active pediocin AcH is thus considered to be a secondary metabolite (Figure 13).

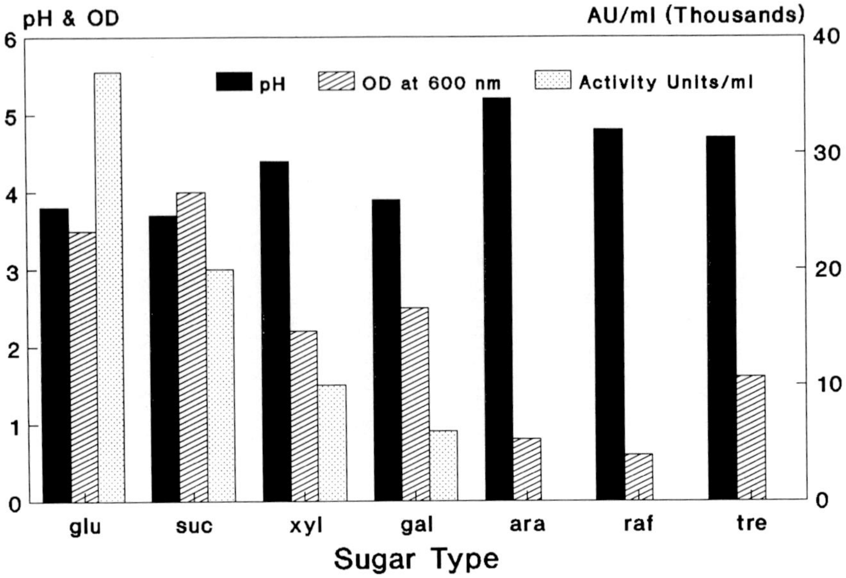

FIGURE 12. Effect of various carbohydrates on final pH, growth (OD_{600nm}), and pediocin activity (AU/ml) by *Pediococcus acidilactici* H. The carbohydrates used were: glucose, sucrose, xylose, galactose, arabinose, raffinose, and trehelose at 1% level. (From Biswas, S. R., Ray, P., Johnson, M. C., and Ray, B., *Appl. Environ. Microbiol.*, 57, 1265, 1991. With permission.)

2. Incubation Temperature and pH

P. acidilactici H has a growth range of about 20 to 50°C with optimum growth at 37°C. When the strain was grown in TGE broth at temperatures between 30 and 40°C, the rate of pediocin AcH production was higher at 37°C up to 16 h. Prolonged incubation can result in reduction in its level at higher temperature. We suspect that cell death and release of intracellular proteolytic enzymes from the cells might be the cause of this loss in activity (Table 13).

The influence of pH on pediocin AcH production during growth was studied by two methods. In one the media pH was adjusted to be between 4.0 and 7.0 before inoculation of cells. After 16 h at 37°C the final pH of all the media was the same (about 3.6), but the maximum pediocin AcH production was only in the medium with an initial pH of 6.5 (Table 14). In the second study the initial pH was 6.5 in all media, but the final pH was controlled automatically in the fermentor between 4 and 6 (Table 15). Under this condition maximum pediocin was produced when the final pH was 4.0. More surprisingly, no pediocin AcH was detected in the amounts tested (5 to 20 μl), at pH above 5.0, although considerable cell growth (OD_{600nm}) occurred.

We suspect that the requirement of a low pH needed for a high production level of pediocin AcH is due to the influence of low pH in posttranslation enzymatic modification.[59a] This probably involves structural modifications of

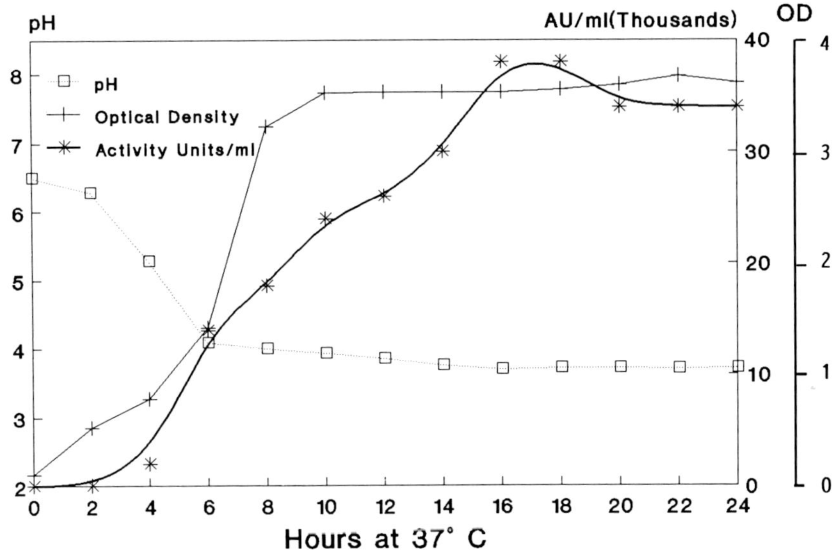

FIGURE 13. Changes in pH, cell mass (OD_{600nm}), and pediocin activity (AU/ml) due to growth of *Pediococcus acidilactici* H during incubation at 37°C for 24 h.[25]

TABLE 13
Effects of Incubation Temperature on Growth, pH, and Pediocin AcH Production by *P. Acidilactici* H

Incubation time (h)	Parameter	AU (× 10^4)/ml at incubation temperature 30°C	37°C	40°C
8	OD at 600 nm	1.4	3.3	2.3
	pH	5.0	3.8	3.9
	AU[a] (× 10 K)/ml	0.0	2.4	0.8
16	OD at 600 nm	3.6	3.8	2.7
	pH	3.7	3.6	3.9
	AU (× 10 K)/ml	4.0	4.0	2.6
24	OD at 600 nm	3.4	3.6	3.5
	pH	3.6	3.6	3.6
	AU (× 10 K)/ml	3.2	3.6	3.0

[a] AU: Activity units.

From Biswas, S. R., Ray, P., Johnson, M. C., and Ray, B., *Appl. Environ. Microbiol.*, 57, 1265, 1991. With permission.

TABLE 14
Influence of Initial pH of the Broth on Growth and Pediocin AcH Production by *P. Acidilactici* H at 37°C for 16 h

pH of broth			
Initial	Final	OD (600 nm)	AU[a] (x 10^4)/ml
7.0	3.6	3.3	2.4
6.5	3.6	3.5	3.6
6.0	3.6	3.4	3.2
5.5	3.7	3.2	3.2
5.0	3.6	3.2	3.0
4.0	3.6	2.8	2.0

[a] AU: Activity units.

From Biswas, S. R., Ray, P., Johnson, M. C., and Ray, B., *Appl. Environ. Microbiol.*, 57, 1265, 1991. With permission.

TABLE 15
Effect of Terminal pH Control on Cell Mass (OD) and Pediocin AcH Production by *P. Acidilactici* H After Growth at 37°C for 16 h

pH of the broth[a]			
Initial	Terminal	OD (600 nm)	10^4 AU/ml
6.5	6.0	3.0	0.0
6.5	5.5	3.1	0.0
6.5	5.0	4.0	1.0
6.5	4.5	4.0	3.0
6.5	4.0	4.0	3.2

[a] As soon as the pH was reduced from the initial (pH 6.5) to a desired terminal pH, it was maintained by the addition of sterile NaOH automatically by the fermenter.

From Biswas, S. R., Ray, P., Johnson, M. C., and Ray, B., *Appl. Environ. Microbiol.*, 57, 1265, 1991. With permission.

the inactive proprepediocin AcH to active pediocin AcH inside the cells and the transportation of the active molecules outside the cells. But the influence of low pH could also be at the transcription and/or translation level instead of at the posttranslation level. Our initial studies have indicated the influence of low pH to be at the posttranslation level.

C. Location of Pediocin in Producer Cells

The limited information so far abvailable has suggested that at the translation level pediocin PA-1 is synthesized as a 62-amino acid molecule. It then undergoes posttranslation modifications that involve removal of an 18-amino acid leader sequence at the NH_2-end and formation of two "S" rings. As detailed information is not available now, it can be speculated that these modifications are enzymatic and result in the conversion of the inactive molecules to active molecules. Also, due to the hydrophobic nature of the active molecules, they have to be transported outside the cell to protect the cytoplasmic membrane immediately following modification or during modification. Similar mechanisms have been speculated about for nisin.[48a] It is probable that active pediocin cannot be present in the cytoplasm or near the cytoplasmic membrane. Our studies indicated that pediocin AcH can be present both in the supernatant and with the producer cells. In the producer cells the molecules remained adsorbed on the cell surface.[20,24] However, when the cells (even after washing) are used to assay against a lawn of sensitive indicator bacteria, zone of growth inhibition occurs. This suggests that the adsorbed pediocin AcH molecules are released from the host cell surface and come in contact with the indicator cells to inhibit growth. Several studies have indicated that use of producer cells at temperatures where the cells do not grow produced antibacterial effects.[33,38,39] This was due to the release of the adsorbed pediocin from the cell surface, and was not due to synthesis of pediocin AcH.[33] The presence of pediocin in both the broth and the cells needs to be considered while preparing pediocin from a broth culture.

D. Purification of Pediocin

Several methods have been used for the purification of pediocin from broth cultures of *P. acidilactici*. Gonzalez and Kunka,[20] after growing *P. acidilactici* PAC 1.0 in MRS broth for 18 h at 35°C, adjusted the pH to 6.0 and removed the cells by centrifugation. The supernatant was filtered through 0.45 μm pore-size membrane filter, and NH_4-sulfate was added to 30% (w/v) and stirred gently for 18 h at 4°C. The supernatant was then centrifuged at 16,300 × g for 20 min, separated from the precipitate and the liquid portion, and brought to 60% NH_4-sulfate by adding more NH_4-sulfate. After 18 h, the precipitate was collected by centrifugation, reconstituted in 50 mm Tris-maleate (TM) buffer (pH 6.5) then dialyzed in the same buffer at 5°C. The dialyzed precipitate was further purified first by DEAE-sephadex A25 and then by CM-Sephadex C-25 by eluting with the TM buffer. The active fractions were dialyzed, concentrated with Carbowax 20, and designated as partially purified pediocin PA-1. This preparation was used to determine molecular weight by gel filtration. Pediocin PA-1 was reported to have a molecular weight of 16,500 Dal.

Bhunia et al.17 initially reported the growth of *P. acidilactici* H in CG broth (casein hydrolysate, 1%; yeast extract, 0.5%; glucose, 1%; Tween 80, 0.1%; Na-acetate, 0.5%; MgSO4, 0.01%, MnSO4, 0.005%; K2HPO4, 0.2%). The broth was prepared at 10 times concentration and dialyzed in a bag (molecular weight cut off 1000 Dal) with deionized water (100 ml in 900 ml); pH was adjusted to 6.8, it was sterilized and then used for growth. After 18 h at 37°C the cells were removed, the supernatant was filtered through membrane filter and mixed with 70% NH_4-sulfate (w/v) for 2 h. The precipitate was collected

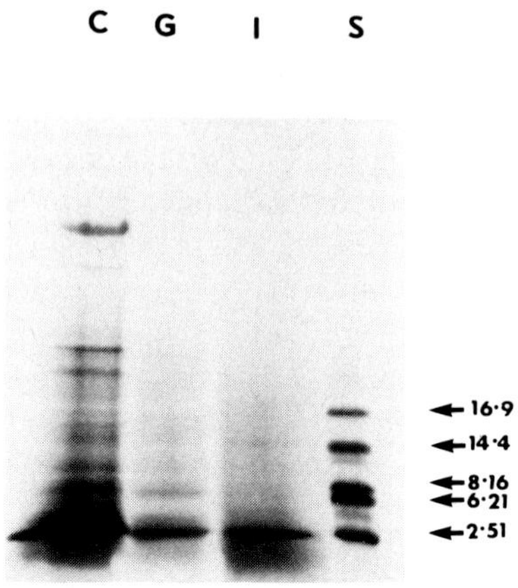

FIGURE 14. Sodium dodecyl sulfate polyacrylamide gel electrophoresis of pediocin AcH preparations after various stages of purification. The gel was silver stained. Lane C crude pediocin AcH; lane G, pediocin AcH after elution from a gel filtration column; lane I, pediocin AcH eluted from ion exchange chromatography; lane S, Molecular Weight Standards (× 1000 Dal) (From Bhunia, A. K., Johnson, M. C., and Ray, B., *J. Appl. Environ. Microbiol.*, 65, 261, 1988. With permission.)

by centrifugation and then dialyzed extensively against deionized water in a bag of 1000 molecular weight cut off. This preparation, designated as crude pediocin AcH, was run through 10 to 25% continuous gradient SDS-PAGE gel with molecular weight standards. Staining with Coommassie blue produced three major and several minor bands. Examination of a similar unstained gel with an indicator bacterial strain indicated that the lower band was the antibacterial pediocin AcH, and was about 2700 Dal molecular weight (Figure 7). The crude preparation was further purified on a 12 HR 10/30 Shepharose 12 Fast Protein Liquid Chromatography (FPLC) column against ammonium acetate buffer. The active fractions were further purified by FPLC anion exchange chromatography using a Mono Q HR 5/5 column against increasing concentration of Tris-HCl buffer. The active fractions, when examined on a 10 to 25% SDS-PAGE gel, gave a single band that had the antibacterial activity and was calculated to be 98.8% pure (Figure 14).[18] This was designated as partially purified pediocin AcH. Further trials to purify this protein by FPLC in C-18 column and by several other methods did not produce fractions that were 100% pure, and thus could not be used for either amino acid composition or amino acid sequence determination. Active fractions from C-18 column however, when analyzed, indicated the presence of the following amino acids: Asp, Glu, Ser, His, Gly, Thr, Arg, Ala, Tyr, Met, Val, Phe, Ile, Leu, Lys, and three unknown amino acids. Efforts to determine the sequence produce confusing data.[41] It has been very difficult to purify pediocin AcH for sequencing studies. Recently, pediocin AcH has been purified by transferring the protein band from SDS-PAGE gel and used for partial amino acid sequencing.[47]

A recent report has indicated that pediocin PA-1 has been purified through gel filtration, ion exchange, and HPLC.[47] These researchers have indicated that purification of PA-1 is extremely difficult. However, they have succeeded in determining the amino acid sequence of pediocin PA.[49a,b]

VI. ANTIBACTERIAL EFFICIENCY OF PEDIOCINS OF *P. ACIDILACTICI* IN FOOD SYSTEMS AGAINST SPOILAGE AND PATHOGENIC BACTERIA

Pure culture studies previously discussed have shown that the pediocins of *P. acidilactici* are bacteriocidal against several spoilage and pathogenic bacteria that are of concern in many foods. However, like nisin and other bacteriocins of lactic acid bacteria, they have several limitations. The most important limitation is that they are not effective against Gram-negative bacteria and fungi. Also, cells and spores of all Gram-positive bacterial species and strains are not killed by pediocins, and even in a most sensitive strain there will be some resistant cells. As pediocins do not have bacteriostatic effect, the resistant cells can multiply under the proper environment. The other important limitation is the possible influence of the food environment in reducing the antibacterial property of pediocins. They are hydrophobic molecules, proteinaceous in nature, and need contact with a sensitive bacterial cell to produce bacteriocidal effect. Thus, special considerations have to be made in the use of pediocins in fatty foods, foods with proteolytic enzymes, foods with high concentrations of NaCl and phosphates, and in solid foods. Finally, they cannot be effective in all types of foods, especially those in which Gram-negative bacteria are predominant, or in foods that have even a high population of Gram-positive bacteria i.e., foods handled under poor sanitary conditions. However, there are foods in which pediocins can be effectively used as biopreservatives. Vacuum-packaged refrigerated raw and semiprocessed foods are probably most suited for the use of pediocins. These foods are expected to have a relatively long shelf life, and are produced under good sanitary conditions to have a low initial microbial load. The availability of low-oxygen-permeable plastic bags and high-efficiency vacuum-packaging machinery have enabled the foods to be vacuum packaged very effectively. Also, there is a current trend to keep many of these products at or close to 0°C during storage. Under these conditions limited types of psychrotrophic anaerobic and facultative anaerobic bacteria can grow, and many of these are Gram-positive. These are mainly lactobacilli, leuconostocs, *Clostridium laramie, Listeria,* and Gram-negative *Yersinia enterocolitica.* Under proper storage temperatures, a low initial population level in a food produced under proper sanitary conditions will take a long time to become a major safety concern, except in the case of some pathogens. However, temperature abuse from the time of production to the time of consumption can change this situation quite radically.

Use of pediocins in these foods will further reduce the initial load of Gram-positive psychrotrophic anaerobic and facultative anaerobic bacteria so that during controlled refrigeration storage, or during brief temperature abuse, the survivors do not reach alarming levels. It can thus function as a secondary barrier coupled with the efficient manufacturing and storage practices to enhance the safety and shelf life of vacuum-packaged refrigerated raw and semi-processed foods. However, very limited studies have been conducted to determine the effectiveness of pediocins in the food system.

A. Effectiveness Against Spoilage Bacteria

1. Sterile Food: Meat and Dairy Products

Holla[30] compared the bacteriocidal effectiveness of pediocin AcH preparations from *P. acidilactici* H against a *Leuconostoc* spp., originally isolated from a spoiled refrigerated meat sample. She used five sterile foods, namely ground beef, sausage mix, cottage cheese, ice cream, and milk for testing purposes. The reduction in cfu was rapid, and at any given pediocin AcH concentration, no large differences were observed in the reduction of cfu amongst the five foods. Viability loss increased from <25% at 300 AU/ml to a maximum of 99% at 1350 AU/ml of pediocin AcH. The survivors multiplied during storage at 4°C, but after 3 weeks the population in the treated samples remained 3 logs lower than in the untreated controls in all five foods (Table 16). More importantly, the cfu in all treated samples remained well below the "spoilage detection level" (about 1-5 $\times$ 10^6/g).

2. Nonsterile Products

a. Fresh Beef

The influence of pediocin AcH in controlling the growth of inoculated meat-spoilage bacteria consisting of *C. laramie*,[28,39,60,61] *Lactobacillus* spp.,[28] *Leuconostoc* spp.,[28] as well as of normal meat flora,[29,61] has been studied during refrigeration storage of vacuum-packaged fresh beef. In all studies the cfu of specific spoilage bacteria, as well as associated bacteria, were much lower in the treated samples than in the untreated samples. Semitendenous muscle (100 g) produced under sanitary conditions was inoculated with about 5 $\times$ 10^3 cells of a *Leuconostoc* spp. isolated from a spoiled refrigerated beef sample and either treated with a pediocin AcH preparation (1400 AU/g) or not treated (control), vacuum packaged and stored at 3 to 4°C. At selected intervals duplicate samples were enumerated for cfu by recommended procedures for aerobic plate count and psychrotrophic lactic acid bacteria. The results presented in Table 17 indicate that in the pediocin AcH-treated samples initially there was a reduction in cfu of 2 logs. During storage the survivors grew but remained below 2 logs in the treated samples up to 8 weeks and after 12 weeks there was only 1 log difference between the treated and untreated samples for

Table 16
Effect of Pediocin AcH Treatment on the Viability Loss of *Leuconostoc Mesenteroides* Ly in Five Sterile Food Systems During Storage at 4°C[a]

Storage time at 4°C	Ground beef		Sausage		Cottage cheese		Ice cream		Milk	
	Ped^-	Ped^+	Ped^-	Ped^+	Ped^-	Ped^+	Ped^-	Ped^+	Ped^-	Ped^+
1 h	28×10^5	23×10^3	36×10^5	29×10^3	24×10^5	20×10^3	33×10^5	27×10^3	24×10^5	24×10^3
1 d	36×10^5	25×10^3	46×10^5	28×10^3	29×10^5	21×10^3	32×10^5	26×10^3	33×10^5	25×10^3
7 d	14×10^6	26×10^3	18×10^6	32×10^3	13×10^6	22×10^3	16×10^6	25×10^3	15×10^6	25×10^3
14 d	68×10^6	84×10^3	81×10^6	93×10^3	56×10^6	74×10^3	74×10^6	86×10^4	64×10^6	82×10^3
21 d	30×10^7	40×10^4	34×10^7	45×10^4	23×10^7	34×10^4	30×10^7	38×10^4	28×10^7	37×10^4

Note: Pediocin AcH preparation was used at 1350 AU/ml. Samples retained residual pediocin activity during storage.

[a] Table adapted from References 30 and 31.

Table 17
Influence of Pediocin AcH Treatment of Fresh Beef Inoclulated with *Leuconostoc* Cells on Bacterial Level During Storage in Vacuum Package at 3°C[a]

Storage Time (weeks)	Log_{10} APC/g		Log_{10} Psy/g	
	Ped⁻	Ped⁺	Ped⁻	Ped⁺
0	3.3	1.0	3.4	1.0
1	3.3	1.9	3.2	1.4
4	3.4	2.2	3.2	2.2
6	4.1	3.1	4.0	3.2
8	5.3	3.3	5.3	3.3
12	7.3	6.3	7.4	6.3

Note: Semitendinous muscle in 100 g portions was inoculated with a Leu*conostoc mesenteroides* Ly at about 5×10^3 cfu/g and pediocin AcH (1400 AU/g), vacuum packaged, and stored at 3°C. Pediocin-AcH untreated samples were also prepared similarly. At different storage times duplicate samples were enumerated for anaerobic plate count (APC/g, 35°C, 2 d) and psychrotrophic counts (Psy per gram; 7°C, 10 d in plate count agar). Average results are presented in Log_{10}. *L. mesenteroides* Ly, originally isolated from spoiled meat sample is capable of growing in plate count agar at 7°C for 10 d (Psy pergram).

[a] Table adapted from References 28 and 29.

both groups of bacteria. Also, the cfu in the treated samples at the end of 12 weeks were not above the "spoilage detection level." Similar studies with commercial samples containing normal microflora indicated no great initial reduction in cfu in the pediocin AcH-treated and control samples. It is probable that this is due to the predominance of Gram-positive resistant and Gram-negative bacteria. Whatever small fractions of psychrotrophic Gram-positive anaerobic and facultative anaerobic bacteria sensitive to pediocin that were present initially were killed (Table 18). During storage in the treated samples only the pediocin-resistant bacteria capable of growing under the storage conditions grew. At the end of 10 weeks the population level of all two bacterial groups remained below spoilage level. Also, the organoleptic quality of the treated samples remained acceptable at the end of storage.

b. Salad Dressing and Salads

Gonzalez[54] reported that a partially purified preparation of pediocin PA-1 was effective in preventing spoilage of salads and salad dressing caused by *Lactobacillus bifermentum*. Pediocin PA-1 was stable in salad dressing environment and was found to extend shelf life of coleslaw, macaroni salad, and potato salad for several days.

TABLE 18
Influence of Pediocin AcH Treatment of Commercial Beef Samples on Bacterial Levels During Vacuum-Packaged Storage at 3°C[a]

Storage time (wks)	Log_{10} APC/g		Log_{10} Psychrotrophs/g	
	Ped^-	Ped^+	Ped^-	Ped^+
0	2.8	2.3	2.9	2.4
2	5.0	3.8	5.1	3.2
4	5.4	5.0	5.5	5.1
6	5.6	5.0	5.7	4.8
8	6.3	5.3	6.2	5.3
10	6.5	5.5	6.4	5.6

Note: Commercial beef (semitendinous muscle) in 100 g portions was treated with a pediocin AcH preparation (2000 AU/g), vacuum packaged, and stored at 3°C. Untreated controls were also similarly stored. At selected intervals the samples in duplicates were enumerated for APC, psychrotrophic counts.

[a] Tabel adapted from Reference 28.

B. Effectiveness Against Pathogenic Bacteria

1. Sterile Food: Meat and Dairy Products

Bacteriocidal efficiency of pediocin AcH on three strains of *L. monocytogenes* cells in sterile ground beef, sausage mix, cottage cheese, ice cream, and milk has been studied.[24,30,31] The results indicate that at a given pediocin AcH concentration, the viability loss of a given population of a *L. monocytogenes* strain did not vary greatly in any of the five food systems, and was not much different from that in the nonfood system (peptone water). However, for a particular strain with a given population level, viability loss increased as the AU/ml of pediocin AcH increased. Conversely, at a given pediocin AcH concentration viability loss decreased as the population level increased. Also, the strains differed greatly in their sensitivity to pediocin AcH. The results of the study on ground beef with four strains of *L. monocytogenes* are presented in Table 19. In general, strains CA and Scott A were more resistant than the strains $Ohio_2$ and $Camp^+/Beta^-$ at all population levels.[31] When the *L. monocytogenes* cells at relatively high concentrations (about 1.3×10^7/ml) were inoculated in the five foods and treated with 1350 AU/ml pediocin AcH, within 1 h there was about 1 log viability loss in all foods, with a slightly higher loss in cottage cheese (Table 20). In the pediocin AcH-untreated control samples the cells multiplied within 7 d at 4°C, while in the treated samples, no increase in cfu was observed in 21 d in cottage cheese and 14 d in other foods. At the end of 28 d there was an average of 3 logs difference in cfu

TABLE 19
Effect of Pediocin AcH Treatment on Viability Loss of Four *L. Monocytogenes* Strains in Milk at Different Population Levels[a]

	Log$_{10}$ cfu reduction with initial cfu/ml			
***L. monocytogenes* strains**	**10^4**	**10^5**	**10^6**	**10^7**
CA	2.1	1.8	1.4	0.9
Scott A	2.8	2.6	1.7	1.3
Camp+ Beta$^-$	3.1	2.4	2.4	2.1
Ohio$_2$	3.3	2.8	2.5	1.9

Note: Cells at about 10^4 to 10^7/ml levels were inoculated in milk along with 1350 AU/ml of pediocin AcH, incubated for 1 h at 4°C and enumerated for cfu in tryptic soy agar. The viability loss as determined from the differences in log$_{10}$ cfu/ml between the pediocin-treated and untreated control are presented. For each strain viability loss decreased as the population level increased. At any population level strains CA and Scott A are more resistant to pediocin AcH than strains Camp$^+$ Beta$^-$ and Ohio$_2$.

[a] Table adapted from References 30 and 31.

between the pediocin AcH-treated and untreated samples.[30] In a separate study with initial *L. monocytogenes* inoculation level of about 100 cells per milliter in pediocin AcH-treated ground beef, the counts after 4 weeks at 4°C remained $<1 \times 10^0$ for strain Ohio$_2$ and 4×10^3 for CA as compared to 2 to 6×10^6/g in the pediocin AcH-untreated beef.[27] In the event of temperature abuse, however, the lag phase in the pediocin AcH-treated samples was greatly reduced; after 21 d at 10°C, the difference between the pediocin-treated and untreated samples was reduced to about 1 log as opposed to about 3 logs in samples stored at 4°C (Table 20). Even at much higher levels pediocin AcH could not extend the lag phase but increased the differences in log$_{10}$ cfu between the pediocin-treated and untreated samples (Table 21). The initial viability loss increased as the pediocin AcH concentrations increased. This was because at about 10^7 cfu/ml about 5000 to 6000 AU/ml are necessary to produce maximum viability loss in the strain Scott A.[26,31]

2. *Nonsterile Products*

a. *Fresh Beef*

Nielsen et al.[32] used a MRS broth preparation of pediocin PA-1 from *P. acidilactici* PAC.1 to study its efficiency in reducing *L. monocytogenes* cells inoculated in irradiated fresh beef. The beef cubes were soaked in a high (about 10^7/ml) and a low (about 10^4/ml) cell suspension of *L. monocytogenes* cells and then treated with 500 to 5,000 AU/ml of pediocin PA-1. At a low population level 1 to 2 logs reduction in cell viability occurred within 2 min at all pediocin PA-1 concentrations. However, at high population levels reduction of about 2 logs was obtained at 5000 AU/ml, but <1 log was observed with 500 AU/

ml. During storage at 5°C, the surviving cells started multiplication after 21 d in samples treated with 5000 AU/ml pediocin PA-1. After 28 d the difference in cfu between the pediocin-treated and untreated control was about 1.2 logs.

In our laboratory we have studied the efficiency of pediocin AcH preparation in reducing the cfu of *Listeria* strains in fresh beef and fresh beef byproducts such as heart, kidney, liver, and cheek muscles. In all types of samples, except in liver, pediocin AcH treatment rapidly reduced the *Listeria* population. The reduction in *Listeria* population was dependent upon strain, cell concentration, and pediocin AcH concentration. In liver samples, the level of cfu reduction under a given set of conditions was less. This was due to the inactivation of pediocin AcH by the protein-hydrolyzing enzymes in fresh liver.[62]

b. Processed Meat

Berry et al.[38] used a pediocin-producing strain of *P. acidilactici* (strain JD1-23) as a starter culture to determine its ability to control the growth of inoculated *L. monocytogenes* cells during fermentation of sausage. They reported that cfu of *L. monocytogenes* were 2 logs lower in products made with pediocin-producing strains as compared to products made with a nonpediocin-producing strain. In a later study they investigated the efficiency of strain JD1-23 in controlling post-heat-treatment contamination of *L. monocytogenes* cells and cells of strain JD1-23 in frankfurters. They vacuum packaged the products and enumerated the population of *L. monocytogenes* during storage at 4°C and 15°C. In all trials, cells of JD1-23 reduced population levels and growth rate of *L. monocytogenes* more than a pediocin non-producing variant.

Yousef et al.[33] used both pediocin AcH and an antibiotic resistant variant of *P. acidilactici* H (designated as JBL 1095) to control the growth of *L. monocytogenes* in wiener exudates. They observed both a rapid reduction followed by a gradual reduction in *L. monocytogenes* cfu during storage at 4°C and 25°C. Studies in our laboratory have also indicated that in heat-processed pork products rapid viability loss of *L. monocytogenes* occurred initially following pediocin AcH treatment. However, depending on the *L. monocytogenes* strain, the cell concentration and pediocin AcH concentration, the survivors grew slowly during storage at 4°C.[62]

c. Dairy Products

Pucci at al.[23] reported the use of dried powder of a preparation of pediocin PA-1 from *P. acidilactici* PAC.1 to control the growth of *L. monocytogenes* in cottage cheese, half-and-half cream, and cheese sauce. *L. monocytogenes* cells were added to the foods at a low level (about 10^2 to 10^3/g or ml) with or without the dry pediocin PA-1 at 10 AU/g or ml. In cottage cheese cfu was reduced after 7 d both in the treated and the untreated samples. In the other two products survivors in the treated samples grew within 7 d at 4°C; the pediocin-treated samples had about 2 to 4 logs cfu less than the untreated samples after 14 d (Table 22).

The results of the limited studies have suggested that pediocins from *P.*

Table 20
Viability Loss and Growth of *L. Monocytogenes* Scott A Cells in Food Treated with Pediocin AcH[a]

Food with and without pediocin AcH at 4 or 10°C	cfu/ml during storage for					
	1 h	1 d	7 d	14 d	21 d	28 d
Ground beef (4°C)						
Ped$^-$	1.3×10^7	1.4×10^7	5.1×10^7	2.1×10^8	9.4×10^8	5.2×10^9
Ped$^+$	1.2×10^6	1.3×10^6	1.2×10^6	1.2×10^6	2.0×10^6	6.7×10^6
Sausage mix (4°C)						
Ped$^-$	1.5×10^7	1.6×10^7	7.2×10^7	3.0×10^8	1.3×10^9	7.0×10^9
Ped$^+$	1.4×10^6	1.3×10^6	1.2×10^6	1.2×10^6	2.5×10^6	8.2×10^6
Cottage cheese (4°C)						
Ped$^-$	1.0×10^7	9.3×10^6	4.2×10^7	1.8×10^8	8.2×10^8	4.3×10^9
Ped$^+$	8.7×10^5	8.2×10^5	7.7×10^5	7.5×10^5	7.5×10^5	1.3×10^5
Ice cream (4°C)						
Ped$^-$	1.5×10^7	1.5×10^7	7.0×10^7	2.9×10^8	1.3×10^9	5.7×10^9
Ped$^+$	1.4×10^6	1.3×10^6	1.2×10^6	1.2×10^6	2.5×10^6	7.8×10^6
Milk (4°C)						
Ped$^-$	1.3×10^7	1.4×10^7	6.1×10^7	2.5×10^8	1.0×10^9	5.0×10^9
Ped$^+$	1.3×10^6	1.2×10^6	1.2×10^6	1.2×10^6	2.1×10^6	7.3×10^6
Sausage mix (10°C)						
Ped$^-$	1.7×10^7	NT	1.3×10^8	8.0×10^8	4.4×10^9	NT
Ped$^+$	1.5×10^6	NT	8.9×10^6	5.1×10^7	2.6×10^8	NT

Cottage cheese (10°C)						
Ped^-	1.2×10^7	NT	7.8×10^7	5.4×10^8	3.2×10^9	NT
Ped^+	9.4×10^5	NT	5.0×10^6	3.2×10^7	1.9×10^8	NT

Note: Sterile food samples were inoculated with *L. monocytogenes* Scott A cells and treated with (Ped^+, 1350 AU/ml) or without (Ped^-) pediocin AcH, stored at 4°C or 10°C, and enumerated during storage up to 28 d.

[a] Table adapted from Reference 30.

TABLE 21
Influence of Pediocin AcH Concentrations on the Viability Loss and Growth of *L. Monocytogenes* in Milk[a]

Pediocin AcH (AU/ml)	cfu/ml after storage at 10°C for		
	1 h	7 d	14 d
0 (control)	1.4×10^7	8.6×10^7	6.1×10^8
1350	1.3×10^6	5.3×10^6	3.5×10^7
3000	1.0×10^6	2.8×10^6	8.3×10^6
4500	4.0×10^5	1.1×10^6	2.7×10^6
6000	2.0×10^5	5.4×10^5	1.9×10^6

Note: About 1.4×10^7 cells per milliter of *L. monocytogenes* were inoculated in sterile 10% nonfat dry milk, treated with 0 to 6000 AU/ml pediocin AcH, and stored at 10°C. During storage up to 14 d the samnples were enumerated for cfu. Cell death increased as pediocin AcH concentration increased. Surviving cells multiplied rapidly during storage at 10°C in all pediocin AcH concentrations.

[a] Table adapted from Reference 30.

acidilactici can be used in many different ways in raw and semipreserved foods to reduce the load of sensitive Gram-positive spoilage and pathogenic bacteria. However, before it can be used in foods more well-designed studies are needed. It appears that pediocin will have good application in both raw and minimally processed foods that are handled under proper sanitary conditions, vacuum packaged, and refrigerated to extend their shelf life. In these products pediocin of *P. acidilactici* in one form or another can be used as an additional barrier to ensure their safety and extend acceptance quality.

VII. LEGAL STATUS

Pediocins of *P. acidilactici* have not yet been approved for use as a food biopreservative by the regulatory agencies. It is not known if any commercial company has petitioned for the use of pediocin as a food preservative. However, a patent involving the use of pediocin PA-1 to prevent spoilage of salad and salad dressings by Gram-positive spoilage bacteria has been approved by the U.S. patent office.[54]

Pediocin-producing *P. acidilactici* strains have been isolated from naturally fermented meat products.[18,19] Many commercial strains have been in use in the controlled fermentation of meat for the last few decades.[20,21] In addition, a commercial strain was used to develop a patented method of preparing bacon in which P. *acidilactici* PO_2 was incorporated to reduce the nitrite level so as to effectively control *C. botulinum* growth.[55] This strain later was found to

TABLE 22
Addition of Pediocin PA-1 in Dairy Foods on the Viability Loss and Growth of *L. Monocytogenes*

Food type	cfu/g of *L. monocytogenes* after storage for[a] 1 d	7 d	14 d
Cottage cheese			
Ped–	7.5×10^3	1.0×10^2	NT[b]
Ped+	$<1.0 \times 10^2$	$<1.0 \times 10^2$	NT
Half-and-Half			
Ped–	1.6×10^4	5.4×10^6	2.0×10^8
Ped+	$<1.0 \times 10^2$	5.9×10^3	9.0×10^5
Cheese sauce			
Ped–	2.6×10^3	1.7×10^7	2.0×10^8
Ped+	$<1.0 \times 10^2$	1.0×10^2	3.0×10^4

[a] The food samples were inoculated with low levels of *L. monocytogenes* cells, and treated (Ped^+) with a dry preparation of pediocin PA-1 (10 Au/g), and stored at 4°C. Untreated samples (Ped^-) were used as controls. Pediocin treatment greatly reduced the cfu. During storage the surviving cells multiplied both in control and treated samples, except in cottage cheese up to 7 d.

From Pucci, M. J., Vedamuthu, E. R., Kunka, B. S., and Vandenberg, P. A., *Appl. Environ. Microbiol.*, 54, 2349, 1988, With permission.

be a producer of pediocin.[21] No adverse effect from the consumption of foods containing the *P. acidilactici* cells and metabolites, including pediocins of these strains, have been recorded. Also, consumption of fermented sausage produced with *P. acidilactici* H did not cause any problem.[51]

Animal injection studies and studies with hybridoma cells have also indicated no undesirable effect on pediocin AcH.[49,52] The limited evidence strongly suggests that pediocins of *P. acidilactici* strains, if consumed by humans at a specific level in the form of a preservative in foods, will not cause health hazards. Toxicological data that will result from feeding laboratory animals will probably be necessary before the regulatory agencies approve this as a food preservative. Nisin has been approved in many countries for use as a food preservative.[50] There is no reason why pediocin will not be approved for the same purpose, if the required information is available.

VIII. CONCLUDING REMARKS

Pediocins of *P. acidilactici* strains can be regarded to be a wide-spectrum bacteriocin against Gram-positive bacteria. Limited information, particularly with pediocin AcH, has suggested that it is effective against some bacterial spores and in sublethally injured Gram-negative bacteria. In addition, the characteristics of pediocin AcH have been well studied, providing especially the information that is necessary in its application as a food preservative. These include physical and chemical properties, stability, mode of action, some toxicological and antigenic properties, economic production, genetic control, and effectiveness in food systems. There is some confusion about the two names pediocin PA-1 from strain *P. acidilactici* PAC.1, and pediocin AcH from strains *P. acidilactici* H, E, and F isolated from fermented meat products, and strain M, obtained from Marshall Laboratories. Restriction enzyme cleavage patterns of the plasmid encoding for pediocin in strain H, E, F, and M have strongly suggested that they are the same. A comparison of the base sequences of the genes as well as the amino acid sequences of the two pediocins (PA-1 and AcH) will be important to resolve this confusion. The genes for pediocin PA-1 and pediocin AcH are present in small plasmids. It will be much easier to understand the mechanisms involved in the transcription, translation, and posttranslation processing of pediocin than that of similar bacteriocins of lactic acid bacteria such as nisin which is located in a polycistronic transposon.[48,48a] It will also be possible to design more effective pediocins through protein engineering as well as to develop desirable strains through transfer of plasmid in strains to produce more pediocin.

Several factors need to be recognized when using pediocin of *P. acidilactici* as a food biopreservative. It is inactivated by many proteolytic enzymes. The possible presence of proteolytic enzymes in raw foods as well as heat-stable proteases in heat-treated foods should be recognized. The presence of some anions such as Cl^-, PO_4^{3-} and HCO_3^- can reduce the effectiveness of antibacterial action of pediocin. However, this can be overcome by increasing the level of pediocin. As pediocins are hydrophobic molecules, uniform distribution in a food system containing large amounts of lipids may be difficult, especially in solid foods. However, heat treatment and proper dispersion of fat in food can be used to reduce this problem. Technological innovations will be necessary to overcome some of the problems, including uniform distribution of pediocin in food systems. This is more important in solid and semisolid foods such as in vacuum-packaged fresh beef.

Finally, in using pediocins as well as other bacteriocins of lactic acid bacteria as food biopreservatives, one should recognize that they produce a bacteriocidal effect by destabilizing the functional integrity of the cytoplasmic membrane of a sensitive bacterial cell. This effect is produced as soon as the molecule(s) come(s) in contact with a sensitive cell irrespective of the metabolic state of the cell. They do not have a bacteriostatic function. Theoretically, in a closed system without any chance of recontamination, the bacteriocidal effect will be produced within a very short time with a uniform dispersion of pediocin in sufficient concentrations that enables every sensitive bacterial cell to be in contact with pediocin. The cells surviving under these conditions are resistant to pediocin and are capable of multiplying even in the presence of residual pediocin in the system. This is also true for other bacteriocins, and is quite different from a metabolic-inhibitor type of preservative. A metabolic inhibitor produces its antibacterial effect only when a cell is metabolizing and thus it needs to be present in the environment in a stable form all the time during storage. Whenever a cell starts metabolic activity in a suitable environment, a metabolic inhibitor should be present to inhibit the function of the cell. However, when used against sensitive spores, residual pediocin will be important to kill the cell as soon as it emerges from a spore.

Pediocins and other bacteriocins under normal conditions are effective only against some Gram-positive bacteria among the different types of microorganisms that can be normally present in many foods. Benefits from their use as a food biopreservative can be enhanced in foods having low initial microbial load due to proper processing sanitation, and in foods that are stored in a controlled refrigerated atmosphere and are not recontaminated.

REFERENCES

1. **Bhunia, A. K., Kim, W. J., Johnson, M. C., and Ray, B.,** Partial purification and characterization of antimicrobial substances from *Pediococcus acidilactici* and *Lactobacillus plantarum*, Institute of Food Technologists Annual Meeting, June 16–19, Las Vegas, 1987.
2. **Collins, M. D., Willimas, A. M., and Wallbanks, S.,** The phylogeny of *Aerococcus* and *Pediococcus* as determined by 16 S rRNA sequence analysis: description of *Tetragenococcus Gen. Nov., FEMB Microbiol. Lett.*, 70, 255, 1990.
3. **Garvie, E. I.,** Genus *Pediococcus* Clanssen, 1903, in *Bergeys's Manual of Systemic Bacteriology*, Vol. 2, Sneath, P. H. A., Ed., William & Wilkins, Baltimore, 1986, 1075.
4. **Pederson, C. S.,** The genus *Pediococcus, Bacteriol. Rev.*, 13, 225, 1949.
5. **Bacus, J. N. and Brown, W. L.,** Use of microbial cultures: meat products, *Food Technol.*, 35(1), 74, 1981.
6. **Dacre, J. C.,** A note on pediococci in New Zealand Cheddar cheese, *J. Dairy Res.*, 25, 414, 1958.
7. **Law, B. A., Castanton, M., and Sharpe, M. E.,** The effect of nonstarter bacteria on the chemical composition and the flavor of Cheddar cheese, *J. Dairy Res.*, 43, 117, 1976.
8. **Bhowmik, T. and Marth, E. H.,** Role of *Micrococcus* and *Pediococcus* species in cheese ripening: a review, *J. Dairy Sci.*, 73, 859, 1989.
9. **Bhowmik, T., Riesterer, R., Van Boekel, M. A. J. S., and Marth, E. H.,** Characteristics of low fat Cheddar cheese made with added *Micrococcus* or *Pediococcus* species, *Milchwissenschaft*, 45, 230, 1990.
10. **Dellaglio, F., Trovatelli, L. D., and Sara, P. G.,** DNA/DNA homology among representative strains of the genus *Pediococcus, Zentralbl. Bakteriol. Hyg.*, 1 (Abstr.), Orig. C2, 140, 1981.
11. **Whittenbury, R.,** Hydrogen peroxide formation and catalase activity in the lactic acid bacteria, *J. Gen. Microbiol.*, 35, 13, 1964.
12. **Etchells, J. L., Costilow, R. N., Anderson, T. E., and Bell, T. A.,** Pure culture fermentation of brined cucumbers, *Appl. Microbiol.*, 12, 523, 1964.
13. **Fleming, H. P., Etchells, J. L., and Costilow, R. N.,** Microbial inhibition by an isolate of *Pediococcus* from cucumber brines, *Appl. Microbiol.*, 30, 1040, 1975.
14. **Rueckert, P. W.,** Studies on a Bacteriocin-Like Activity Produced by *Pediococcus* and Effective Against Gram-Positive Organisms, M.S. thesis, Michigan State University, East Lansing, 1979.

15. **Daeschel, M. A. and Klaenhammer, T. R.,** Association of a 13.6 MDal plasmid in *Pediococcus pentosaceus* with bacteriocin activity, *Appl. Environ. Microbiol.,* 50, 1538, 1985.
16. **Graham, D. C. and McKay, L. L.,** Plasmid DNA in strains of *Pediococcus cerevisiae* and *Pediococcus pentosaceus, Appl. Environ. Microbiol.,* 50, 532, 1985.
17. **Bhunia, A. K., Johnson, M. C., and Ray, B.,** Direct detection of an antimicrobial peptide of *Pediococcus acidilactici* in sodium-dodecyl sulfate-polyacrylamide gel electrophoresis, *J. Ind. Microbiol.,* 2, 319, 1987.
18. **Bhunia, A. K., Johnson, M. C., and Ray, B.,** Purification, characterization and antimicrobial spectrum of a bacteriocin produced by *Pediococcus acidilactici, J. Appl. Bacteriol.,* 65, 261, 1988.
19. **Ray, S. K., Johnson, M. C., and Ray, B.,** Bacteriocin plasmids of *Pediococcus acidilactici, J. Ind. Microbiol.,* 4, 163, 1989.
20. **Gonzales, C. F. and Kunka, B. S.,** Plasmid-associated bacteriocin production and sucrose fermentation in *Pediococcus acidilactici, Appl. Environ. Microbiol.,* 53, 2534, 1987.
21. **Hoover, D. G., Walsh, P. M., Kolaetis, K. M., and Daly, M. M.,** A bacteriocin produced by *Pediococcus* species associated with a 5.5 MDal plasmid, *J. Food Prot.,* 51, 29, 1988.
21a. **Hoover, D. G., Dishart, K. J., and Hermes, M. A.,** Antagonistic effect of *Pedioсiccus* spp. against *Listeria monocytogenes, Food Biotechnol.,* 3(2), 183, 1989.
22. **Scherwitz, K. M., Baldwin, K. A., and McKay, L. L.,** Plasmid linkage of a bacteriocin-like substance in *Streptococcus lactis* subsp. *diacetylactis* strain WM4: transferability of *Streptococcus lactis, Appl. Environ. Microbiol.,* 45, 1506, 1983.
23. **Pucci, M. J., Vedamuthu, E. R., Kunka, B. S., and Vandenbergh, P. A.,** Inhibition of *Listeria monocytogenes* by using bacteriocin PA-1 produced by *Pediococcus acidilactici* PAC 1.0 *Appl. Environ. Microbiol.,* 54, 2349, 1988.
24. **Bhunia, A. K., Johnson, M. C., Ray, B., and Kalchayanand, N.,** Mode of action of pediocin AcH from *Pediococcus acidilactici* on sensitive bacterial strains, *J. Appl. Bacteriol.,* 70, 25, 1991.
25. **Biswas, S. R., Ray, P., Johnson, M. C., and Ray, B.,** Influence of growth conditions on the production of a bacteriocin, pediocin AcH by *Pediococcus acidilactici* H, *Appl. Environ. Microbiol.,* 57, 1265, 1991.
26. **Motlagh, A. M., Johnson, M. C., and Ray, B.,** Viability loss of foodborne pathogens by starter culture metabolites, *J. Food Prod.,* 54, 873, 1991.
27. **Motlagh, A. M., Holla, S., Johnson, M. C., Ray, B. and Field, R. A.,** Inhibition of *Listeria* spp. in sterile food systems by pediocin AcH, a bacteriocin from *Pediococcus acidilactici* H, *J. Food Prot.,* accepted, 1992.
28. **Kalchayanand, N.,** Extension of Shelf-Life of Vacuum-Packaged Refrigerated Fresh Beef by Bacteriocins of Lactic Acid Bacteria, Ph.D. thesis, University of Wyoming, 1990.
29. **Rozbeh, M.,** Extension of Shelf-Life in Vacuum-Packaged Fresh Beef, M.S. thesis, University of Wyoming, Laramie, 1990.
30. **Holla, S.,** Efficiency of Pediocin AcH on Viability Loss of Pathogenic and Spoilage Bacteria in Food, M.S. thesis, University of Wyoming, Larmie, 1990.
31. **Motlagh, A. M.,** Antimicrobial Efficiency of Pediocin AcH and Genetic Studies on the Plasmid Encoding for Pediocin AcH, Ph.D. thesis, University of Wyoming, Laramie, 1991.
32. **Nielson, J. W., Dickson, J. S., and Crouse, J. D.,** Use of bacteriocin produced by *Pediococcus acidilactici* to inhibit *Listeria monocytogenes* associated with fresh meat, *Appl. Environ. Microbiol.,* 56, 2142, 1990.
33. **Yousef, A. E., Luchansky, J. G., Degnan, A. J., and Doyle, M. P.,** Behavior of *Listeria monocytogenes* in wiener exudates in the presence of *Pediococcus acidilactici* H, *Appl. Environ. Microbiol.,* 57, 1265, 1991.
34. **Kim, Wang-June,** Development of Plasmid Transfer System in Lactic Acid Bacteria Used in Meat Fermentation, Ph.D. thesis, University of Wyoming, Laramie, 1990.

35. **Lewis, C. B., Kaiser, A., and Montville, T. J.,** Inhibition of foodborne bacterial pathogens by bacteriocins from lactic acid bacteria isolated from meat, *Appl. Environ. Microbiol.*, 57, 1683, 1991.
36. **Ray, B., Kalchayanand, N., and Field, R. A.,** Isolation of a *Clostridium* spp. from spoiled vacuum-packaged refrigerated beef and its susceptibility to bacteriocin from *Pediococcus acidilactici, Proc. 35th Int. Congr. of Meat Science and Technol.*, Copenhagen, Denmark, Danish Meat Research Instititue, Roskilde, Denmark, 1989, 285.
37. **Harris, L. J., Daeschel, M. A., Stilles, M. E., and Klaenhammer, T. R.,** Antimicrobial activity of lactic acid bacteria against *Listeria monocytogenes, J. Food Prot.*, 52, 3784, 1989.
38. **Berry, E. D., Liewen, M. B., Mandigo, R. W., and Hutkins, R. W.,** Inhibition of *Listeria monocytogenes* by bacteriocin producing *Pediococcus* during the manufacture of fermented semidry sausage, *J. Food Prot.*, 53, 194, 1990.
39. **Berry, E. D., Hutkins, R. W., and Mandigo, R. W.,** The use of bacteriocin-producing *Pediococcus acidilactici* to control postprocessing *Listeria monocytogenes* contamination of frankfurters, *J. Food Prot.*, 54, 681, 1991.
39a. **Ray, B.,** unpublished data, University of Wyoming, Laramie.
40. **Okereke, A. and Montville, T. J.,** Bacteriocin inhibition of *Clostridium botulinum* spores by lactic acid bacteria, *J. Food Prot.*, 54, 349, 1991.
41. **Bhunia, A. K.,** Characteristics and Mode of Action of an Antimicrobial Peptide Produced by *Pediococcus acidilactici* Strain H, Ph.D. thesis, University of Wyoming, Larmie, 1989.
42. **Barefoot, S. F. and Klaehammer, T. R.,** Detection and activity of lactacin B, a bacteriocin produced by *Lactobacillus acidophilus, Appl. Environ. Microbiol.*, 45, 1988, 1983.
43. **Ramseir, N. R.,** The action of nisin on *Clostridium butyricum, Arch. Microbiol.*, 37, 57, 1960.
43a. **Wicken, A. J. and Knox, K. W.,** Lipoteichoic acid: a new class of bacterial antigen, *Science*, 187, 1161, 1975.
44. **Johnson, M., Ray, B., and Speck, M. L.,** Freeze-injury in cell wall and its repair in *Lactobacillus acidophilus, Cryo Lett.*, 5, 171, 1984.
44a. **Ray, B. and Johnson, M. C.,** Freeze-drying injury of surface layer protein and its protection in *Lactobacillus acidophilus, Cryo. Lett.*, 7, 210, 1986.
44b. **Brennen, M., Wanismail, B., Johnson, M. C., and Ray, B.,** Cellular damage in dried *Lactobacillus acidophilus, J. Food Prot.*, 49, 47, 1986.
45. **Ray, B.,** *Injured Index and Pathogenic Bacteria,* CRC Press, Boca Raton, FL, 1989, chap. 1.
46. **Kalchayanand, N., Ray, B., Field, R. A., and Johnson, M. C.,** Spoilage of vacuum-packaged refrigerated beef by *Clostridium, J. Food Prot.*, 52, 424, 1989.
47. **Motlagh, A. M., Buhunia, A, K., Szostek, F., Hansen, T. R., Johnson, M. C. and Ray, B.,** Nucleotide and amino acid sequence of *pap* gene (peiocin AcH production in *Pediococcus acidilactici* H, *Letters Appl. Microbiol.*, accepted, 1992.
48. **Buchman, G. W., Banerjee, S., and Hansen, J. N.,** Structure, expression and evaluation of a gene encoding the precursor of nisin, a small protein antibiotic, *J. Biol Chem.*, 263, 16260, 1988.
48a. **Steen, M. T., Chung, Y. J., and Hansen, J. N.,** Characterization of the nisin gene as a part of a polycistronic operon in chromosome of *Lactococcus lactis* ATCC 11454, *Appl. Environ. Microbiol.*, 57, 1181, 1991.
49. **Bhunia, A.,** Studies on antigenic property and amino acid sequence of pediocin AcH from *P. acidilactici* H, personal communication, 1991.
49a. **Van Belkum, M. J.,** Lactococcal Bacteriocins: Genetics and Mode of Action, Ph.D. thesis, University of Gorningen, The Netherlands, 1991, 22.
49b. **Vandenburgh, P. A.,** Pediocin PA-1, produced by *Pediococcus acidilactici,* in *Proc., Annual Meeting Society Industrial Microbiology,* Philadelphia, 1991, S101.

50. **Delves-Broughton, J.,** Nisin and its uses as a food preservative, *Food Technol.,* 44(11), 100, 1990.
51. **Wu, W. H., Rule, D. C., Busborm, J. R., Field, R. A., and Ray, B.,** Starter culture and time/temperature of storage influences on quality of fermented mutton sausage, *J. Food Sci.,* 56, 916, 1991.
51a. **Wu, Wen-Hsin,** Evaluation of Fermented Mutton Sausages Made with the Different Lactic Acid Starter Cultures, M.S. thesis, University of Wyoming, Laramie, 1989.
52. **Bhunia, A. K., Johnson, M. C., Ray, B., and Belden, E. L.,** Antigenic property of pediocin AcH produced by *Pediococcus acidilactici* H, *J. Appl. Bacteriol.,* 69, 21, 1990.
53. **Gonzalez, C. F. and Kunka, B. S.,** Plasmid transfer in *Pediococcus* spp.: intergeneric and intragenic transfer of pIP501, *Appl. Environ. Microbiol.,* 46, 81, 1983.
54. **Gonzalez, C. F.,** Methods for Inhibiting Bacterial Spoilage and Resulting Composition, U.S. Patent 4883673, 1989.
55. **Tanaka, N., Meske, L., Doyle, M. P., Traisman, E., Thayer, D. W., and Johnson, R. W.,** Plant trials of bacon made with lactic bacteria, sucrose and lowered sodium nitrite, *J. Food Prot.,* 48, 679, 1985.
56. **Kim, W. J., Ray, B., and Johnson, M. C.,** Plasmid transfer by conjugation and electroporation in *Pediococcus acidilactici, J. Appl. Bacteriol.,* 72, 201, 1992.
57. **Ray, B., Nie, H., and Menke, S.,** Mapping and cloning of pSMB74, a plasmid encoding pediocin AcH production in *Pediococcus acidilactici* H, *Annual Meeting, American Society Microbiology,* Dallas, 0.31, (Abst.) 263, 1991.
58. **Ray, B., Motlagh, A. M., and Bozoglu, F.,** Mapping of pSMB74, a plasmid encoding bacteriocin pediocin AcH production (Pap^+) phenotype in *Pediococcus acidilactici* H, *Lett. Appl. Microbiol.,* submitted, 1992.
59. **Ray, S. K., Kim, W. J., Johnson, M. C., and Ray, B.,** Conjugal transfer of a plasmid encoding bacteriocin production and immunity in *Pediococcus acidilactici* H, *J. Appl. Bacteriol.,* 66, 393, 1989.
59a. **Johnson, M. C., Hanlin, M. B., and Ray, B.,** Low pH and lactate are necessary for conversion of prepediocin to active pediocin AcH in *Pediococcus acidilactici* H, *American Society Microbiology Annual Meeting,* (Abstr.) 1992.
60. **Kalchayanand, N., Ray, B., and Field, R. A.,** Characteristics of *Clostridium laramie,* a new psychrotrophic species associated with spoilage of vacuum-packaged fresh beef, *Proc., Western Section, American Society Animal Science Annual Meeting,* University of Wyoming, Laramie, 42, 26, 1991.
61. **Kalchayanand, N., Ray, B., Field, R. A., and Johnson, M. C.,** Spoilage of vacuum-packaged refrigerated beef by a *Clostridium* spp., *J. Food Prot.,* 52, 424, 1989.
62. **Owens, J., Kalchayanand, N., and Ray, B.,** Influence of biopreservatives in reducing spoilage and pathogenic bacteria in fresh beef byproducts and processed meat, unpublished data, University of Wyoming, Laramie.

Chapter 11

BACTERIOCINS OF LACTIC ACID BACTERIA

Mark A. Daeschel

TABLE OF CONTENTS

I. INTRODUCTION

Many lactic acid bacteria (LAB), and more recently members of the propionibacteria, have been observed to synthesize and excrete bacteriocins. Bacteriocins are protein-containing macromolecules that can either kill susceptible bacteria by membrane lysis or by inhibition of essential cell processes such as protein synthesis or DNA synthesis. Many bacteriocins, such as the colicins produced by *Escherichia coli,* have been thoroughly characterized in terms of structure, synthesis, mode of action, and genetic determinants.[1] Information regarding bacteriocins from bacteria used as starter cultures in foods is beginning to accumulate in sufficient detail and depth, to the point where potential applications are being investigated. The bacteriocin nisin was the first to be characterized from food bacteria, and it continues to be the most studied in terms of chemistry, molecular biology, and applications. However, interest in food-grade bacteriocins other than nisin is strong and continues to gain momentum. The potential application of these macromolecules in science and food preservation is enormous and exciting. Molecular biology has the tools available to custom tailor these molecules to a vast array of application requirements. In the coming decade it is likely that progress in bacteriocin research will result in a whole generation of food preservatives and processing aids that provide economic and safe alternatives to the current cadre of chemical preservatives. It is the objective of this chapter to provide a brief review of some of the investigations of the last decade on bacteriocins from food bacteria. It is not the intent to cover all investigations and information, but rather to provide an overview of a cross section of representative bacteriocins from the four genera of the LAB. A complete listing of described bacteriocins from LAB is given previously, in Chapter 8, and information regarding nisin and pediocin AcH is also treated elsewhere in separate chapters.

II. *LACTOCOCCUS*

The newly established genus name *Lactococcus* is reserved for species formerly referred to as the lactic streptococci, dairy streptococci, or the group N streptococci. It should be noted that literature prior to 1988 describing bacteriocins from these microorganisms will have the old nomenclature. In the following discussion the author will use the current nomenclature exclusively. Bacteriocinogenicity has been described for the three most economically important *Lactococcus* species: *L. lactis* subsp. *lactis, L. lactis* subsp. *diacetylactis,* and *L. lactis* subsp. *cremoris*. The importance of these microorganisms to the dairy industry cannot be overstated. The marketplace identity of most cultured dairy products is directly dependent on the action of these microorganisms. The preservative effect these microorganisms instill to cultured dairy products is well established and may in some instances be due to the presence of bacteriocins. Nisin produced by *L. lactis* subsp. *lactis* is intentionally used as a permissible additive to prevent the outgrowth of

Clostridium botulinum spores in dairy foods. Other bacteriocins produced by the Lactococci, such as diplococcin may also have application in food preservation.

A. *Lactococcus Lactis* Subsp. *Cremoris*

The initial recognition of the antimicrobial properties of *L. lactis* subsp. *lactis* in 1928 by Rogers[2] was followed five years later by reports of antagonistic properties exhibited by *Lactococcus lactis* subsp. *cremoris*. The report of Whitehead[3] focused on describing how certain Lactococci could inhibit the acid-producing capability of New Zealand cheesemaking strains. Two subsequent reports[4–5] from New Zealand ascribed the antagonistic activity to a proteinaceous metabolite produced by *L. lactis* subsp. *cremoris*. The inhibitor, termed diplococcin, was partially purified by Oxford,[6] who described it as being a water-soluble protein and as being heat-stable under acidic conditions. Comparisons of diplococcin and nisin established important differences between the two. Diplococcin was discussed by Hirsch[7] as being active against only closely related species within its genus, whereas nisin displayed activity against many different Gram-positive bacteria. Another significant difference was that nisin caused cell membrane lysis in some susceptible cells whereas diplococcin rapidly killed susceptible cells, without lysis. Interest in diplococcin waned after World War II, likely because of its narrow spectrum of activity, and like nisin it had little application as a clinical antibiotic. The next investigation into diplococcin[8] appeared 37 years after Oxford's report and provided characterization as to the molecular weight and amino acid composition of the molecule. In contrast to nisin, diplococcin did not contain the sulfur-containing amino acids, lanthionine, and β–methyllanthionine that are diagnostic of nisin. A molecular weight of 5300 was estimated for Diplococcin, as compared to 3350, their estimate for nisin. This investigation[8] also confirmed the narrow spectrum of activity of diplococcin as being restricted to that within the *Lactococcus* genus. The investigators[8] discussed an approach for designing desirable cheese-making strains of *L. lactis* subsp. *cremoris* by making them free or cured of the diplococcin-producing ability so they would not dominate over or inhibit other cultures in multiple starter systems. Nondiplococcin-producing variants of *L. lactis* subsp. *cremoris* 346 were isolated following growth at elevated temperatures.[9] Subsequently, these variants were demonstrated to be suitable for cheese production, having retained desirable properties in that regard.[9] Genetic conjugal transfer of diplococcin-producing ability into other *L. lactis* subsp. *cremoris* strains was demonstrated by Davey et al,[10] however plasmid analysis did not suggest a plasmid-borne determinant for diplococcin. In a subsequent study by Davey[11] using two different plasmid free-conjugation recipients, the acquisition of a 54 MDl plasmid was correlated with subsequent diplococcin production. Additional proof[11] of plasmid involvement was seen as the loss of the plasmid from transconjugants coupled with loss of diplococcin production, and the appearance of diplococcin sensitivity.

Workers in Poland described a series of bacteriocins called lactostrepcins from *Lactococcus* spp. These bacteriocins all had in common the properties of being active only under acidic conditions (pH< 5), of being inactive at neutral pH, and of regenerating activity when acid conditions were restored. The lactostrepcins were also described as having a narrow spectrum of activity, generally being limited to other *Lactococcus* spp., with some exceptions being observed. Lactostrepcin 5, produced by *Lactococcus lactis* subsp. *cremoris* 202, was purified approximately 170-fold and used in experiments to determine the mode of killing. The bacteriocidal effect of lactostrepcin 5 was shown with survival curves to exhibit multihit kinetics where it appeared that more than one molecule of bacteriocin was necessary to kill each cell. The effect of lactostrepcin 5 on cell processes was seen to slow down and then stop DNA, RNA, and protein synthesis. Other observed effects included a decrease in intercellular ATP content to less than 1% within 30 min after exposure, an inhibition of uridine transport, and a leakage of potassium ions. These investigators concluded that lactostrepcin 5 acts by modifying the integrity of the cell membrane, causing disruption of ionic gradients, which lead to disruption of synthetic functions and cellular death. Information regarding chemical composition of lactostrepcin 5 or genetic determinants for production have not to date appeared in the literature.

Geis et al.[14] conducted a comprehensive survey of a large number of dairy lactococci for bacteriocin-mediated antagonistic activity. Eight strains of *L. lactis* subsp. *cremoris* were observed to be bacteriocinogenic and were categorized into four types based on heat stability and inhibition spectrum. Some bacteriocins from the cremoris strains resembled Diplococcin based on heat sensitivity at alkaline pH. The spectrum of activity of the strains was limited to other lactic acid bacteria with the exception that all were active against some of the clostridia strains tested. All bacteriocin producers possessed one or more plasmids that were investigated in a later study as possible genetic determinants for bacteriocin production. Two bacteriocinogenic strains of *L. lactis* subsp. *cremoris* (9B4 and 4G6) were used as donors in conjugal matings with plasmid-free recipients. Transconjugants that acquired bacteriocin-producing ability also acquired a 39.6-MDl (60 Kb) plasmid. Direct correlation of the loss and regaining of bacteriocinogenicity with disappearance and appearance of the plasmid provided proof that bacteriocin production in these strains was plasmid mediated and transmissible.[15]

Plasmid 9B4-6, containing genetic determinants for bacteriocin production and immunity in *L. lactis* ssp. 9b4 was analyzed with restriction enzymes and fragments cloned into shuttle vectors.[16] Two different fragments (1.8 kb and 15.0 kb) each expressed antagonistic activity in *L. lactis* subsp. *lactis* IL1403. Each activity had a different inhibition specificity, however hybridization experiments indicated a high degree of homology between a 4-kb fragment of the 15-kb fragment with the 1.8-kb fragment. Both immunity and bacteriocin production were localized to the original 1.8-kb fragment and to a 1.3-kb subfragment of the 4.0-kb fragment. This was the first observation that two distinct antagonistic activities, and associated immunities, were determined by a single plasmid in Lactococcal species. In a subsequent study[17] the same research group sequenced and analyzed, by deletion and frameshift mutation analyses, the regions of p9B4-6 which encoded bacteriocin production and immunity. Sequence data on the 1.8-kb fragment indicated three open reading frames (ORFs), with the first two being responsible for bacteriocin production and the other one for immunity. The 1.3-kb fragment had two ORFs, one apparently for bacteriocin production and the other for immunity. The sequence of the two immunity ORFs did not have significant similarity, however, a 20-base sequence of the ORFs responsible for bacteriocin production from the 1.8-kb and 1.3-kb fragments were identical. These observations prompted the authors to suggest that transposition-like rearrangements may be occurring within the cells' genome. It is hoped that further elucidation of the genetic organization and regulation of this bacteriocin plasmid is forthcoming shortly. Information on the mode of action of the p9B4-6 bacteriocins, how the immunity products function, and the mechanism and range of conjugal transmission are areas yet to be explored.

B. *Lactococcus Lactis* Subsp. *Diacetylactis*

Antagonistic activities associated with *L. lactis* subsp. *diacetylactis* have been reported in the literature by several workers.[18–20] However, studies indicating the production of bacteriocin-like substances did not appear until the report of Scherwitz et al.[21] who observed the transfer of antagonistic activity from *L. lactis* subsp. *diacetylactis* WM4 to *L. lactis* subsp. *lactis*. The described inhibitor had a different protease sensitivity pattern then that of nisin, but similar to that of diplococcin. The inhibitor had a narrow spectrum of activity being limited to other Lactococcal species. Genetic studies suggested that bacteriocin production by the WM4 strain was mediated by an 88-Mdal plasmid (131.1 kb) referred to as pNP2. Plasmid curing experiments using novobiocin generated one colony that was bacteriocin negative and that was missing the 131.1-kb plasmid. This observation, coupled with the appearance of bacteriocinogenicity and pNP2 in transconjugants, gave strong evidence of plasmid involvement. In a subsequent study, Harmon and McKay [22] cloned restriction fragments of pNP2. Three *Bcl*I fragments cloned into the vector pGB301 and transformed into *L. lactis* subsp. *lactis* LM0230 yielded a bacteriocinogenic transconjugant having a recombinant plasmid pKSH1. Apparent differences between the expression level of the bacteriocin genes were observed between pNP2- and PKSH1-containing strains, however it was not determined why this occurred.

A recent report[23] describes a bacteriocin-producing strain of *L. lactis* subsp. *diacetylactis*. The bacteriocin designated S50 was sensitive to proteases, stable to heating (110°C/60 min), and like other bacteriocins from *L. lactis* subsp. *diacetylactis*, was inhibitory only toward other lactococcal spp. The inhibitor was active against a nisin-producing strain, providing evidence that it itself was not nisin. Evidence was not presented that would rule out that the inhibitor was diplococcin. The activity spectrum and protease sensitivity pattern were characteristic of diplococcin, however diplococcin was characterized by Davey[8] as losing 75% of its activity after 1 min at 100°C, which is in sharp contrast to the heat stability reported for bacteriocin S50. More complete characterizations of this inhibitor would be useful for definitive comparisons.

III. *LACTOBACILLUS*

Lactobacillus is the largest of the four genera that constitute the lactic acid bacteria. Metabolically they are also the most diverse genus, consisting of species that are obligate homofermenters, heterofermeters, and facultative heterofermenters with respect to glucose dissimilation. The lactobacilli are ubiquitous in nature and occupy a diverse range of environmental niches, including the surfaces of plants, decaying vegetation, sewage and manure, and the gastrointestinal tracts of various mammals.[24] They are also able to grow over a wide range of temperatures (5–55°C), are more acid tolerant than the other LAB genera, and are able to degrade a wide variety of organic acids and sugars.[24] They are important commercially and are used for the production of lactic acid, are employed in vitamin bioassays, and are used as starter cultures for fermented meats, fermented vegetables, sourdough breads, silage, and cultured dairy products. They are also implicated as spoilage agents in certain commodities such as packaged meats, beers, and wines.

Bacteriocinogenicity has been described for several of the obligate homofermenters *(L. acidophilus, L. helveticus,* and *L. fermentum),* for the facultative heterofermenters *(L. casei, L. plantarum, L. sake),* and most recently for the heterofermentative *L. brevis.* It is likely that more lactobacilli species will be discovered to be bacteriocinogenic, considering that there are more then 40 species within the genus.

A. *Lactobacillus Acidophilus*

Antimicrobial activity associated with *Lactobacillus acidophilus* has been observed and reported since the 1960s. The earliest reports[25–28] attributed the antagonism by *L. acidophilus* either to hydrogen peroxide, to acid production, or to what was termed antibiotic substances. A study by Gilliland and Speck[28] concluded that the antimicrobial effect of a strain of *L. acidophilus* against a group of bacteria including *Salmonella typhimurium* and *Staphylococcus aureus* was partially due to hydrogen peroxide, but also concluded that other undefined substances were responsible for inhibition. These undefined substances were not systematically characterized to the point that conclusive statements could be made as to the chemical nature of the inhibitors. The report of Vincent et al.[25] described an inhibitory substance that appeared to have the characteristics of a bacteriocin as we know them today. The authors described the inhibitor as being nondialyzable, active at neutral pH, and unaffected by catalase. Mehta et al.[29,30] reported on the properties and purification of an inhibitory protein from *L. acidophilus* AC_1. The inhibitor was described as being active over a pH range of 4.0 to 7.5 , but also as being heat labile, with complete inactivation occurring within 20 min at 50°C. A partially purified inhibitor preparation subjected to gel filtration gave a molecular weight estimate of 5400 for the active component. Like Vincent, et al.,[25] they reported the inhibitor as being active against both Gram-positive and Gram-negative bacteria and obeservation which has not been made for other bacteriocins from lactobacilli.

More definitive characterizations of bacteriocins from *L. acidophilus* were first reported in 1983[31] with a description of lactacin B, a bacteriocin produced by *L. acidophilus* N2. The investigators observed antagonistic activity in 63% of 57 strains of *L. acidophilus* examined. An inhibitor from *Lactobacillus acidophilus* N2 was further examined and found to be bacteriocidal without lysis, degraded by proteolytic enzymes, and stable to heating (100°C/60 min) at pH 5. The inhibitor was shown to be active only against certain *Lactobacillus* species. In the same study, ultrafiltration experiments provided a molecular weight of 100,000 for the inhibitor, however, further purification in a later study [32] using ion-exchange chromotography and gel filtration gave a molecular weight in the range of 6000 to 6500 for the purified inhibitor. Biological activity was destroyed by proteinase K and the purified compound gave maximum absorbance at 211 nm, indicative of peptide bonds. These characteristics provided further confirmation that the inhibitor was a bacteriocin. Genetic studies[32] did not provide evidence that lactocin B was a plasmid-determined gene product, which prompted the investigators to suggest that a chromosomal determinant may encode for the bacteriocin.

In the survey of Barefoot and Klaenhammer,[31] which screened 52 strains of *L. acidophilus* for antagonistic activity, a strain designated 11088 was observed to produce an inhibitor that was broader then that of the Lactocin-B producing *L. acidophilus* N2. The inhibitory activity of *L. acidophilus* 11088 (88) was examined by Muriana and Klaenhammer[33] and found to be characteristic of bacteriocins. The inhibitor, designated as lactacin F, was degraded by proteolytic enzymes, but was unaffected by sterilization conditions (121°C/15 min). Production of lactacin F was very pH independent with minimal production at pH 5.0 or 7.5 and maximal production at pH 7.0. Lactacin F was progressively purified by ammonium sulfate precipitation, gel filtration, and reverse phase HPLC. A key purification step was recovery of almost all activity in a floating pellicle upon saturation of culture supernatants with ammonium sulfate (35–40%). These observations indicated that native lactacin F was part of a large macromolecular complex, even visible under electron microscopy as particles having an approximate size of 25 to 75 nm.[34] Lactacin F purified from the large complexes was sized by SDS-Page as a 2500 dalton peptide, whereas amino acid compositional analysis indicated a peptide of 56 amino acids (6200 molecular weight).[34] The size discrepancy was discussed by the investigators as probably being due to abnormal electrophoretic behavior of small hydrophobic peptides. In a subsequent study,[35] cloning experiments and sequencing provided evidence that lactacin F was a 57-amino acid peptide.

The initial genetic studies[33] of lactacin F production focused on documenting the genetic conjugal transfer of plasmids-encoding production and producer-cell immunity. Selection of bacteriocinogenic transconjugants was achieved by using a preparation of lactacin F in media as a selective agent for lactocin F immunity. Conjugal transfer of lactacin F and immunity determinants were concluded to be facilitated by a mobilization event in which the chromosomal determinants for lactocin F and immunity integrated into plasmids that, upon transfer, reintegrated the determinants into the recipient genome. A 63-mer probe based on the amino acid sequence of lactacin F was synthesized and found to be adequate to sufficiently detect lactocin F DNA sequences that are conjugally transferred. A shuttle vector was constructed and used to transfer by electroporation a clone containing a fragment encoding for lactacin F and immunity into nonbacteriocinogenic recipients. Expression of lactacin F production and immunity in transformants was documented, and characteristics of the expressed bacteriocin (heat stability and activity spectrum) were identical to lactacin F. This study was the first to sequence the DNA and clone the genes responsible for bacteriocin production and immunity from a *Lactobacillus* species.

The most recent description of antagonistic activity associated with *L. acidophilus* due to a proteinaceous substance was reported by Toba et al.,[36] who termed the inhibitor acidophilucin A. Culture supernatants of *L. acidophilus* LAPT 1060 were used in agar well diffusion assays to demonstrate antagonism. Antagonistic activity was lost when proteolytic enzymes (actinase E and trypsin) were added to culture supernatants. In a limited survey to determine the spectrum of activity, antagonism was evident toward several strains of *L. delbruckii* subsp. *bulgaricus* and *L. helveticus,* but not against *Escherichia coli*, *Staphylococcus aureus,* and *Bacillus subtilis.* The authors considered acidophilucin A to be a new, previously undescribed *L. acidophilus* bacteriocin based on its different heat sensitivity (inactivated at 70°/10 min) in contrast to the greater stability of lactacin B and F.

B. *Lactobacillus Brevis* and *Lactobacillus Casei*

Rammelsberg and Radler [37] screened 79 strains of *Lactobacillus* species isolated from plants and food fermentations and observed antagonistic activity in 12 strains displayed against other lactic acid bacteria. The inhibitory activities of two of the strains *Lactobacillus brevis* B37 and *Lactobacillus casei* B80, were further characterized in terms of sensitivity to heat, proteolytic enzymes, chloroform, filtration, and activity spectrum. The activity spectrum of the inhibitor from *L. brevis* 37 (brevicin 37) was deduced from testing its activity against 145 Gram-positive bacteria, 42 Gram-negative bacteria, and 34 yeasts. *L. brevis* B37 culture supernatants were observed to be active only against lactic acid bacteria, with the exception of *Nocardia corollina,* which was also sensitive. A much more narrow activity spectrum was observed for *L. casei* supernatants (casecin 80), which were restricted to being inhibitory to only 1 of 24 *L. casei* strains tested.The activity of brevicin 37 and casecin 80 were degraded by pronase E and trypsin, in contrast to nisin which was insensitive to trypsin. In terms of heat stability, brevicin 37 was stable at 121°C for 1 h, while casecin 80 was no longer active when held for 10 min at 60°C. The activity of brevicin 37 was lost upon exposure to chloroform and during filtration through 0.2 µm cellulose nitrate membranes. These observations led the investigators to suggest that the inhibitor may have a lipid component essential for activity. The inhibitors described above clearly fall within the criteria for bacteriocins and each appeared to be distinctly different from each other and from nisin.

Rammelsberg et al.[38] further pursued the characterization of casecin 80 in preference to brevicin 37 probably because of its superior stability during laboratory manipulations. Casecin 80 production was optimized to obtain inhibitor suitable for purification by growing *L. casei* 80 in a synthetic medium supplemented with peptone. Mytomycin C (0.5–1.0 µg/ml) was used to increase casecin 80 production by 5 to 7-fold, presumably by an induction-type event. Casecin 80 was initially concentrated by ultrafiltration and then purified by ion exchange (cellulose SE-23) chromotography. Molecular sizing of purified casecin 80 was accomplished with gel filtration that gave an apparent molecular mass (mr) of between 40,000 and 42,000. The isoelectric point of casecin 80 determined by isoelectric focusing in polyacrylamide was given as 4.5. The inhibitory effect of casecin 80 was observed to be more than bacteriostatic, but not strongly bacteriocidal. In an experiment in which 10^3 cells of a sensitive strain of *L. casei* were added to a tomato juice medium containing casecin 80, growth was halted with a decrease in population of about 60% after 24 h exposure. Experiments to determine if casecin 80 killed susceptible cells via a lytic mechanism were not included in their study. Both casecin 80 and brevicin 37 represent the first described bacteriocins from *L. casei* and *L. brevis,* and the information presented regarding these inhibitors will be valuable in understanding bacteriocinogenicity within this commercially important genus.

C. *Lactobacillus Helveticus*

Lactobacillus helveticus is an obligate homofermentative species closely related to *L. acidophilus*[24] and is used commercially as a starter for the production of Swiss cheese and other cultured dairy products. Two bacteriocins have been characterized from the species: lactocin 27,[39,40] and heveticin J.[41,42]

Upreti and Hinsdill[39] observed antagonistic activity in 7 of nearly 100 lactobacilli strains screened and went on to characterize the inhibitory activity associated with *L. helveticus* LP27. The inhibitory activity, isolated from culture supernatants, was described as a protein-lipopolysaccharide complex. Denaturing of the complex with sodium dodecyl sulfate resulted in the isolation (gel filtration) of a 12,400-molecular weight glycoprotein identified as the native inhibitor. Evidence suggesting that the inhibitor could be categorized as a bacteriocin was its sensitivity to proteolytic enzymes (trypsin and pronase), its being nondialyzable, its and having a narrow spectrum of activity (restricted to *L. helveticus* and *L. acidophilus*). The purified inhibitor (lactocin 27) was described as very heat stable, with no loss in activity being measured after heating for 1 h at 100°C, similar to tthe results reported for lactacin B[31] and F[33]. The amino acid composition of lactacin 27 was determined and found to contain few, if any, sulfur-containing residues. Treatment of lactocin 27 with reductive S-carboxymethylation, to disrupt any potentially existing disulfide bonds, did not diminish the activity of inhibitor. This line of evidence suggests in retrospect, that lactocin does not have the thioether linkages that nisin has. With further study of lactocin 27[40] it was concluded that the inhibitor manifests bacteriostatic rather then bacteriocidal effects on sensitive bacteria. The mode of action was described as inhibition of protein synthesis, but with no effect on DNA or RNA synthesis or ATP generation. Lysis of cell membranes was not observed, however leakage of potassium ions was noted. Evidence was not obtained to suggest specific cellular receptors for lactocin 27 on the surface of sensitive cells. The inhibitor was reported to bind indiscriminately to sensitive or insensitive cells. Genetic studies (plasmid-curing experiments) failed to implicate plasmids as determinants for lactocin 27 production.

The second bacteriocin (helveticin J) characterized from *L. helveticus* was identified by Joerger and Klaenhammer[41,42] who successfully purified the inhibitor, sequenced the structural gene, and cloned and expressed the gene in a different species.

The bacteriocinogenic properties of helveticin J were described as: sensitivity to proteolytic enzymes (pronase, trypsin, pepsin, ficin, and subtilisin), bacteriocidal mode of action, and active against closely related species. In contrast, lactocin 27 was not deactivated by ficin and had greater stability to heating the helveticin J (inactivated within 30 min at 100°C). Similar to many of the lactobacilli bacteriocins previously described in this review, helveticin J was isolated as a large aggregate with a molecular weight greater than 300,000. Subsequent purification by gel chromotography and SDS-PAGE identified a 37,000 molecular weight protein as responsible for antibacterial activity. *L. helveticus* was found to harbor one plasmid designated pMJ1008 that was evaluated as a potential genetic determinant for helveticin J. Plasmid-curing experiments generated plasmid-free variants that still maintained bacteriocinogenicity. To ensure that curing procedures did not cause pMJ1008 to integrate into the chromosome, thus masking its extrachromosomal state as a determinant for helveticin J, pMJ1008 probes were constructed, but did not hybridize to genomic DNA from the plasmid-free variants. This evidence, coupled with identical phenotypic characteristics between *L. helveticus* 481 and the plasmid-free variants, provided conclusive proof that helveticin J production and producer immunity were chromosomally determined.

A subsequent study[42] confirmed this conclusion when lambda phage libraries of *L. helveticus* 481 chromosomal DNA were prepared and screened for fusion proteins that would react to polyclonal antihelveticin J IgG. Two phage clones produced protein that gave positive reactions when treated with antibody. The complete helveticin J gene was isolated, sequenced, and two complete open reading frames were identified, the second of which could encode a protein the size of helveticin J. An immunity protein was suggested for the first open reading frame, although such a protein has not been identified. A recombinant plasmid was constructed to contain a DNA fragment encompassing both open reading frames. The plasmid was transformed into, and expressed by, *L. acidophilus* NCK 48. The protein(s) expressed displayed an inhibition spectrum and heat sensitivity characteristic of helveticin J.

D. *Lactobacillus Plantarum* and *Lactobacillus Sake*

Lactobacillus plantarum and *L. sake* are both facultative heterofermenters and are isolated from similar environments such as sauerkraut, fermenting plant material, meats, and pickled foods.[24] They are able to grow at 15°C and they ferment a broad spectrum of pentoses, hexoses, and disaccharides. However, they have fundamental taxonomic differences in terms of peptidoglycan composition and electrophoretic mobilities of D and L lactate dehydrogenases.[24].

The first identification of an "antibiotic" substance from *L. plantarum* was related by Kodama[43] in 1953, who termed the substance lactolin. A crude powder preparation was compared to diplococcin and nisin and was found to have significantly different properties. Lactolin was not characterized in terms of bacteriocinogenic properties, as is the case with more recent investigations.

A description of bacteriocin activity in *L. plantarum* was first communicated by Daeschel, et al.,[44,45] who described an inhibitor that was bacteriocidal toward some species of the four genera of lactic acid bacteria, but not active against other Gram-positive or Gram-negative bacteria or yeasts. The observed spectrum of activity was intermediate in breadth as compared to the narrow spectrum of lactocin B and F, helveticin J and casecin 80, and the broad spectrum of nisin and pediocin A.[46] The inhibitor described in the study was designated plantaricin A, produced by *L. plantarum* C-11, a cucumber fermentation isolate. Crude preparations of plantaracin A were shown to be stable to heating (>30 min at 100°C) , nondialyzable (>8000 molecular weight), and active from pH 4.0 to 6.5. *Lactobacillus plantarum* C-11 was observed to harbor at least two plasmids which were evaluated as potential genetic determinants for plantaricin A. Plasmid-curing procedures employing ethidium bromide, acridine orange, and novobiocin were unsuccessful in generating non-bacteriocinogenic (Bac–) variants. Bac– variants were obtained from chemostat cultures, but had plasmid profiles identical to that of the parent culture. Thus, the genetic determinants for plantaricin A were not conclusively identified.

West and Warner[47] described an inhibitor (plantacin B) from *L. plantarum* NCDO 1193, a strain originally isolated from ensiled vegetable matter.[48] Detection of antagonism was evaluated using the deferred-antagonism method of spotting cultures onto agar surfaces and overlayering with indicator-seeded agar. In a limited activity spectrum survey, antagonism was observed against two strains of *L. plantarum,* a strain of *Leuconostoc mesenteroides,* and of *Pediococcus damnosus.* Sensitivity to proteolytic enzymes (pronase, pepsin, trypsin, and chymotrypsin) was demonstrated by their incorporation into agar plates described above. Activity could not be demonstrated in liquid broth, which precluded determining if the inhibitor had a bacteriocidal mode of action.

Schillinger and Lucke[49] surveyed 221 strains of lactobacilli isolated from raw and processed meats for antimicrobial activity against other lactobacilli and *Listeria monocytogenes*. Dialyzed and pH adjusted (pH 6.5) cell-free supernatants from six strains of *L. sake* were observed to possess antagonistic factors. The inhibitory activity of *L. sake* Lb706 was further characterized and found to be destroyed by trypsin or pepsin, stable to heating (20 min at 100°C). A bacteriocidal mode of action was demonstrated where sensitive cells (3×10^5 cfu/ml) were reduced to 30 cfu after 5 h exposure in 10-fold concentrated, pH-neutralized supernatant. The *L. sake* Lb706 inhibitor was designated sakacin A, and was determined to have an intermediate activity spectrum with antagonism displayed against *Lactobacillus, Carnobacterium, Enterococcus,* and *Listeria,* but not against *S. aureus, Salmonella typhimurium,* and other Gram-positive and Gram-negative bacteria. A single plasmid was observed to be native to *L. sake* Lb706, and with curing experiments the plasmid was implicated as a possible genetic determinant for sakacin A production and associated producer cell immunity. Schillinger [50] conducted applied trials using *L. sake* Lb 706 in minced meats and fresh mettwurst products to evaluate the effectiveness of sakacin A in inhibiting intentionally added *L. monocytogenes.* In contrast to liquid media studies, *L. sake* Lb 706 did not dramatically decrease populations of *L. monocytogenes* in the meat products. An effect, more bacteriostatic than bacteriocidal, was noted in the meat products, which led the investigator to suggest that sakacin A may be less active or may be bound to components in the meat environment. Nevertheless, the activity of the sakacin A in meat products was such that the potential for controlling *Listeria* in these products exists and warrants further investigation.

A second plasmid-associated bacteriocin produced by a *L. sake* strain was identified by Mortvedt and Nes[51] and designated Lactocin S. The bacteriocinogenic strain, *L. sake* L 45 was identified from a group of 43 lactobacilli isolated from Norwegian fermented dry sausage. Antimicrobial activity, evaluated by deferred antagonism methods, indicated a spectrum of activity for lactocin S that included other lactobacilli, *L. mesenteroides,* and *Pediococcus* species. Interestingly, lactic acid bacteria associated with dairy environments were insensitive to lactocin S. In addition, the enteric Gram-negative species and bacilli species tested were insensitive. In contrast to sakacin A, lactocin S was not active against *L. monocytogenes,* indicating they are most likely different bacteriocins. Bacteriocinogenic properties of lactocin S included heat stability (50% loss after 1 h at 100°C) and degradation by proteolytic enzymes. *Lactobacillus sake* was observed to contain two native plasmids, pCIM1 (50kb) and pCIM2 (34kb). Curing experiments provided evidence linking pCIM1 to lactocin S production and immunity. Hybridization studies showed that the DNA in pCIM2 had a homologous sequence in pCIM2 that suggested that pCIM2 could be a form of pCIM1 that had undergone deletion. DNA sequence rearrangements of pCIM1 were observed in bacteriocin-negative immunity-positive variants that led the authors to suggest that a transposition-like event may have occurred.

In a subsequent study,[52] lactocin S was purified by sequential treatment with ion exchange chromotography, hydrophobic interaction(Phenyl sepharose), gel filtration, hydrophobic interaction, (phenyl superose) and reverse-phase (C_2/C_{18}) chromotography. A 40,000-fold increase in activity was achieved with a 3% yield. Molecular weight determination of lactocin S was inconclusive, but purified preparations gave estimates of a molecular weight of less the 13,700. An amino acid analysis (33 residues) and hydropathic index (1.37) for lactocin S suggested a hydrophobic protein that may contain modified cysteine residues reminiscent of those that occur with nisin.

In another survey of lactic acid bacteria for antagonistic activity, this time from strains isolated from Spanish fermented dry sausage, 119 of 720 isolates displayed antagonism toward an indicator strain *(Lactobacillus fermentu*m). The antimicrobial activity of one of the isolates, *L. sake* 148, was further characterized and described to be somewhat heat stable (60% activity remaining after 20 min at 100°C) and to be inactivated by papain and protease XIV. Concentrated culture supernatants were inhibitory toward strains of *L. brevis, L. curvatus, Carnobacterium divergens, L. mesenteroides, S. aureus, Clostridium botulinum, Clostridium perfringens,* and *L. monocytogenes,* but not against enteric Gram-negative food-borne pathogens. Inhibition studies indicated a bacteriostatic rather then bacteriocidal effect in broth cultures. A name or designation was not given for the inhibitory agent.

IV. *PEDIOCOCCUS PENTOSACEUS*

The pediococci are ubiquitous in fermented foods and beverages and are commonly isolated from dairy products, meats, fresh and fermented vegetables, silage, wine, beer, and soy sauce. In commodities such as alcoholic beverages and dairy foods they are considered spoilage agents, whereas in other foods such as fermented meats and pickles they are intentionally used as starter cultures. They are homofermentative on glucose, and some species are tolerant of inhibitory environments such as high-salt concentrations > 10% NaCl, >10% ethanol, and <pH 3.5. The pediococci are unique among the lactic acid bacteria in that cell division occurs alternately in two planes at right angles.[54] This feature allows rapid presumptive identification with microscopic observation. Bacteriocinogenicity has been described with two species: *P. pentosaceus* and *P. acidilactici*. It should be noted that *P. cerevisiae* is no longer a valid designation and that strains previously referred to as *P. cerevisiae* are either *P. pentosaceus* or *P. damnosus*.[54] Bacteriocins of *P. acidilactici* have been reviewed in a previous chapter, thus the ensuing discussion will be limited to bacteriocinogenicity in *P. pentosaceus*.

The first observation that *Pediococcus* spp. possess antagonistic activity other than that associated with acid production was discussed by Etchells et al [55] in interpreting the observed inhibition of *L. plantarum* in mixed culture with *P. pentosaceus* in experimental cucumber fermentations. They concluded: "The rapid acid production by the *Pediococcus* species is the only possible reason evident from our data for this initial inhibition of *L. plantarum* growth. However, *L. plantarum* is so acid tolerant that it does not appear likely that this is the sole reason for the inhibition, and that there may be other factors more important." This observation prompted further investigation by Fleming et al.[56] who screened a group of *Pediococcus* strains for inhibitory activity against bacteria and yeast associated with fermented cucumbers. Two strains of *P. pentosaceus,* FBB 61 and L-7230, originally isolated from two geographically distant pickle-processing plants, possessed identical inhibitory spectra against Gram-positive bacteria with neither displaying antagonism toward Gram-negative bacteria, yeast, or each other. Further characterization[56] eliminated the possibility of hydrogen peroxide as being the antagonistic agent produced by the pediococci.

Chemical characterization of the inhibitor produced by *Pediococcus pentosaceus* FBB-61 was the focus of an M.S. thesis (P. Reuckert, Michigan State University, East Lansing, 1979) that described the inhibitor as stable to heating (100°C for 60 min), as possessing a bacteriocidal mode of action, as being nondialyzable, and as being deactivated upon exposure to the enzyme pronase. These characteristics are representative of the bacteriocins of which Reuckert concluded the inhibitor to be a member.

Daeschel and Klaenhammer[46] elaborated on the activity spectra of *P. pentosaceus* FBB-61 and L-7230 and recorded inhibition of the food-borne pathogens, *Clostridium botulinum, Clostridium perfringens, Staphylococcus aureus,* and of several species of lactic acid bacteria. In a study by Spelhaug and Harlander,[57] the antagonistic activity of *P. pentosaceus* FBB-61 was demonstrated with strains of *Bacillus cereus, C. perfringens,* and *S. aureus.* In the same study,[57] 14 strains of *Listeria monocytogenes* and four strains of *Listeria innocua* were inhibited by *P. pentosaceus* FBB-61. Harris et al.[58] also demonstrated the sensitivity of *L. monocytogenes* (eight strains) to the inhibitory activity of *P. pentosaceus* FBB-61 and L 7230.

The sensitivity of *C. botulinum* to the inhibitor of *P. pentosaceus* FBB-61 and L 7230, first observed by Daeschel and Klaenhammer,[46] was further investigated by Okereke and Montville,[59] who determined the minimum inhibitory cell concentration (MICC) of the inhibitor-producer strain required for observing inhibition in deferred antagonism assays. They evaluated 11 strains of *C. botulinum,* representing both type A and B strains, all of which were inhibited by *P. pentosaceus* FBB-61 and L7230. The MICC (spot inoculum) required to produce an inhibition zone against a 10^4 composite cell population of *C. botulinum* strains seeded into agar was 1×10^5 cfu/ml for *P. pentosaceus* FBB-61 and 4×10^6 for *P. pentosaceus* L-7230. In comparison, an MICC of 7×10^5 was required for the nisin-producer *L. lactis* 11454. The investigators conclude that *P. pentosaceus* FBB-61 was the "most promising strain for further research on bacteriocin-mediated protection against *C. botulinum* hazards," The inhibitor from *P. pentosaceus* FBB-61 and L7230 designated pediocin A in the study of Daeschel and Klaenhammer[46] can be categorized as broad spectrum similar to nisin in that it is active against many Gram-positive genera. Most significant is the effectiveness of pediocin A against Gram-positive bacteria of public health significance. *P. pentosaceus* FBB-61 and L 7230 were deposited with the American Type Culture Collection and have accession numbers 43200 and 43201 respectively.

The studies reviewed above all entailed using deferred antagonism assays to demonstrate the antagonistic activity of pediocin A. Attempts to isolate and purify pediocin A in the author's laboratory have been uniformly unsuccessful, probably because routine laboratory manipulations are destabilizing or denaturing it in some manner.

Two genetic studies[46,60] have addressed bacteriocin production and producer cell immunity in *P. pentosaceus*. Graham and McKay[60] observed bacteriocin-like activity in *Pediococcus cerevisiae* FBB-63, and with curing experiments using novobiocin, linked bacteriocin production to a 10.5 Mdl plasmid. Comparison of the activity spectra of *P. cerevisiae* FBB-63 and *P. pentosaceus* FBB-61 suggests that the inhibitory agent in these strains is the same. *Pediococcus cerevisiae* and the variants cured of bacteriocin activity (Bac–) were not evaluated with respect to their native immunity to the inhibitor. Plasmid curing studies[46] with *P. pentosaceus* FBB-61 generated Bac– variants when the strain was cultivated in chemostat culture at 42°C, 4 degrees higher then the optimum growth temperature. Two variants were examined for reversion to the Bac+ phenotype. A total of 37,000 and 28,000 colonies from each Bac– culture were evaluated and found to remain Bac–. In addition all Bac– variants were inhibited by the parent strain (FBB-61), and by strain L-7230, in deferred antagonism assays; this suggested a concurrent loss in immunity with bacteriocin production. Plasmid analysis of strains L 7230 and FBB 61 revealed the presence of a 13.6 Mdl plasmid in both strains. Strain FBB-61 also possessed a 5.0 Mdl plasmid. Variants of FBB-61 that were cured of the Bac+ phenotype and associated immunity were missing the 13.6 Mdl (pMD136) plasmid, but not the 5.0 Mdl plasmid. Curing of strain L7230 of bacteriocin activity or plasmids was not successful. Restriction enzyme analysis indicated that pMD 136 in FBB-61 and L-7230 were similar, and both determined Pediocin A production and immunity. More direct evidence could be provided by genetic transfer of pMD136, however attempts with electroporation and conjugal transfer protocols have not been successful (author's laboratory).

In the study of Daeschel and Klaenhammer[46] described above it was observed that the stability of pMD136 in strain FBB-61 was affected by cultivation at higher then optimum temperatures. At 10 d of continuous culture (chemostat) at 42°C, 17% of the population were Bac–, and at 14 d 53% were Bac–. With further study.[61] It was observed that populations of strain FBB-61 harbor Bac– cells (pMD136 absent) at a frequency of approximately 0.5% when cultivated

TABLE 1
Effect of Temperature on the Stability of the Pediocin A phenotype in *P. Pentosaceus* FBB-61

Temperature, °C.	Bac^-/Bac^+ cells	% Bac^- cells
37a	15/2754	0.54
	9/1632	0.55
42	1341/1342	99.99
	1061/1062	99.90
37	111/504	22.02
	99/566	17.49
37	2/2202	0.09
	0/2138	>0.001

[a] Replicate cultures were transferred twice at a 1% inoculum level in MRS broth at each indicated temperature, and allowed to grow overnight (approximately 16 h) before plating with deferred antagonism assay for pediocin A activity.

at an optimum temperature of 37 to 38°C (Table 1). After two subcultures of the strain at 42°C, the cells present were almost entirely Bac–. When the incubation temperature was decreased back to 37°C and transferred twice again, the populations began to contain a greater proportion of Bac+ cells. After two more transfers at 37°C , the populations were almost entirely Bac+. This data suggested that 37°C is a permissive temperature, whereas 42°C may be termed restrictive. The ability of Bac– populations to regain a Bac+ phenotype may illustrate a self-selecting pressure that may result from Bac– cells being suppressed or killed. This is particularly impressive when one considers that Bac+ populations had much slower specific growth rates when compared to Bac– populations (Table 2), regardless of temperature or medium. This suggested that Bac+ cells, although handicapped by slower growth rates, can out-compete the faster growing Bac– cells by the action of pediocin A. The reasons why Bac+ populations grow so much slower is not known, however, one can speculate that perhaps a slower growth rate results from the physical burden of maintaining a plasmid or perhaps it is an overall result from cells spontaneously cured of pMD136 being killed, coupled with plasmid-bearing cells growing at normal rates. Based upon these observations with pMD136, it is conceivable that the inclusion of bacteriocin-immunity genes in the construction of food-grade cloning vectors may serve to stabilize the introduced plasmid as well as provide a directly selectable marker .

The only other description of bacteriocinogenicity in *P. pentosaceus* was a brief report communicated by Hoover et al.[62] who observed that bacteriocin

TABLE 2
Effect of Harboring the Plasmid pMD 136 on the Specific Growth Rates (K) of *P. pentosaceus* FBB-61 in Different Media and Temperatures

	(K) in TSB broth		(K) in MRS broth	
Temperature (°C)	FBB-61[a]	FBB-61-2[b]	FBB-61	FBB-61-2
30	0.31	0.40	—	—
37	0.41	0.68	0.771.01	
40	—	—	0.58	0.65

[a] Harbors pMD136.
[b] Cured of pMD136.

activity was linked to a 5.5 Mdl plasmid that was resident in three strains of *P. acidilactici* and one strain of *P. pentosaceus.*

V. *LEUCONOSTOC*

Leuconostocs are found in many of the same environments as lactobacilli, pediococci, and lactococci. The genus consists of four species *L. oenos, L. lactis, L. paramesenteroides,* and *L. mesenteroides.* The latter has three subspecies, *L. mesenteroides* subsp. *mesenteroides,* subsp. *dextranicum,* and subsp. *cremoris.*[63] Two recently described new species of *Leuconostoc, L. gelidium* and L. *carnosum,* isolated from meats, have been proposed for inclusion in the genus.[64] Leuconostocs are obligate heterofermenters and generally have optimum growth at temperatures (20–30°C) lower then the optimum for other LAB. They are found in many food environments, including dairy products, meat, and fruits and vegetables. They are used commercially in cultured dairy products *(L. cremoris)* to generate desirable flavor components by dissimilating citrate. They are also important in the wine industry where *L. oenos* is used to reduce the acidity of wine by decarboxylating malic acid. Vegetable fermentations, such as in sauerkraut manufacture, rely on the growth of *L. mesenteroides* to generate the sufficient amounts of volatile acid necessary for a quality product. The leuconostocs have been observed to produce metabolites including acetic acid and diacetyl (Reviewed in Chapters 5 and 6), which are recognized to be inhibitory toward certain microorganisms, particularly Gram-negative species. Babel[65] reviewed the antimicrobial properties of dairy cultures including *Leuconostoc* spp., and discussed the practical applications of antibiosis.

In a study addressing the occurrence of plasmid DNA in *Leuconostoc* spp., Orberg and Sandine[66] also described inhibition of *Leuconostoc* spp. and *Lactococcus lactis* by three strains of *L. dextranicum* and two *Leuconostoc* spp.

using the Kekessey-Piquet[67] procedure for bacteriocin detection. The inhibitory action of these strains was not characterized further.

A. *Leuconostoc Gelidium*

Two recent reports [68,69] addressed bacteriocinogenicity in different strains of the newly proposed species *L. gelidium*. Harding and Shaw[68] described the antagonistic activity of *L. gelidium* IN139 as heat stable (100°C/60 min), sensitive to protease, and nondialyzable, indicating a molecular weight greater then 10,000 daltons. A bacteriocidal mode of action was demonstrated against *L. monocytogenes* in which a decrease in optical density of a log-phase culture occurred upon addition of inhibitor extract. Adsorption studies indicated that the inhibitor could not adsorb to the producer cells, but 50 to 100% of activity adsorbed to sensitive cells under the same conditions. The sum of these observations clearly suggests that this unnamed inhibitor is likely to be a bacteriocin. The activity spectra included species of *Carnobacterium, Lactobacillus, Enterococcus, Leuconostoc, Listeria,* but not *S. aureus, C. botulinum,* or Gram-negative pathogens. Shortly after the report of Harding and Shaw appeared, Hastings and Stiles[69] published observations of a bacteriocin-like inhibitor from a strain of *L. gelidium* (UAL187) isolated from meat. The inhibitor was sensitive to proteolytic enzymes and had an activity spectrum similar to that of *L. gelidium* IN139. The inhibitor was described as being stable to heating at 62°C for 30 min, however, since stability at 100°C was not reported, comparison to the heat stability of the inhibitor from *L. gelidium* IN139 cannot be made. Hastings and Stiles concluded that the substance was more likely to be bacteriostatic then bactericidal in its effect, based on their results from killing kinetics experiments. Plasmid-curing studies with *L. gelidium* UAL 187 suggested that a 7.6-Mda plasmid may determine inhibitor production and immunity. A novel observation was that the inhibitor was synthesized and excreted at temperatures of 1 and 5°C, which suggested possible applications in protecting chilled foods like meat from *L. monocytogenes*.

B. *Leuconostoc Mesenteroides*

The first report of bacteriocinogenicity within the species *L. mesenteroides* was communicated by Daba, et.al.[70]. *Leuconostoc mesenteroides* UL5 isolated from Cheddar cheese produced an inhibitor that appeared to have a unique activity spectrum that was limited essentially to *Listeria* spp. Twenty-two strains representing the genera *Lactobacillus, Lactococcus, Leuconostoc,* and *Pediococcus* (except one strain) were not sensitive to the inhibitor. Besides the 12 strains of *Listeria* that were sensitive, only 4 other bacterial strains were sensitive and these were described as being only partially so. Gram-negative bacteria were also reported as insensitive. The inhibitor from this study, referred to as mesenterocin 5, was sensitive to proteolytic enzymes, was partially denatured by chloroform, was heat stable (100°C/30 min), and it was concluded that it was bacteriostatic in action. Direct detection of the inhibitor using SDS-PAGE indicated an apparent molecular mass of 4.5k Da. Variants of *L. mesenteroides* UL5 deficient in mesenterocin 5 production but not in associated immunity were obtained by cultivation with acriflavin. Plasmid profiles of the parent strain and variants were not presented as part of the study, thus the genetic determinants for mesenterocin 5 and immunity are yet to be identified. The authors dicussed the uniqueness of the activity spectra in terms of possible applications in foods. The point made was that it is possible to consider using mesenterocin 5 along with insensitive lactic acid bacteria to provide a synergistic antagonism against pathogens such as *L. monocytogenes*.

VI. CONCLUDING REMARKS

Since the review of Klaenhammer[71] in 1988 on bacteriocins of lactic acid bacteria, there has been a very significant increase in information on existing bacteriocins and on new, previously undescribed, bacteriocins. The breadth of investigations have encompassed fundamental studies on the chemistry and molecular biology of bacteriocins to developing applications using bacteriocins to protect the food supply. Recent notable accomplishments in the field include the purification of bacteriocins from *L. acidophilus* and *L. helveticus,* and the subsequent cloning, sequencing, and expression of genes encoding for bacteriocin production. A unique plasmid in *Lactococcus lactis* subsp. lactis was found to contain genes that encode for two distinct antagonistic activities and associated immunities. Bacteriocin-like substances have been characterized from genera (Leuconostoc) and species that have not been recognized previously as bacteriocinogenic; including *L. casei*, *L. sake, L., brevis, L. gelidium,* and *L. mesenteroides.* These are but a few examples that illustrate the progress in the field that has occurred in the last four years.

We are coming closer to realizing the potential applications of bacteriocins as food preservation agents. Many LAB bacteriocins have been observed to be antagonistic to *L. monocytogenes*, a pathogen that frequently contaminates refrigerated foods. Strategies to protect such foods include the use of LAB that are capable of producing bacteriocins under refrigeration temperatures. Other studies have focused on how to make bacteriocins more effective against food-borne pathogens. An example is the investigation of Stevens et al.,[72] who described making *Salmonella* spp. sensitive to the action of nisin by the addition of EDTA, which is thought to destabilize the outer membrane of Gram-negative bacteria. Another example is the study of Jung et al.,[73] which demonstrated that an emulsifier such as tween 80 could significantly increase the effectiveness of nisin in inhibiting *L. monocytogenes* in fluid milk.

It is envisioned that future investigations will provide information on the mode of action of bacteriocins, on producer associated immunity, on genetic regulation of bacteriocin synthesis and export, and on protein engineering of bacteriocins. These types of fundamental investigations are essential to take full advantage of the potential applications of bacteriocins in food science.

REFERENCES

1. **Konisky, J.,** Colicins and other bacteriocins with established modes of action, *Ann. Rev. Microbiol.,* 36, 125, 1982.
2. **Rogers, L. A.,** The inhibiting effect of *Streptococcus lactis* on *Lactobacillus bulgaricus, J. Bacteriol.,* 16, 321, 1928.
3. **Whitehead, H. R.,** A substance inhibiting bacterial growth by certain strains of lactic streptococci, *Biochem. J.,* 27, 1793, 1933.
4. **Whitehead, H. R. and Riddet, W.,** Slow development of acidity in cheese manufacture. Investigation of a typical case of "non-acid" milk, *N.Z. J. Agric.,* 46, 225, 1933.
5. **Cox, G. A. and Whitehead, H. R.,** Streptococci which produce a substance inhibiting the growth of lactic streptococci, *N.Z. J. Agric.,* 52, 38, 1936.
6. **Oxford, A. E.,** Diplococcin, an antibacterial protein elaborated by certain milk streptococci, *Biochem. J.,* 38, 178, 1944.
7. **Hirsch, A.,** Some general aspects of the inhibitory streptococci, *Proc. Soc. Appl. Bacteriol.,* 26 ,1946.
8. **Davey, G. P. and Richardson, B. C.,** Purification and some properties of diplococcin from *Streptococcus cremoris* 346, *Appl. Environ. Microbiol.,* 41, 84, 1981.
9. **Davey, G. P. and Pearce, L. E.,** The use of *Streptococcus cremoris* strains cured of diplococcin production as cheese starters, *N.Z. J. Dairy Sci. Technol.,* 15, 51, 1980.
10. **Davey, G. P. and Pearce, L. E.,** Production of diplococcin by *Streptococcus cremoris* and its transfer to nonproducing group N streptococci, *Microbiology,* Schlessinger, D., Ed., American Society for Microbiology, Washington, D.C., 1982, 221.
11. **Davey, G.,** Plasmid associated with diplococcin production in *Streptococcus cremoris, Appl. Environ. Microbiol.,* 48, 895, 1984.
12. **Kozak, W. J., Bardowski, J., and Dobrzanski, W. T.,** Lactostrepcin — acid bacteriocins produced by lactic streptococci, *J. Dairy Res.,* 45, 247, 1978.
13. **Zajdel, J. K., Ceglowski, P., and Dobrzanski, W. T.,** Mechanism of action of Lactostrepcin 5, a bacteroicin produced by *Streptococcus cremoris* 202, *Appl. Environ. Microbiol.,* 49, 969, 1985.
14. **Geis, A., Singh, J., and Teuber, M.,** Potential of lactic streptococci to produce bacteriocin, *Appl. Environ. Microbiol.,* 45, 205, 1983.
15. **Neve, H., Geis, A., and Teuber, M.,** Conjugal transfer and characterization of bacteriocin plasmids in group N (lactic acid) streptococci, *J. Bacteriol.,* 157, 833, 1984.
16. **van Belkum, M. J., Hayema, B. J., Geis, A., Kok, J., and Venema, J.,** Cloning of two bacteriocin genes from a *Lactococcus* bacteriocin plasmid, *Appl. Environ. Microbiol.,* 55, 1187, 1989.
17. **van Belkum, M. J., Hayema, B. J., Jeeninga, R. E., Kok, J., and Venema, G.,** Organization and nucleotide sequences of two Lactococcal operons, *Appl. Environ. Microbiol.,* 57, 492, 1991.
18. **Collins, E. B.,** Domination among strains of lactic streptococci with attention to antibiotic production, *Appl. Microbiol.,* 9, 200, 1961.
19. **Daly, C., Sandine, W. E., and Elliker, P. R.,** Associative growth and inhibitory properties of *Streptococcus diacetylactis, J. Dairy Sci.,* 53, 637, 1970.
20. **Daly, C., Sandine, W. E., and Elliker, P. R.,** Interaction of food starter cultures and food-borne pathogens: *Streptococcus diacetylactis* versus food pathogens, *J. Milk Food Technol.,* 35, 349, 1972.
21. **Scherwitz, K. M., Baldwin, K. A., and McKay, L. L.,** Plasmid linkage of a bacteriocin-like substance in *Streptococcus lactis* subsp. *diacetylactis* strain WM4: transferability to *Streptococcus lactis, Appl. Environ. Microbiol.,* 45, 1506, 1983.
22. **Harmon, K. S. and McKay, L. L.,** Restriction enzyme analysis of lactose and bacteriocin plasmids from *Streptococcus lactis* subsp. *diacetylactis* WM4 and cloning of *Bcl*I fragments coding for bacteriocin production, *Appl. Environ. Microbiol.,* 53, 1171, 1987.

23. **Kojic, M., Svircevic, J., Banina, A., and Topisirovic, L.,** Bacteriocin-producing strain of *Lactococcus lactis* subsp. *diacetylactis* S50, *Appl. Environ. Microbiol.,* 57, 1835, 1991.
24. **Kandler, O. and Weiss, N.,** Regular, nonsporing Gram-positive rods, Section 14, in *Bergey's Manual of Systematic Bacteriology,* Vol. 2, Sneath, P. A., Ed., Williams & Wilkins, Baltimore, 1986, 1208.
25. **Vincent, J. G., Veonett, R. C., and Riley, R. G.,** Antibacterial activity associated with *Lactobacillus acidophilus, J. Bacteriol.,* 78, 477, 1959.
26. **Sabine, D. B.,** An antibiotic-like effect of *Lactobacillus acidophilus, Nature (London),* 199, 811, 1963.
27. **Hamadan, I. Y. and Mikolajcik, E. M.,** Acidolin: an antibiotic produced by *Lactobacillus acidophilus, J. Antibiot.,* 27, 631, 1974.
28. **Gilliland, S. E. and Speck, M. L.,** Antagonistic action of *Lactobacillus acidophilus* toward intestinal and foodborne pathogens in associative cultures, *J. Food Prot.,* 40, 820, 1977.
29. **Mehta, A. M. and Dave, P. J.,** Inhibition of curd contaminants by an inhibitory protein from *Lactobacillus acidophilus, Microbios,* 37, 37, 1983.
30. **Mehta, A. M. , Patel, K. A., and Dave, P. J.,** Purification and properties of the inhibitory protein from *Lactobacillus acidophilus, Microbios,* 38, 37, 1983.
31. **Barefoot, S. F. and Klaenhammer, T. R.,** Detection and activity of Lactacin B, a bacteriocin produced by *Lactobacillus acidophilus, Appl. Environ. Microbiol.,* 45, 1808, 1983.
32. **Barefoot, S. F. and Klaenhammer, T. R.,** Purification and characterization of the *Lactobacillus acidophilus* bacteriocin Lactacin B, *Antimicrob. Agents Chemother.,* 26, 328, 1984.
33. **Muriana, P. M. and Klaenhammer, T. R.,** Conjugal transfer of plasmid-encoded determinants for bacteriocin production and immunity in *Lactobacillus acidophilus* 88, *Appl. Environ. Microbiol.,* 53, 553, 1987
34. **Muriana, P. M. and Klaenhammer, T. R.,** Purification and partial characterization of lactacin F, a bacteriocin produced by *Lactobacillus acidophilus* 88, *Appl. Environ. Microbiol.,* 57, 114, 1991
35. **Muriana, P. M. and Klaenhammer, T. R.,** Cloning, phenotypic expression, and DNA sequence of the gene for Lactacin F, an antimicrobial peptide produced by *Lactobacillus* spp., *J. Bacteriol.,* 173, 1779, 1991.
36. **Toba, T., Yoshiko, E., and Itoh, T.,** Acidophilucin A, a new heat labile bacteriocin produced by *Lactobacillus acidophilus* LAPT 1060, *Lett. Appl. Microbiol.,* 12, 106, 1991.
37. **Rammelsberg, M. and Radler, F.,** Antibacterial polypeptides of *Lactobacillus* species, *J. Appl. Bacteriol.,* 69, 177, 1990.
38. **Rammelsberg, M., Müller, E., and Radler, F.,** Casecin 80: purification and characterization of a new bacteriocin from *Lactobacillus caseii, Arch. Microbiol.,* 154, 249, 1990.
39 **Upreti G. C. and Hinsdill, R. D.,** Isolation and characterization of a bacteriocin from a homofermentative *Lactobacillus, Antimicrob. Agents Chemother.,* 4, 487, 1973.
40. **Upreti G. C. and Hinsdill, R. D.,** Production and mode of action of lactocin 27: bacteriocin from a homofermentative *Lactobacillus, Antimicrob. Agents Chemother.,* 7, 139, 1975.
41. **Joerger, M. C. and Klaenhammer, T. R.,** Characterization and purification of helveticin J and evidence for a chromosomally determined bacteriocin produced by *Lactobacillus helveticus* 481, *J. Bacteriol.,* 167, 439, 1986.
42. **Joerger, M. C. and Klaenhammer, T. R.,** Cloning, expression, and nucleotide sequence of the L*actobacillus helveticus* 481 gene encoding the bacteriocin helveticin J, *J. Bacteriol.,* 172, 6339, 1990.
43. **Kodama, R.,** Studies in lactic acid bacteria, II, Lactolin, a new antibiotic substance produced by lactic acid bacteria, *J. Antibiol.,* 72, 1952.
44. **Daeschel, M. A., McKenney, M. C., and McDonald, L. C.,** *ASM Bacteriol. Proc.,* P-13 (Abstr.), 277, 1986.

45. **Daeschel, M. A., McKenney, M. C., and McDonald, L. C.,** Bacteriocidal activityof *Lactobacillus plantarum* C-11, *Food Micro.*, 7, 91, 1990.
46. **Daeschel, M. A. and Klaenhammer, T. R.,** Association of a 13.6-megadalton plasmid in *Pediococcus pentosaceus* with bacteriocin activity, *Appl. Environ. Microbiol.*, 50, 1538, 1985.
47. **West, C. A. and Warner, P. J.,** Plantacin B, a bacteriocin produced by *Lactobacillus plantarum* NCDO 1193, *FEMS Microbiol. Lett.*, 49, 163, 1988.
48. **ARFC Institute of Food Research**, National Collection of Food Bacteria, Catalogue of strains, 3rd ed., Reading, England, 1986.
49. **Schillinger, U. and Lucke, F. C.,** Antibacterial activity of *Lactobacillus sake* isolated from meat, *Appl. Environ. Microbiol.*, 55, 1901, 1989.
50. **Schillinger, U.,** Bacteriocins of lactic acid bacteria, in *Biotechnology and Food Safety*, Bills, D. and Kung, S., Eds., Butterworths, London, 1990, 55.
51. **Mortvedt, C. I. and Nes, I. F.,** Plasmid associated bacteriocin production by a *Lactobacillus sake* strain, *J. Gen. Microbiol.*,136, 1601, 1989.
52. **Mortvedt, C. I. and Nes, I. F.,** Purification and amino acid sequence of lactocin S, a bacteriocin produced by *Lactobacillus sake* L45, *Appl. Environ. Microbiol.*, 57, 1829, 1991.
53. **Sobrino, O. J., Rodriguez, J. M., Moreira, W. L., Fernandez, M. F., Sanz, B., and Hernandez, P. E.,** Antibacterial activity of *Lactobacillus sake* isolated from dry fermented sausages, *Intl. J. Food Microbiol.*, 13, 1, 1991.
54. **Garvie, E. I.,** Genus Pediococcus, in *Bergey's Manual of Systematic Bacteriology*, Vol. 2, Sneath, P. A., Ed., Williams & Wilkens, Baltimore, 1986, 1075.
55. **Etchells, J. L., Costilow, R. N., Anderson, T. E., and Bell, T. A.,** Pure culture fermentation of brined cucumbers, *Appl. Microbiol.*, 12, 523, 1964.
56. **Fleming, H. P., Etchells, J. L., and Costilow, R. N.,** Microbial inhibition by an isolate of *Pediococcus* from cucumber brines, *Appl. Microbiol.*, 30, 1040, 1975.
57. **Spelhaug, S. R. and Harlander, S. K.,** Inhibition of foodborne bacterial pathogensby bacteriocins from *Lactococcus lactis* and *pediococcus pentosaceus*, *J. Food Prot.*, 52, 856, 1989.
58. **Harris, L. J., Daeschel, M. A., Stiles, M. E., and Klaenhammer, T. R.,** Antimicrobial activity of lactic acid bacteria against *Listeria monocytogenes*, *J. Food Prot.*, 52, 3784, 1989.
59. **Okereke, A. and Montville, T. J.,** Bacteriocin inhibition of *Clostridium botulinum*, *J. Food Prot.*, 54, 349, 1991.
60. **Graham, D. C. and McKay, L.,**Plasmid DNA in strains of *Pediococcus cerevisiae* and *Pediococcus pentosaceus*, *Appl. Environ. Microbiol.*, 50, 532, 1985.
61. **Daeschel, M. and Mormino, R. E.,** Stability of the bacteriocin plasmid pMD 136 and its effect on growth of *Pediococcus pentosaceus*, *ASM Bacteriol. Proc.*, P-36, (Abst.), 280, 1987.
62. **Hoover, D. G., Walsh, P. M., Kolaetis, K. M., and Daly, M. M.,** A bacteriocin produced by *Pediococcus* species associated with a 5.5-megadalton plasmid, *J. Food Prot.*, 51, 29, 1988.
63. **Garvie, E. I.,** Genus Leuconostoc, in Bergey'*s Manual of Systematic Bacteriology*, Vol. 2, Sneath, P. A., Ed., Williams & Wilkens, Baltimore, 1986, 1071.
64. **Shaw, B. G. and Harding, C. D.,** *Leuconostoc gelidium* sp. nov. and *Leuconostoc carnosum* sp. nov. from chill-stored meats, *Int. J. Syst. Bacteriol.*, 39, 217, 1989.
65. **Babel, F. J.,** Antibiosis by lactic culture bacteria. *J. Dairy Sci.*, 60, 815, 1977.
66. **Orberg, P. K. and Sandine, W. E.,** Common occurrence of plasmid DNA and vancomycin resistance in *Leuconostoc* spp., *Appl. Environ. Microbiol.*, 48, 1129, 1984.
67. **Kekessey, D. A. and Piquet, J. D.,** New method for detecting bacteriocin production, *Appl. Microbiol.*, 20, 282, 1970.
68 **Harding, C. D. and Hardy, B. G.,** Antimicrobial activity of *Leuconostoc gelidum* against closely related species and *Listeria monocytogenes*, *J. Appl. Bacteriol.*, 69, 648, 1990.

69. **Hastings, J. W. and Stiles, M. E.,** Antibiosis of *Leuconostoc gelidium* isolated from meat, *J. Appl. Bacteriol.,* 70, 127, 1991.
70. **Daba, H., Pandian, S., Gosselin, J. F., Simard, R. E., Huang, J., and Lacroix, C.,** Detection and activity of a bacteriocin produced by *Leuconostoc mesenteroides, Appl. Environ. Microbiol.,* 57, 3450, 1991.
71. **Klaenhammer, T. R.,** Bacteriocins from lactic acid bacteria, *Biochimie,* 70, 337, 1988.
72. **Stevens, K. A., Sheldon, B. W., Klapes, N. A., and Klaenhammer, T. R.,** Nisin treatment for inactivation of *Salmonella* species and other Gram-negative bacteria, *Appl. Environ. Microbiol.,* 57, 3613, 1991.
73. **Jung, D. S., Bodyfelt, F. W., and Daeschel, M. A.,** Influence of fat and emulsifiers on the efficacy of nisin in inhibiting *Listeria monocytogenes* in fluid milk, *J. Dairy Sci.,* 75, 387, 1992.

Chapter 12

METABOLITES OF YEASTS AS BIOPRESERVATIVES

Alan T. Bakalinsky

TABLE OF CONTENTS

I. INTRODUCTION

While the role of yeasts in the production of fermented beverages was only recognized little more than a century ago by Louis Pasteur, the making of bread, beer, wine, and related products dates back thousands of years. The making of these must surely be considered among the earliest and most successful applications of biotechnology. Continual refinement of these products has led to the domestication of *Saccharomyces cerevisiae,* because nearly all yeasts used in their manufacture are recognized as strains of this species. In addition to playing premier roles in the production of fermented foods, and more recently serving as hosts for large-scale production of recombinant proteins, genetically defined derivatives enjoy wide use as model eukaryotes in the study of diverse problems in biology.

Because of the central importance of *Saccharomyces cerevisiae* in the production of fermented foods and beverages, this chapter will be confined to a discussion of its metabolites known to possess antimicrobial activity — ethanol, sulfite, and killer toxins. Antimicrobial metabolites produced by other food-grade or industrial yeasts will be discussed only in light of their relevance to S. *cerevisiae,* and unless stated otherwise, the term "yeast" will refer exclusively to this species.

II. ETHANOL

A. Biosynthesis

The formation of ethanol as an endproduct of glycolysis in *S. cerevisiae* (Figure 1) serves the same purpose as in other fermentative organisms: regeneration of an oxidized cofactor, usually nicotinamide adenine dinucleotide (NAD^+). Yeast does not form lactic acid from pyruvate but can catalyze the reverse reaction when grown aerobically on lactate due to the activity of a mitochondrial lactate dehydrogenase (flavocytochrome b2). Interestingly, even a mutant lacking functional copies of all known genes encoding alcohol dehydrogenase still produces ethanol from glucose. The acetaldehyde-

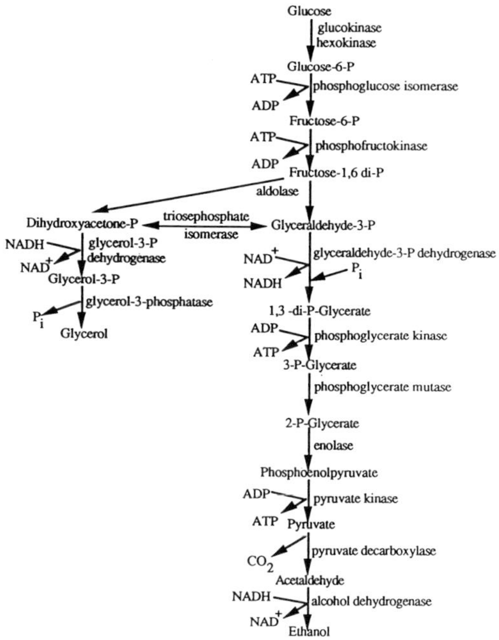

FIGURE 1. Glycolysis in *Saccharomyces cerevisiae.*

reducing activity requires mitochondrial electron transport and is apparently linked to an acetaldehyde-oxidizing activity that results in significant acetate production.[1] Reviews of glycolysis and related carbohydrate metabolism in yeast are available.[2,3]

B. Chemical and Physical Properties

Ethanol is a clear, colorless, flammable liquid miscible with water and many organic solvents in all ratios. It is hygroscopic and relatively nontoxic, exhibiting an oral LD50 in rats of 13.7 g/kg. When ethanol is added to water, a rise in temperature occurs due to the heat of solvation, with a consequent increase in volume. When the mixture is returned to the original temperature, the volume decreases, becoming slightly less than the sum of the individual water and ethanol volumes. Maximum contraction occurs at a molar ratio of eight parts water to one part ethanol. At a pressure of one atmosphere, a mixture of 95.6% ethanol and 4.4% water on a mass basis forms a constant boiling point mixture called an azeotrope. In practice, this means that the concentration of ethanol cannot be increased beyond 95.6% by simple distillation. Some physical properties of ethanol are listed in Table 1.

C. Mode and Spectrum of Action

In ethanol-producing microorganisms, the rate of production generally slows as a function of increasing concentration, suggesting that in some way ethanol inhibits its own synthesis. Understanding the nature of this inhibition has

TABLE 1
Physical Properties of Ethanol[4]

Formula	CH_3CH_2OH
Molecular weight, daltons	46.07
Boiling point, °C	78.32
Freezing point, °C	–114.1
Density, , g/ml	0.7893
Refractive index, n_D^{20}	1.361
Viscosity at 20° C, cP	1.17
Dielectric constant at 20°C	25.7
Heat of fusion, J/g	104.6
Heat of vaporization at 78.32°C, J/g	839.31

practical ramifications. The recognized limit to the supply of petroleum-based fuels, and their attendant expense, has led to development of fuel-ethanol-production facilities, most notably in Brazil. Worldwide interest in ethanol as a renewable fuel source has generated research and development programs in many other countries. An obvious focus for research has been the nature of ethanol inhibition, because alcohol production rates decline and plant utilization becomes correspondingly less efficient as higher concentrations are achieved.

Although ethanol has been observed to be more or less inhibitory to all microorganisms examined, most studies on mechanisms of inhibition have been conducted in yeast because of the obvious application of any findings to the development and use of strains possessing greater ethanol tolerance. Part of the complexity of studying and describing ethanol toxicity is lack of general agreement on defining the endpoint of the effect. As discussed in reviews of the topic,[5,6] ethanol has been shown to have different and separable effects on yeast growth, viability, and fermentation. Fermentation appears to be the least sensitive parameter, which explains the paradoxical observation that while growth ceases and cell viability begins to decrease during fermentation, ethanol continues to increase.

1. *Effects on Enzymes*

One explanation for the inhibitory effect of ethanol on fermentation is that it is an inhibitor of one or more of the enzymes involved in its own biosynthesis. In vitro studies[7] that examined the stability and activity of the glycolytic enzymes of yeast and *Zymomonas mobilis* in the presence of ethanol found insignificant denaturation at physiologically relevant concentrations. Sensitivity of the enzymes to inhibition varied, but none was found to be more sensitive than pyruvate decarboxylase which exhibited 50% inhibition at an ethanol concentration of slightly more than 10% (v/w). Since decline in fermentation rates is commonly observed at ethanol concentrations well below this level, it would appear that ethanol inhibition of the glycolytic enzymes is not a significant contributing factor. Reports that intracellular ethanol concentrations greatly exceeded that of the surrounding medium formerly supported the possibility that inhibitory levels were reached, until subsequent work demonstrated that the earlier observations were artifacts of the assay procedures used. Ethanol does not accumulate intracellularly during the course of fermentation.[8,9]

2. *Effects on Membranes*

The ability of alcohols to inhibit cellular functions is directly related to their hydrophobicity.[10] Ethanol is less polar than water and its addition to aqueous solutions increases their dielectric strength. This increases the strength of interactions between dissolved ionic species at the expense of interactions with the solvent. Because interactions between ionic species and ethanol are weaker than with water, the solubility of ions in ethanolic solutions is less than in non-ethanolic aqueous systems. Hydroxyl groups of ethanol compete with water for hydrogen bond formation with polar solutes and other water molecules as a function of the molar ratios of the two solvents. Because ethanol partitions into the hydrophobic regions of membranes more efficiently than water, it is likely that it effectively replaces water bound to macromolecules within the membrane. Replacement of water with relatively bulky molecules of ethanol is likely to alter macromolecular interactions and folding, and weaken the important hydrophobic interactions that assure membrane integrity.[10]

The damage caused to the plasma membrane by ethanol has been shown to result in altered membrane permeability and organization. Ethanol increases the permeability of the membrane to protons and in acidic media the cell must expend one ATP/proton to reverse the flow through theactivity of the plasma membrane-bound, proton-pumping ATPase.[11] Energy formerly available for cell growth is thus diverted to maintain normal intracellular pH. The increased permeability of the membrane is not limited to protons. Ethanol causes leakage of cofactors, ions, and other polar metabolites that normally cannot cross the plasma membrane passively.[12] Loss of magnesium and nucleotides has been shown to be sufficient to explain the inhibitory effect of ethanol on fermentation in both *Z. mobilis* and yeast.[13,14] Medium supplementation with magnesium has been shown to have significant effects on prolonging yeast growth and fermentative activity.[15]

3. *Effects on Solute Transport*

An indirect but critical inhibitory effect of ethanol on nutrient accumulation in yeast cells relates to the damage done to membrane integrity. Active transport of many nutrients is coupled to transmembrane ion gradients which are dissipated by the ethanol-induced increase in membrane permeability. Ethanol has been shown to be a noncompetitive inhibitor of several transport systems: glucose[16] — whose transport is not linked to ion gradients,[17] maltose,[18] amino acids via the general amino acid permease,[19] and ammonia,[20,21] but not fructose[21]. Fructose uptake was found to be inhibited in a noncompetitive manner by ethanol in the yeast K*luyveromyces marxianus*.[22] This effect of ethanol may explain why addition of a readily available nitrogen source such as ammonia to an incompletely fermented, nitrogen-deficient grape must is rarely effective if done too late. In the presence of a significant ethanol concentration, the yeast is unlikely to be able to import the added nitrogen.

4. *Other Aspects*

The toxicity of ethanol increases synergistically in the presence of sugar, presumably due to the resultant increase in osmotic pressure. This partially explains the observation that in saké fermentations, final ethanol concentrations may reach over 20% (v/v). During the fermentation, hydrolysis of rice starch by amylases of A*spergillus oryzae* occurs at the same time that sugar is converted to ethanol by yeast. Thus, the steady-state sugar concentration never becomes significant. If a saké fermentation were begun with an initial sugar concentration sufficient to yield 20% (v/v) ethanol, the yeast would likely not carry out a complete conversion. The other significant factor that promotes high ethanol yields in saké is the presence of a proteolipid complex produced by A*spergillus oryzae,* which confers greater ethanol tolerance to the yeast.[23,24] Theprotein component apparently serves a passive role as a carrier for the lipid since high ethanol yields were still obtained when it was replaced by methylcellulose. The lipid fraction was found to contain significant amounts of phosphatidylcholine, linoleic acid, and ergosterol, and in addition to promoting higher ethanol yields, the complex was shown to protect the integrity of the yeast cell membrane against ethanol-induced leakage of UV-absorbing constituents.[25]

D. Genetic Aspects

Although little is known about the genetic control of ethanol tolerance (toxicity), the complexity of the inhibitory effects of ethanol on yeast suggests that a large number of genes are involved. This was demonstrated by Ismail and Ali[26] who observed great variation in the tolerance exhibited by meiotic progeny of a diploid and by hybrids constructed by crossing different meiotic segregants. Although it has been extremely difficult to isolate ethanol-resistant mutants by conventional techniques, a turbidostat-based continuous culture system has been successfully employed.[27] In contrast, ethanol-sensitive mutants representing numerous complementation groups have been isolated at high frequency,[28] mirroring the results of Ismail and Ali.[26]

Ethanol has been found to be mildly mutagenic relative to known chemical mutagens. At 24 and 30% (v/v), a 10- and 40-fold increase in the frequency of "petite" mutants was observed, respectively.[29] "Petites" are yeast mutants that are deficient in respiratory metabolism due to nuclear mutations, or more commonly, to mutations in the mitochondrial genome.

E. Studies in Food Systems

As a sole agent of preservation in beverages, minimum concentrations ranging from 18 to 21% (v/v) are required to assure microbiological stability.

Table wines that contain significantly less alcohol are stable because of additional factors: their naturally high acidity, low pH, high content of phenolic compounds, and lack of sugar. Sugar and ethanol in combination have been observed to exert a synergistic inhibitory effect which explains the stability of a class of sweet dessert wines that contains several percent sugar and only 7 to 10% (v/v) ethanol. Beers, which contain very little sugar and 2 to 8% (v/v) ethanol are not microbiologically stable unless pasteurized or subjected to sterile filtration. The most ethanol-tolerant microorganisms known, Lacto*bacillus homohiochii* and *L. heterohiochii* were isolated from saké, and are capable of growth in ethanolconcentrations exceeding 16% (v/v).[30] To prevent their growth, saké and dessert wines such as port and sherry are routinely bottled at an alcohol concentration of 18 to 21% (v/v).

A special type of fruit preserve, "brandied fruit," relies largely on the use of alcohol as a preservative. The intact fruit (usually cherries, plums, apricots, or pears) is steeped in a mixture of sugar and brandy for a period of months to acquire the desired taste and texture.

F. Advantages and Disadvantages

Ethanol is not particularly toxic and is required at high molar concentrations to be an effective preservative. As already noted, in order to be effective as a sole agent of preservation, concentrations in the range of 18 to 21% (v/v) are necessary. This severely limits its use because such high concentrations drastically alter the nature of the food or beverage to which the alcohol is added. In the U.S., beverages containing greater than 0.5% (v/v) are taxed on the basis of their alcohol content. Governmental prohibition in one form or another in many countries forbids consumption of alcoholic foods and beverages by all or some citizens.

For a combination of the above considerations, ethanol is not widely used as a preservative, except in flavor extracts, certain medicines, and distilled beverages. One might reasonably argue that in the case of distilled beverages, the inhibitory concentration of ethanol is appreciated for reasons other than its preservative capacity.

III. SULFITE

A. Biosynthesis

1. Reductive Assimilation from Sulfate

In yeast, sulfite is produced primarily through the reductive assimilation of sulfate (Figure 2). The first step in the formation of sulfite from sulfate is transport of the sulfate anion across the cell membrane. Although two transport systems were suggested through the isolation of mutants resistant to the sulfate analogs chromate and selenate,[31] the actual number is still unresolved.[32] Once inside the cell, sulfate is activated in two ATP-requiring steps. First, sulfate is converted to adenylylsulfate (APS) by ATP-sulfurylase, and APS is subsequently converted to 3′-P- adenylylsulfate (PAPS) by APS kinase. PAPS

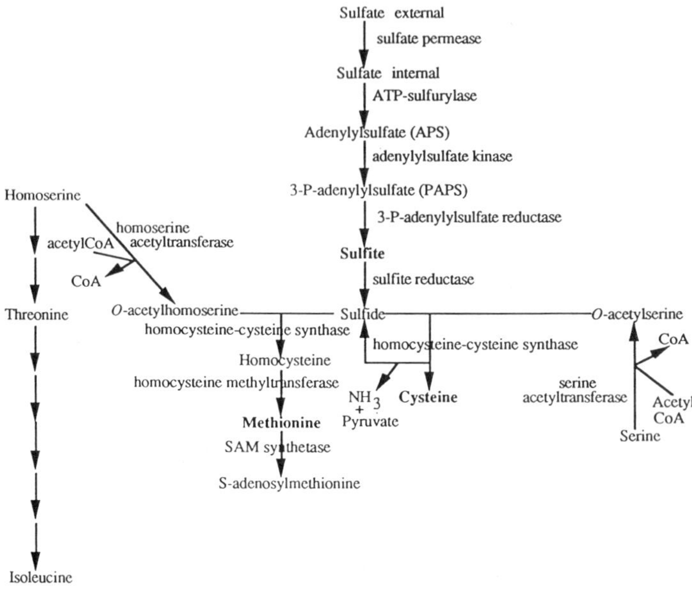

FIGURE 2. Reductive assimilation of sulfate and simplifed scheme of related metabolism in *Saccharomyces cerevisiae.*

reductase converts PAPS to adenosine-3′,5′- diphosphate and free sulfite in a thioredoxin-dependent reduction. Recent work suggests a lack ofa bound sulfite intermediate.[33] Free sulfite is subsequently reduced to sulfide by the activity of sulfite reductase. The sulfur-containing amino acids cysteine and methionine are formed through condensation of sulfide with O-acetylserine and *O*-acetylhomoserine, respectively. The extent to which free sulfite is excreted by cells depends in part on the availability of these carbon skeletons.

2. Catabolism of Cysteine

Although yeast cannot grow on cysteine as a sole nitrogen source,[34] it is capable of metabolizing it to pyruvate, ammonia, and sulfide.[35] This activity may be significant under conditions of nitrogen deprivation. Because excretion of sulfide during wine (or beer) fermentation is extremely undesirable, supplementation of ammonia or urea to nitrogen-deficient grape musts has been recommended as a preventive measure. The extent to which yeast is capable of oxidizing sulfide to sulfite is not clear. The reverse reaction catalyzed by sulfite reductase involves a transfer of six electrons mediated by thioredoxin but ultimately derived from NADPH. The extent to which sulfur compounds of intermediate oxidation state and stability accumulate has not been determined. Suggestive evidence that yeast is capable of oxidizing sulfite — but not necessarily sulfide — is based on the fate of labeled sulfite added prior to fermentation. Both low- and high-sulfite-forming strains (discussed below) produced labeled sulfate and labeled reduction compounds.[35–37] Whether the oxidation was enzyme-catalyzed or not would appear to be of critical importance because nonenzymatic oxidation of sulfite occurs spontaneously. An analogous experiment using labeled cysteine would clarify whether cysteine can indeed serve as a source of sulfite.

B. Chemical and Physical Properties

Sulfur dioxide in various forms (collectively referred to as sulfites) is widely used as a preservative in foods, beverages, and medicinal products. It has both antioxidant and antimicrobial properties, and is a weak acid in aqueous solution:

$$SO_2 + H_2O_2 = H_2SO_3 \quad (1)$$

$$H_2SO_3 = H+ + HSO_3 \quad (2)$$

$$HSO_3 = H+ + SO_3 \quad (3)$$

The pKa values for the dissociation equilibria for Equations 2 and 3 are 1.77 and 7.20, respectively.[38] The relative concentration of each species is pH-dependent; at high pH, the sulfite ion (SO_3^{-2}) predominates; at intermediate pH values, the bisulfite ion (HSO_3^-) predominates; and at low pH, undissociated sulfurous acid is the dominant species. Although SO_2 is volatile, evaporative loss only becomes significant at acid pH values when SO_2 is present at a significant concentation relative to the other species. Sulfur dioxide is a relatively unstable form of sulfur that is potentially reactive with a broad class of biomolecules including nucleic acids, vitamins, sugars, and proteins.

C. Mode and Spectrum of Action

The reactions of sulfite with different classes of biomolecules have been reviewed elsewhere.[39] A few are described here to illustrate potential causes of cytotoxicity.

1. pH Dependence of Toxicity

The toxicity of sulfite to microorganisms has been shown to be a pH-dependent phenomenon because it is only the undissociated sulfurous acid that is able to cross the cell membrane. In yeast,[40] and presumably in most other microorganisms, sulfite is not actively transported, but rather crosses the membrane by passive diffusion. Since uncharged sulfurous acid is only present at a significant concentration (relative to sulfite and bisulfite ions) at low pH, it follows that sulfite is most effective as a preservative in acid foods and beverages.

2. Carbonyl Compounds

Sulfites react with aldehydes and ketones to form hydroxysulfonates.[41] Reaction rates between carbonyls and sulfites are fast, and below pH 8, the equilibrium greatly favors hydroxysulfonate formation.[42] The ability of sulfites to inhibit enzymatic and nonenzymatic browning is largely due to their reactions with carbonyl compounds produced in these reactions.[43] The stability and rate of formation of the hydroxysulfonates depends in large measure on thecarbonyl species and the reaction pH. In general, reducing sugars such as glucose do not form stable hydroxysulfonates.[41] In contrast, acetaldehyde hydroxysulfonate is an extremely stable reaction product and its formation has been exploited in the use of yeast to produce significant quantities of glycerol during fermentation. The process was developed by Neuberg (Neuberg's "second form" of fermentation) in Germany in response to a shortage of glycerol during World War I. Under appropriate conditions, addition of sulfite and subsequent reaction with acetaldehyde blocks formation of ethanol and regeneration of oxidized NAD+. Instead, dihydroxyacetone phosphate is reduced to glycerol phosphate to regenerate the oxidized cofactor. Free glycerol is formed following enzymatic hydrolysis of the phosphate group.[44]

3. *Sulfitolysis*

*D*isulfide bonds can be cleaved by sulfite resulting in formation of a thiol (R-SH) and an S-sulfonate ($R\text{-}SSO_3^-$).[45] The reaction with cystine goes essentially to completion at physiologic pH. Susceptibility of protein disulfides to sulfitolysis is greatly enhanced upon denaturation. Reaction with glutathione disulfide may be a physiologically relevant mechanism of sulfite detoxification.[46] Intracellular glutathione contents of UV-generated sulfite-resistant mutants of yeast were found to be higher than in the wildtype parent, although the statistical significance of the difference was not indicated in the report.[47] Glutathione contents of ethyl methanesulfonate- generated sulfite-resistant and sulfite-sensitive mutants examined in a different study were significantly different (and lower) only in two of the latter mutants, representing two of five complementation groups.[48]

4. *Thiamine and NADH*

The reaction of sulfite with thiamine has been known for many years and is considered deleterious because it destroys the vitamin's nutritive value. Irreversible nucleophilic attack of sulfite on the quaternary nitrogen of the thiazole ring results in formation of pyrimidine sulfonic acid and 4-methyl-5-β-hydroxyethylthiazole.[39]

Sulfite can form adducts with NADH by addition to the 3,4 double bond. The stability of the addition product is enhanced if NADH is associated with an enzyme.[49]

5. *Selected Enzymes*

It has been reported that millimolar concentrations of sulfite cause a rapid depletion of the ATP content of yeast at low pH values. Studies on the effects of sulfite on energy metabolism have shown that the glycolytic pathway is effectively impaired. The most important enzymes affected are glyceraldehyde-3-phosphate dehydrogenase and alcohol dehydrogenase.[50–52] *a. Glyceraldehyde-3-Phosphate Dehydrogenase*

Inhibition of glycolysis at the step of glyceraldehyde-3-phosphate dehydrogenase is thought to be largely responsible for the decrease in ATP generation. Among enzyme activities assayed in sulfite-treated yeast cells, glyceraldehyde-3-phosphate dehydrogenase was shown to be the most sensitive to sulfite.[50–52] Glyceraldehyde-3-phosphate dehydrogenase catalyzes the conversion of glyceraldehyde-3-phosphate to 1,3-diphosphoglycerate, which in a subsequent step is converted to 3-phosphoglycerate with the concomitant production of ATP. Inactivation of the enzyme blocks the glycolytic pathway and causes a depletion of ATP, since the two ATPs expended earlier in the pathway cannot be recovered.

b. Alcohol Dehydrogenase

Inhibition of alcohol dehydrogenase was also observed.[52] The enzyme catalyzes the reduction of acetaldehyde to ethanol by NADH during alcoholic fermentation. While this step does not involve the production of ATP directly, the NAD+ regenerated by acetaldehyde reduction is an obligate electron acceptor in the earlier reaction catalyzed by glyceraldehyde-3-phosphate dehydrogenase. Thus, when alcohol dehydrogenase is inhibited, NAD+ is no longer produced, and this in turn inhibits the oxidation of glyceraldehyde-3-phosphate, indirectly causing depletion of ATP. The most obvious result of this effect is the blocking of ethanol formation.

6. Effects on Intracellular pH

When sulfite crosses the cell membrane as undissociated sulfurous acid, it is converted into bisulfite and sulfite ions because of the near-neutral intracellular pH. At the same time, the intracellular pH decreases due to the dissociation, which in turn lowers the transmembrane pH gradient, assuming an acidic growth medium. This would tend to retard or inactivate processes that require energy derived from proton-motive force such as active transport.[53]

Maier et al.[52] measured the effect of sulfite on the intracellular proton concentration of glucose-starved yeast cells and found that in 1 m*M* sulfite the average intracellular proton concentration increased about 100-fold. This intracellular acidification specifically stimulated the F1-ATPase resulting in ATP depletion. Yeast cells treated with formic acid underwent a similar internal acidification analogous to the sulfite-treated cells. However, compared to sulfite the effect on ATP content was much less marked. This finding suggested that dissipation of the proton-motive force during intracellular acidification of itself was of little or secondary importance with respect to sulfite-induced depletion of ATP.[52]

D. Genetic Aspects

1. Control of Formation

The ability of yeasts to form sulfite during alcoholic fermentation has long been known. Since sulfite is regularly added to wine as an antioxidant and antimicrobial agent, and since health concerns have led to efforts to restrict its use as an additive, the production of sulfite by yeast itself has become a matter of some importance. Yeast strains have been shown to differ in their ability to form sulfite. In addition to genetic factors, the nutritional status of the growth medium plays an important and complex role. Gyllang et al.[54] have attempted to rationalize possible effects of sugar concentration and amino acid availability during fermentation to explain why sulfite formation by a brewing strain increased as sugar levels increased, and why it either decreased or increased in the presence of certain amino acids.

They reasoned that a measured increase in osmotic pressure caused by the addition of glucose stimulated glycerol production. The known inhibitory effect of glycerol on alcohol dehydrogenase caused acetaldehyde and pyruvate to accumulate. The large intracellular pools of these two carbonyl compounds bound sulfite, and the resultant hydroxysulfonates freely crossed the cell membrane, stimulating more sulfite production. A further effect was proposed: excess carbon dioxide (present as bicarbonate) competed with bisulfite for sulfite reductase, effectively blocking sulfite reduction and starving the cell for sulfide, methionine, and S-adenosyl methionine (SAM). Such starvation would stimulate further futile reduction of sulfate to sulfite.

The observed inhibition of sulfite formation by methionine was explained by the known repression of sulfite formation by SAM, which is derived from methionine. The stimulation bythreonine and isoleucine was explained in the following manner: threonine feedback inhibits its own synthesis and that of homoserine, the precursor of O-acetylhomoserine. This is turn causes sulfide and sulfite to accumulate because of lack of a carbon backbone to which the sulfide can condense to form the immediate precursor of methionine, homocysteine. This accounting, however, does not explain why O-acetylserine, the precursor of cysteine, should be simultaneously unavailable for sulfhydrylation by the accumulating sulfide. The stimulatory effect of isole ucine is based on its ability to block utilization of threonine (from which it is derived), by a combination of feedback inhibition and repression. Thus, addition of isoleucine has the same net effect as addition of threonine. The observed stimulation of sulfite formation by addition of serine could not be rationalized.[54]

Most *S. cerevisiae* strains produce between 10 and 30 mg of sulfite per liter when tested under comparable conditions, but some produce less than 10 while others form more than 100, indicating great genetic variability.[35] Eschenbruch[35] has suggested calling the former strains "low-sulfite-forming" and the latter, "high-sulfite-forming". In grape must, a differential response to sulfate levels has been observed. Low-sulfite-forming strains failed to form more sulfite when sulfate levels increased, whereas high-sulfite-forming strains did. Addition of cysteine or methionine to grape must has been shown to reduce or completely suppress sulfate uptake and sulfite formation by both low- and high-sulfite-forming strains. Similar results have been reported for brewing strains.[35]

In a review of genetic aspects of sulfite formation in wine yeasts, Snow[55] suggested that enzymes involved in sulfite biosynthesis may be productive targets for converting high-sulfite- forming strains to low-sulfite formers and vice versa. For example, Heinzel and Trüper[56] showed that although in low- and high-sulfite-forming strains, the same specific actvity of ATP-sulfurylase was found, the enzymes possessed different kinetic and regulatory properties. The enzyme in the former strain was saturated by an ATP concentration of 4 m*M*, whereas in the latter strain, saturation occurred at 30 m*M*. The enzyme from the low-sulfite-forming strain was strongly feedback-inhibited by APS, whereas only slight inhibition was observed in the high-sulfite- forming strain. Differences in the active uptake of sulfate and its regulation by methionine were also found in low- and high-sulfite-forming strains.[57] Clearly, deliberate introduction of the genes encoding the desirable forms of the relevant enzymes or rational in vitro modification of the cloned genes offer one means of exerting better control of sulfite formation. However, it is doubtful that such an integral trait will ever be completely divorced from environmental influence. And because winemakers and brewers cannot predict precisely the nutritional composition of their raw materials,attainment of absolute control over sulfite formation during fermentation will remain a challenge, at least for the near future.

2. Resistance

Saccharomycodes ludwigii[58] and wine strains of *S. cerevisiae* are relatively resistant to sulfite. However, the basis for the resistance has not been established. Zambonelli et al.[59] found great differences in resistance among wine strains, and determined that a single dominant gene was responsible for the resistant phenotype. As noted earlier (Section C.3), elevated levels of glutathione were observed in UV-generated sulfite-resistant mutants of *S. cerevisiae*[47] but not in another collection of resistant mutants for which a single dominant gene was also found responsible.[48]

Rose[60] has proposed three different mechanisms to explain resistance: relative impermeability of the plasma membrane to sulfite, an unusual acidic intracellular pH, and production of sulfite-binding compounds such as acetaldehyde. The first possibility is in conflict with the finding that resistance is dominant, because if sulfite transport is blocked or impaired, one would expect that a resistant x wildtype hybrid would have both wildtype and mutant membrane components, rendering it sensitive to sulfite. Just the opposite has been observed. The yeast cytoplasm has a pH of about 6.5, which means that upon entry into the cell, SO_2 quickly dissociates to bisulfite and sulfite ions, causing more SO_2 to diffuse in. If the intracellular pH were lower than 6.5, one would expect a decreased rate of entry of SO_2 into the cell, and a decreased sensitivity. Finally, production of sulfite-binding compounds such as acetaldehyde would result in formation of hydroxysulfonates which can presumably cross the plasma membrane passively and provide the cell with a means for removal of intracellular sulfite. By this reasoning one might expect that all strains of the vigorously fermentative *S. cerevisiae* would produce sufficient levels of acetaldehyde and pyruvate to scavenge the relatively small amounts of sulfite added as a preservative. However, great variation in the level of resistance has been observed among different strains. Cloning and characterization of the resistance gene(s) ought to help clarify the mechanisms.

E. Studies in Food Systems

The FDA recognizes sulfite as safe for food use, except on fresh fruits and vegetables intended for consumption in the raw state. When residues exceed 10 ppm (expressed as total SO_2), current regulations require that a label on packaged foods and beverages indicate its presence.[61] Sulfite prevents browning reactions, controls microbial spoilage, and is used as a bleaching agent in the manufacture of certain foods, such as maraschino cherries.

Sulfur dioxide is used in gaseous or liquid form (as the sodium or potassium salt of sulfite) on dried fruits, molasses, wines, fruit juices, and other foods. It has been used as a food preservative since ancient times, and its use as a meat preservative in the U.S. dates to at least 1813.[62] Because it is reactive with thiamine, its use is no longer permitted in meats or other foods considered sources of the vitamin. Post-harvest growth of molds such as *Botrytis* on grapes is controlled by gassing with sulfur dioxide.

Direct comparison of the effectiveness of various sulfite concentrations in different foods against different organisms is not meaningful for at least two reasons. One is the known pH dependence of the bacteriocidal effect and the other is the potential reactivity of sulfite with carbonyl and other compounds. The carbonyl-addition products such as acetaldehyde hydroxysulfonate are not bacteriocidal and hence reduce the effective sulfite concentration. In general, among fungi, the obligately aerobic molds are more sensitive to sulfite than yeasts. Among yeasts, *Saccharomycodes ludwigii* appears to be the most resistant species examined, exhibiting greater tolerance than wine strains of *Saccharomyces cerevisiae*. Sulfite is effective against all known bacterial species capable of causing food spoilage. In fruit juices and other beverages, it has been shown to inhibit growth of *Acetobacter* spp. and lactic acid bacteria. In meat products, sulfite-inhibited growth of *C. botulinum,* Salmonellae and other Enterobacteriaceae, *Serratia liquefaciens, S. marcescens,* and *Hafnia alvei*.[62]

F. Advantages and Disadvantages

The greatest advantages of sulfite as a preservative are its GRAS status, its broad specificity against microorganisms at relatively low concentration, its relative nontoxicity to humans, and its extensive history of use in food preservation. As an antioxidant and antimicrobial agent, sulfite has been used since ancient times in winemaking. For many applications including winemaking, sulfite has no obvious replacement. Although control of sulfite production by yeast during growth has potential application for reducing sulfite additions to grape must and wine, our limited understanding of the factors which regulate such production make this a remote possibility in the near future.

Nonetheless, sulfite has disadvantages. The pH dependence of its antimicrobial activity restricts its judicious use to acidic foods and beverages, or to those foods to which acid can be added without deleterious effect. To be effective at higher pH, greater concentrations are required. Of greater importance, at least one million individuals in the U.S are believed to be sulfite-sensitive. In these individuals, exposure to sulfite can trigger life-threatening hypersensitivity reactions. Sulfite-induced deaths caused by consumption of improperly sulfited lettuce in a salad bar and other nonfatal cases of hypersensitivity provoked the current FDA regulations regarding sulfite use and labelling requirements. Because of heightened public awareness of the potential danger of sulfite, much interest has been generated in finding acceptable alternatives.

Because regulation of sulfite formation in yeast is still incompletely understood, rational control of this trait is not yet possible. And without control, deliberate exploitation of sulfite formation for use in food and beverage preservation cannot be expected soon.

IV. KILLER TOXINS

The killer phenomenon among fungi was first observed in certain strains of *Saccharomyces cerevisiae* that were originally classified into three phenotypes: "killer," "sensitive," and "neutral."[63] Killers secrete a proteinaceous toxin to which they and neutral cells are immune, but which kills sensitive cells. Sensitive cells generally belong to the same genus as the killer in the case of *Sacchromyces,* but this is not usually so in other genera in which toxins have been observed to possess significantly broader activity.[64] Since the original finding, many genera of yeast as well as other fungi have been found to possess this trait.[65–68] What distinguishes most killer systems identified thus far in *Saccharomyces* is that the killer character is generally encoded by dsRNA viruses. The phenomenon is widespread, as most laboratory strains that have been tested are K1 killers.

A. Biosynthesis

In yeast, the most thoroughly characterized killer toxins identified thus far are encoded by different dsRNA mycoviruses. Young and Yagiu[69] classified types K1, K2, K3; Pfeiffer and Radler[70] reported a fourth type, KT28; Extremera et al.[71] have described a fifth; and Goto et al.[72] have described a chromosomally encoded toxin, KHR. Young[65] has noted that other *S. cerevisiae* strains have been identifed in which the killer character was inherited in a Mendelianmanner, indicating a chromosomally encoded trait. Killer toxins are secreted from cells in the exponential phase of growth.

In the K1 killer system, the most extensively studied, two types of viral particles have been identified: the M1 dsRNA genome that encodes the killer toxin and immunity factor, and the L dsRNA genome that encodes the capsid protein for both particles. With respect to the K1 toxin, the distinction made

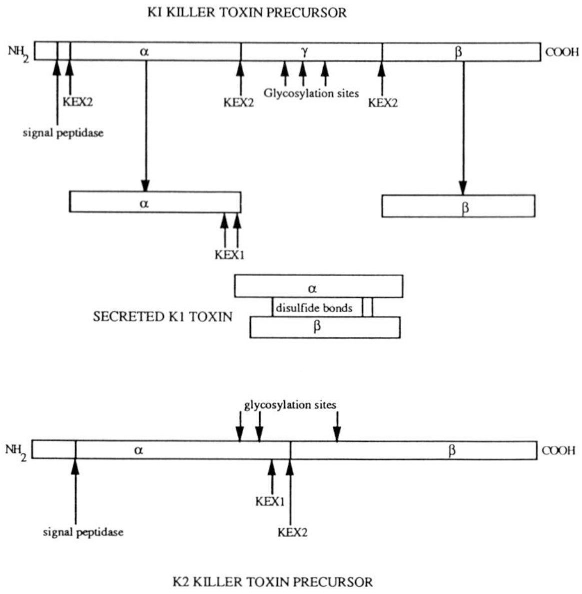

FIGURE 3. Postulated structures and precessing of the K1 and K2 killer toxins of *Saccharomyces cerevisiae* (not drawn to scale).[73,74]

between toxin-insensitive "neutral" and "resistant" strains is that neutral strains contain an M ds RNA genome that produces an immunity factor but that no longer encodes a functional toxin, whereas resistant strains harbor a nuclear gene or genes that confer resistance.

Production of the mature K1 toxin requires extensive processing of a precursor polypeptide[73] (Figure 3). Reading from the amino terminus, the precursor contains a signal peptide of some 26 amino acid residues, and an 18-residue proregion followed by α, γ, and β subunits. Cleavage sites recognized by the chromosomally encoded KEX2 endoprotease separate the subunits from one another, and the α subunit from the proregion. Final proteolytic processing involves removal of arginine residues from the carboxy terminus of the α subunit by another chromosomally encoded protease, KEX1. The mature toxin is an α-β dimer linked by disulfide bridges. Mutational analysis[75] has shown that the N-glycosylated γ subunit plays important roles in precursor processing and secretion, perhaps by promoting correct folding. A similar analysis has shown that α sequences are involved in conferring immunity but likely as part of the intact precursor, since KEX1- and KEX2-processing proteases are not essential for immunity functions.

The K2 toxin is encoded by a dsRNA species termed M2. At the nucleic acid and protein levels, the K2 toxin shows no sequence identity with K1. A cDNA copy of the M2 ds RNA has been expressed in yeast using the yeast alcohol dehydrogenase *(ADH1)* promoter.[74] The precursor toxin and immunity polypeptide appear to be processed similarly to that encoded by K1. While KEX1 and KEX2 processing proteases are required for efficient K2-specific killing, K2- specific immunity is independent of the proteases. Thus, as in the case of the K1 precursor, the intact K2 precursor may confer immunity. The K2 toxin is glycosylated and the putative subunits do not appear to be linked through disulfide bonds based on their separate detection on Western blots of SDS PAGE gels in the absence of mercaptoethanol.

B. Physical Properties

The K1 toxin (an α-β disulfide-bridged dimer) has a calculated molecular weight of 20,600 daltons and a pI of 4.34. The α subunit has two hydrophobic domains likely to be part of a channel-forming, membrane-associated region. The dimer is capable of forming aggregates, but it is not known whether such formation is required for channel formation. The observation that reduction of the disulfide bonds renders the α subunit insoluble in aqueous solution suggests a role for the β subunit in maintenance of toxin solubility.[73] The toxin is not glycosylated.

Dignard et al.[74] have proposed that the K2 toxin consists of two polypeptide components, the major one of which is *N*-glycosylated and has a molecular weight of 21,000 daltons.

The K28 killer toxin, isolated from a wine strain of *S. cerevisiae,* has a molecular weight of 16,000 daltons. Its carbohydrate moiety consists of 10 O-linked D-mannose residues via serines and/or threonines.[76]

The KHR toxin described by Goto et al.[71] consists of a single polypeptide with a molecular weight of 20,000 and a pI of 5.3. The gene encoding the toxin has been localized to chromosome IX.

C. Mode and Spectrum of Action

The K1 killer toxin kills sensitive cells of *S. cerevisiae* and *Torulopsis glabrata*[77] but has no effect on whole cells of other microorganisms tested. In contrast, spheroplasts of cells belonging to the genera *Candida, Kluyveromyces,* and *Schwanniomyces* are sensitive.[78] The killing activity is expressed within a narrow pH range, 4.6 to 4.8, is unstable at temperatures in excess of 25°C, and can be stabilized against shearing by gelatin.[79] Although less stable than K1 with respect to pH, and in spite of a lack of sequence identity, the physiology of killing by the K2 toxin is very similar to that of K1.[73] Killing requires at least two steps: attachment to a cell wall receptor containing a β-1,6-D-glucan, and perturbation of the plasma membrane resulting in ion leakage and cell death. Toxin-resistant mutants lacking the glucan receptor *(kre)* could be sensitized to the toxin if converted to spheroplasts. By use of the patch clamp technique, Martinac et al.[80] have shown that the toxin forms voltage-independent cation channels in sensitive yeast spheroplasts and in artificial liposomes. Extensive mutational analysis implicates both α and β subunits in binding to the β-glucan cell wall receptor, and the hydrophobic α subunit in ion channel formation and immunity.[81]

Although the K28 killer toxin is similar to the K2 toxin in amino acid composition, molecular weight, pI, and in being a glycoprotein, it does not bind to the β-1,6-D-glucan receptor but rather to a cell wall mannoprotein.[82,83] Toxin binding is effectively blocked by concanavalin A, which is known to bind to the D-mannopyranose moiety of the mannoprotein, as well as by antimannoprotein antibodies.[84] Mutants with defined alterations in mannoprotein side chain structure *(mnn2* and *mnn5)* do not bind toxin and are rendered resistant to K28-induced killing. When the mutants are converted to spheroplasts by enzymatic digestion of the cell wall, they regain sensitivity to the toxin just as *kre* mutants lose their resistance to the K1 killer when converted to spheroplasts.

D. Genetic Aspects

In the killer systems encoded by cytoplasmic dsRNA viruses, the killer phenotype is inherited in a non-Mendelian manner, although a large number of nuclear genes have been identified that affect expression of the killer character. Young[65] has summarized the genetics of the K1 system which has been intensively studied by Wickner and co-workers.

In this system, chromosomal mutations have been described that affect killer expression (*kex*), resistance *(rex),* maintenance of the M dsRNA genome (*mak* and mutations *pet18*, *spe2*, and *spe10),* regulation of killer activity *(ski),* and killer replication bypass *(KRB).* Mutations in *KEX1* and *KEX2* impair proteolytic processing of the precursor polypeptide and result in loss of killer expression although immunity is retained. The *rex1* mutation confers a suicide phenotype. Such strains can only be grown under conditions in which the toxin is inactive, such as at pH 5 or higher. Unsuccessful efforts to isolate suicide mutants after hydroxylamine mutagenesis of the toxin-immunity cDNA expression vector have suggested to some workers that it is unlikely that a genuine suicide mutant has been described or is even possible.[75] Thirty mutations have been described that result in loss of the M dsRNA genome, but not the L dsRNA. Some *mak* mutations cause slow growth, *pet18* results in loss of mitochondrial DNA, *MAK1*, *MAK6,* and *MAK8* are allelic with genes for topoisomerase, *SPE2* encodes adenosyl methionine carboxylase, which is involved in synthesis of spermidine and spermine, and *SPE10* encodes ornithine decarboxylase required for putrescine biosynthesis. Additional chromosomal mutations have been isolated that confer a "super-killer" phenotype *(ski)* resulting in increased production of the killer toxin. Thedominant mutation termed *KRB* (killer replication bypass) suppresses *mak7* and *pet18* which otherwise cause loss of the M dsRNA genome. *KRB* does not, however, enable *pet18* mutants to maintain mitochondrial DNA.

Results of the extensive mutational analysis of an M1 dsRNA cDNA expression vector undertaken by Bussey and co-workers have been described in previous sections (IV.A and B). Because this approach has been so fruitful in dissecting the functional elements of the M1 viral genome, it is currently being applied to the other dsRNA-based killer systems.

E. Studies in Food Systems

Wine and beer fermentations are subject to contamination by killer strains of many yeast genera. Their presence in grape musts can no longer be overlooked as a possible cause of stuck or sluggish fermentations that in the past were often ascribed exclusively to nitrogen deficiency.[85–87] Most recent surveys of the yeast microflora of grapes and other natural fermenting substrates have found killer yeasts.[88–90] Killer yeasts belonging to *Debaromyces, Hansenula, Candida,* and *Pichia* have also been isolated from such fermented foods as miso, soy sauce, and salted vegetables.[91]

Heard and Fleet[88] conducted grape fermentations with different combinations of killer and sensitive wine strains of *S. cerevisiae* and observed killing at pH 3.0 and 3.5 and at 15, 20, and 25°C. The sensitive strains were eliminated after 2 to 3 d when the same concentration of killer and sensitive cells were inoculated. No killing effect was observed when the initial ratio of killer-to-sensitive cells was approximately 1:7 or 1:10. Although the killer type was not specified, activity at pH 3.0 to 3.5 suggests the K2 toxin. van Vuuren and Wingfield85 have documented that wild K2 killer wine strains were responsible for stuck fermentations in commercial cellars. In controlled, mixed-strain fermentations in which the initial ratio of K2 killer to nonkiller was 1:500, the time required for completion was significantly prolonged at 20°C, and noticably but less prolonged at 30°C.

Jacobs and van Vuuren[87] investigated the effects of four different K2 killers isolated from grapes on two sensitive wine strains as a function of initial cell ratios. The less sensitive of the two wine strains was hardly affected irrespective of the initial concentration of killer cells. In contrast, the other sensitive strain was eliminated by the end of the fermentation when the initial killer inoculum constituted 0.1% of the total yeast population. An earlier report by Tredoux et al.[86] based on use of various combinations of different sensitive and killer strains indicated that unlessthe starting concentration of killer cells was at least 2.5% of the total yeast population, little effect on fermentation rate was observed. Other than indicating which strain predominated, no quantitative data were presented regarding the relative concentration of killer and sensitive cells isolated at the end of the fermentations. Petering et al.[92] observed that in mixed-culture grape fermentations, killer activity was only detected when the initial ratio of killer-to-sensitive cells exceeded 1:2, and at a ratio of 2:1, sensitive cells were not completely eliminated.

F. Advantages and Disadvantages

The major disadvantages of killer toxins as antimicrobial agents are their comparatively narrow spectra of activity and equivocal effectiveness in practical settings. None is active against bacteria, and the most extensively studied, the K1 and K2 toxins, are limited to killing sensitive strains of S. *cerevisiae* and *Torulopsis glabrata*. In contrast, killer toxins produced by non- *Saccharomyces* species appear to have broader activity, i.e., all kill some *Saccharomyces* strains.[65] Aside from the specificity of killing, the *Saccharomyces* killer toxins are only active in a narrow range of pH and temperature. For example, activity is generally assayed at 20°C, because instability has been observed at temperatures in excess of 25°C. The K2 toxin is active from pH 2.9 to 4.9,[87,93] whereas the K1 toxin retains significant activity only from pH 4.6 to 4.8.[79]

Clearly, these considerations severely limit the potential broad application of yeast killer toxins in foods and beverages. At the same time, because the toxins are unstable proteins, it is highly unlikely that they would survive passage through the human gastrointestinal tract and cause untoward side effects. A number of workers have introduced and evaluated wildtype K1 or K2 toxin production in brewing[94–97] and wine[98–100] strains.

Development of recombinant killer toxins that overcome some of the limitations of the natural forms holds promise as a means of exploiting the killer phenomenon. An engineered yeast toxin with the broad specificity and relative stability of nisin, for example, would have obvious application in brewing and winemaking. Based on the intense interest in fungal killer toxins, it may not be unreasonable to expect that in the near future, recombinant toxins with relatively broad killing activity derived from non-*Saccharomyces* yeasts will be expressed in production strains of *S. cerevisiae*. Use of such strains will have to be reconciled with the caution[85] that their inevitable contamination of wine cellars, breweries, and other fermentation facilities will preclude future use of sensitive but otherwise desirable yeasts.

V. CONCLUDING REMARKS

Among the three yeast metabolites discussed in this review, killer toxins appear to have the greatest practical potential for widespread use as biopreservatives in fermented foods and beverages. Aside from our relatively poor understanding of how environmental and genetic factors control sulfite formation, current industrial and regulatory unease about sulfite use will hamper the significant research efforts required to exploit this trait rationally. Although we know a great deal about ethanol formation, the limited and specialized application of ethanol as a preservative is due mostly to its relative nontoxicity (on a molar basis) and to a lesser extent, to restrictive regulation.

Because natural killer toxins produced by *Saccharomyces* species appear to lack the requisite potency and broad specificity necessary for use as food preservatives, their genetic modification is indicated. Two approaches are conceivable: (1) genetic modification of *Saccharomyces* toxin genes and (2) introduction of non-*Saccharomyces* killer toxin genes into production strains of *Saccharomyces*. Both approaches are ambitious and will engender regulatory complications. At the outset, it might be prudent to assign the killer toxins a more neutral designation, such as "zymocins"[65] to allay inevitable and understandable public alarm. Eventual development of recombinant S*accharomyces* strains possessing zymocidal and bacteriocidal activity promises to have broad application in the fermentation industries and to further illuminate a fascinating natural phenomenon.

REFERENCES

1. **Drewke, C., Thielen, J., and Ciriacy, M.,** Ethanol formation in *adh^o* mutants reveal the existence of a novel acetaldehyde-reducing activity in *Saccharomyces cerevisiae, J. Bacteriol.,* 172, 3909, 1990.
2. **Fraenkel, D. G.,** Carbohydrate metabolism, in *The Molecular Biology of the Yeast Saccharomyces: Metabolism and Gene Expression,* Strathern, J. N., Jones, E. W., and Broach, J. R., Eds., Cold Spring Harbor Laboratory, Cold Spring Harbor, New York, 1982, 1.
3. **Wills, C.,** Regulation of sugar and ethanol metabolism in *Saccharomyces cerevisiae, Crit. Rev. Biochem. Mol. Biol.,* 25, 245, 1990.
4. **Sherman, P. D. and Kavasmaneck, P. R.,** Ethanoic acid, in *Encyclopedia of Chemical Technology,* Vol. 9, 3rd ed., Mark, H. F., Othmer, D. F., Overberger, C. G., and Seaborg, G. T., Eds., John Wiley & Sons, New York, 1980, 338.
5. **Casey, G. P. and Ingledew, W. M.,** Ethanol tolerance in yeasts, *CRC Crit. Rev. Microbiol.,* 13, 219, 1986.
6. **D'Amore, T., Panchal, C. J., Russell, I., and Stewart, G. G.,** A study of ethanol tolerance, *Crit. Rev. Biotechol.,* 9, 287, 1990.
7. **Millar, D. G., Griffiths-Smith, K., Algar, E., and Scopes, R. K.,** Activity and stability of glycolytic enzymes in the presence of ethanol, *Biotechnol. Lett.,* 4, 601, 1982.
8. **Guijarro, J. M. and Lagunas, R.,** *Saccharomyces cerevisiae* does not accumulate ethanol against a concentration gradient, *J. Bacteriol.,* 160, 874, 1984.
9. **Dasari, G., Keshavarz, E., Connor, M. A., and Pamment, N. B.,** A reliable method for detecting the intracellular accumulation of fermentation products: application to intracellular ethanol analysis, *Biotechnol. Lett.,* 7, 541, 1985.
10. **Ingram, L. O.,** Ethanol tolerance in bacteria, *Crit. Rev. Biotechnol.,* 9, 305, 1990.
11. **Leao, C. and van Uden, N.,** Effects of ethanol and other alkanols on passive proton influx in the yeast *Saccharomyces cerevisiae, Biochim. Biophys. Acta,* 774, 43, 1984.
12. **Salgueiro,. S. P., Sá-Correia, I., and Novais, J. M.,** Ethanol-induced leakage in *Saccharomyces cerevisiae*: kinetics and relationship to yeast ethanol tolerance and alcohol fermentation productivity, *Appl. Environ. Microbiol.,* 54, 903, 1988.
13. **Ingram, L. O. and Buttke, T. M.,** Effects of alcohol on microorganisms, *Adv. Microbiol. Physiol.,* 25, 256, 1984.
14. **Osman, Y. A. and Ingram, L. O.,** Mechanism of ethanol inhibition of fermentation in *Zymomonas mobilis* CP4, *J. Bacteriol.,* 164, 173, 1985.
15. **Dombek, K. M. and Ingram, L. O.,** Magnesium limitation and its role in apparent toxicity of ethanol during yeast fermentation, *Appl. Environ. Microbiol.,* 52, 975, 1986.
16. **Leão, C. and van Uden, N.,** Effects of ethanol and other alkanols on the glucose transport system of *Saccharomyces cerevisiae, Biotechnol. Bioeng.,* 24, 2601, 1982.
17. **Eddy, A. A.,** Mechanism of solute transport in selected eukaryotic microorganisms, *Adv. Microbiol. Physiol.,* 23, 1, 1982.
18. **Loureiro-Dias, M. C. and Peinado, J. M.,** Effect of ethanol and other alkanols on the maltose transport system of *Saccharomyces cerevisiae, Biotechnol. Lett.,* 4, 721, 1982.
19. **Ferreras, J. M., Iglesias, R., and Girbés, T.,** Effect of the chronic ethanol action on the activity of the general amino acid permease from *Saccharomyces cerevisiae* var. *ellipsoideus, Biochim. Biophys. Acta,* 979, 375, 1989.
20. **Leão, C. and van Uden, N.,** Effects of ethanol and other alkanols on the ammonium transport system of *Saccharomyces cerevisiae, Biotechnol. Bioeng.,* 25, 2085, 1983.
21. **Iglesias, R., Ferreras, J. M., Arias, F. J., Muñoz, R., and Girbés, T.,** Effect of continued exposition to ethanol on activity of the ammonium and fructose transport systems in *Saccharomyces cerevisiae* var. *ellipsoideus, Biotechnol. Bioeng.,* 37, 389, 1991.
22. **Sá-Correia, I. and van Uden, N.,** Effect of ethanol on the fructose transport system of *Kluyveromyces fragilis, Biotechnol. Lett.,* 5, 413, 1983.

23. **Hayashida, S., Feng, D., and Hongo, M.,** Function of the high concentration alcohol-producing factor, *Agric. Biol. Chem.,* 38, 2001, 1974.
24. **Hayashida, S., Feng, D., Ohta, K., Chaitiumvong, S., and Hongo, M.,** Composition and role of *Aspergillus oryzae* proteolipid as a high concentration alcohol producing factor, *Agric. Biol. Chem.,* 40, 73, 1976.
25. **Hayashida, S. and Ohta, K.,** Cell structure of yeasts grown anaerobically in *Aspergillus oryzae*-proteolipid supplemented media, *Agric. Biol. Chem.*, 42, 1139, 1978.
26. **Ismail, A. A. and Ali, A. M. M.,** Selection of high ethanol yielding *Saccharomyces.* II. Genetics of ethanol tolerance, *Folia Microbiol.*, 16, 350, 1971.
27. **Brown, S. W. and Oliver, S. G.,** Isolation of ethanol-tolerant mutants of yeast by continuous selection, *Eur. J. Appl. Microbiol. Biotechnol.,* 16, 119, 1982.
28. **Aguilera, A. and Benítez, T.,** Ethanol-sensitive mutants of *Saccharomyces cerevisiae, Arch. Microbiol.,* 143, 337, 1986.
29. **Zakharov, I. A. and Bandas, E. L.,** Induction of mutations of respiratory deficiency in yeast by ethyl alcohol, *Sov. Genet.,* 15, 620, 1979.
30. **Ingram, L. O.,** Microbial tolerance to alcohols: role of the cell membrane, *Trends Biotech.*, 4, 40, 1986.
31. **Breton, A. and Surdin-Kerjan, Y.,** Sulfate uptake in *Saccharomyces cerevisiae*: biochemical and genetic study, *J. Bacteriol.*, 132, 224, 1977.
32. **Cartwright, C. P, Rose, A. H., Calderbank, J., and Keenan, M. H. J.,** Solute transport, in *The Yeasts: Metabolism and Physiology of Yeasts,* Vol. 3, 2nd ed., Rose, A. H. and Harrison, J. S., Eds., Academic Press, London, 1989, 5.
33. **Schween, J. D., Krone, F. A., and Husmann, K.,** Yeast PAPS reductase: properties and requirements of the purified enzyme, *Arch. Microbiol,* 150, 313, 1988.
34. **Cooper, T. G.,** Nitrogen metabolism in *Saccharomyces cerevisiae,* in *The Molecular Biology of the Yeast Saccharomyces: Metabolism and Gene Expression,* Strathern, J. N., Jones, E. W., and Broach, J. R., Eds., Cold Spring Harbor Laboratory, Cold Spring Harbor, New York, 1982, 39.
35. **Eschenbruch, R.,** Sulfite and sulfide formation during winemaking — a review, *Am. J. Enol. Vitic.,* 25, 157, 1974.
36. **Eschenbruch, R., Haasbroek, F. J., and de Villiers, J. F.,** On the metabolism of sulphate and sulphite during the fermentation of grape must by *Saccharomyces cerevisiae, Arch. Microbiol.*, 93, 259, 1973.
37. **Eschenbruch, R. and Kleynhans, P. H.,** The influence of copper-containing fungicides on the copper content of grape juice and on hydrogen sulfide formation, *Vitis,* 12, 320, 1974.
38. **King, A. D., Jr., Ponting, J. D., Sanshuck, D. W., Jackson, R., and Mihara, K.,** Factors affecting death of yeast by sulfur dioxide, *J. Food Prot.*, 44, 92, 1981.
39. **Taylor, S. L., Higley, N. A., and Bush, R. K.,** Sulfites in foods: uses, analytical methods, residues, fate, exposure assessment, metabolism, toxicity, and hypersensitivity, *Adv. Food Res.,* 30, 1, 1986.
40. **Stratford, M. and Rose, A. H.,** Transport of sulfur dioxide by *Saccharomyces cerevisae, J. Gen. Microbiol.,* 132, 1, 1986.
41. **Joslyn, M. A. and Braverman, J. B. S.,** The chemistry and technology of the pretreatment and preservation of fruit and vegetable products with sulfur dioxide and sulfites, *Adv. Food Res.,* 5, 97, 1954.
42. **Burroughs, L. F. and Sparks, A. H.,** Sulphite-binding power of wines and ciders. I. Equilibrium constants for the dissociation of carbonyl bisulphite compounds, *J. Sci. Food Agric.*, 24, 187, 1973.
43. **Haisman, D. R.,** The effect of sulphur dioxide on oxidizing enzyme systems in plant tissues, *J. Sci. Food Agric.*, 25, 803, 1974.
44. **Phaff, H. J., Miller, M. M., and Mrak, E. M.,** *The Life of Yeasts,* 2nd ed., Harvard University Press, Cambridge, MA, 1978, 250.

45. **Means, G. E. and Feeney, R. E.,** *Chemical Modification of Proteins,* Holden-Day, San Francisco, 1971, 152.
46. **Kagedal B., Källberg, M., and Sörbo, B.,** A possible involvement of glutathione in the detoxication of sulfite, *Biochem. Biophys. Res Commun.,* 136, 1036, 1986.
47. **Casalone, E., Colella, C. M., Ricci, F., and Polsinelli, M.,** Isolation and characterization of *Saccharomyces cerevisiae* mutants resistant to sulphite, in *Yeasts as a Main Protagonist of Biotechnology,* John Wiley & Sons, New York, 1989.
48. **Xu, X., Wightman, J. D., Geller, B. L., and Baklinsky, A. T.,** Sulfite toxicity in the yeast *Saccharomyces cerevisiae*, Abstract #3123, Proc. Annual Meeting of the American Society for Microbiology, New Orleans, May 26–31, in press, 1992.
49. **Gunnison, A. F.,** Sulphite toxicity: a critical review of in *vitro* and *in vivo* data, *Food Cosmet. Toxicol.,* 19, 667, 1981.
50. **Hinze, H. and Holzer, H.,** Effect of sulfite or nitrite on the ATP content and the carbohydrate metabolism in yeast., Z. *Lebensm. Unters. Forsch.,* 181, 87, 1985.
51. **Hinze, H. and Holzer, H.,** Analysis of the energy metabolism after incubation of *Saccharomyces cerevisiae* with sulfite or nitrite, *Arch. Microbiol.,* 145, 27, 1986.
52. **Maier, K., Hinze, H., and Leuschel, L.,** Mechanism of sulfite action on the energy metabolism of *Saccharomyces cerevisiae, Biochim. Biophys. Acta,* 848, 120, 1985.
53. **Pilkington, B. J. and Rose, A. H.,** Reactions of *Saccharomyces cerevisiae* and *Zygosaccharomyces bailii* to sulphite, *J. Gen. Microbiol.,* 134, 2823, 1988.
54. **Gyllang, H., Winge, M., and Korch, C.,** Regulation of sulfur dioxide formation during fermentation, *Proc 22nd Eur. Brew. Conv.,* Zurich, 1989, 347.
55. **Snow, R.,** Genetic improvement of wine yeast, in *Yeast Genetics, Fundamental and Applied Aspects,* Spencer, J. F. T., Spencer, D. M., and Smith, A. R. W., Eds., Springer-Verlag, New York, 1983, 439.
56. **Heinzel, M. and Trüper, H. G.,** Sulfite formation by wine yeasts. II. Properties of ATP-sulfurylase, *Arch. Microbiol.,* 107, 293, 1976.
57. **Dott, W., Heinzel, M., and Trüper, H. G.,** Sulfite formation by wine yeasts. IV. Active uptake of sulfate by "low" and "high" sulfite producing wine yeasts, *Arch. Microbiol.,* 112, 283, 1977.
58. **Stratford, M., Morgan, P., and Rose, A. H.,** Sulfur dioxide resistance in *Saccharomyces cerevisae* and *Saccharomycodes ludwigii, J. Gen. Microbiol.,* 133, 2173, 1987.
59. **Zambonelli, C., Guerzoni, M. E., and Nanni, M.,** Selecione genetica nei lieviti della fermentazione vinaria. III. La resistenza alla anidride solforosa, *Revista Vitic. Enol.,* 25, 170, 1972.
60. **Rose, A. H.,** Responses to the chemical environment, in *The Yeasts: Yeasts and the Environment,* Vol. 2, 2nd ed., Rose, A. H. and Harrison, J. S., Eds., Academic Press, London, 1987, 32.
61. **Anon.,** Food labelling; declaration of sulfiting agents, *Fed. Reg.,* 51, 25012, 1986.
62. **Jay, J. M.,** *Modern Food Microbiology,* 3rd ed., Van Nostrand Reinhold, New York, 1986, 264.
63. **Bevan, E. A. and Makower, M.,** The physiological basis of the killer character in yeast, in *Genetics Today, Proc. 11th Int. Congr. Genet.,* Gert, S. J., Ed., 1(Abstr.), 203, 1963.
64. **Philliskirk, G. and Young, T. W.,** The occurrence of killer character in yeasts of various genera, *Antonie van Leeuwenhoek,* 41, 147, 1975.
65. **Young, T. W.,** Killer yeasts, in *The Yeasts: Yeasts and the Environment,* Vol. 2, 2nd ed., Rose, A. H. and Harrison, J. S., Eds., Academic Press, London, 1987, 131.
66. **Stark, M. J. R., Boyd, A., Mileham, A. J., and Romanos, M. A.,** The plasmid-encoded killer system of *Kluyveromyces lactis*: a review, *Yeast,* 6, 1, 1990.
67. **Tipper, D. J. and Bostian, K. A.,** Double-stranded ribonucleic acid killer systems in yeasts, *Microbiol. Rev.,* 48, 125, 1984.
68. **Wickner, R. B.,** Double-stranded RNA replication in yeast: the killer system, *Annu. Rev. Biochem.,* 55, 373, 1986.

69. **Young, T. W. and Yagiu, M.,** A comparison of the killer character in different yeasts and its classification, *Antonie van Leeuwenhoek, J. Microbiol. Ser.,* 44, 59, 1978.
70. **Pfeiffer, P. and Radler, F.,** Purification and characterization of extracellular and intracellular killer toxin of *Saccharomyces cerevisae* strain 28, *J. Gen. Microbiol.,* 128, 2699, 1982.
71. **Extremera, A. L., Martin, I., and Montoya, E.,** A new killer toxin produced by *Saccharomyces cerevisae, Curr. Genet.,* 5, 17,1982.
72. **Goto, K., Iwase, T., Kichise, K., Kitano, K., Totuka, A., Obata, T., and Hara, S.,** Isolation and properties of a chromosome-dependent KHR killer toxin in *Saccharomyces cerevisae, Agric. Biol. Chem.,* 54, 505, 1990.
73. **Bussey, H., Boone, C., Zhu, H., Vernet, T., Whiteway, M., and Thomas, D. Y.,** Genetic and molecular approaches to synthesis and action of the yeast killer toxin, *Experientia,* 46, 193, 1990.
74. **Dignard, D., Whiteway, M., Germain, D., Tessier, D., and Thomas, D. Y.,** Expression in yeast of a cDNA copy of the K2 killer toxin gene, *Mol. Gen. Genet.,* 227, 127, 1991.
75. **Vernet, T., Boone, C., Greene, D., Thomas, D. Y., and Bussey, H.,** Yeast killer toxin: site-directed mutations affecting toxin and immunity domains of the precursor gene, in *FEMS Symp. Natural Antimicrobial Systems,* 35, Bath University Press, Claverton Down, England, 1986, 222.
76. **Schmitt, M. J. and Pfeiffer, P. C.,** Immunochemical analysis of the carbohydrate moiety of yeast killer toxin K_{28}, *Antonie van Leeuwenhoek,* 58, 277, 1990.
77. **Bussey, H. and Skipper, N.,** Killing of *Torulopsis glabrata* by *Saccharomyces cerevisiae* killer factor, *Antimicrob. Agents Chemother.,* 9, 352, 1976.
78. **Zhu, H. and Bussey, H.,** The K1 toxin of *Saccharomyces cerevisiae* kills speroplasts of many yeast species, *Appl. Environ. Microbiol.,* 55, 2105, 1989.
79. **Woods, D. R. and Bevan, E. A.,** Studies on the nature of the killer factor produced by *Saccharomyces cerevisiae, J. Gen. Microbiol.,* 51, 115, 1968.
80. **Martinac, B., Zhu, H., Kubalski, A., Zhou, X.-L., Culbertson, M., Bussey, H., and Kung, C.,** Yeast K1 killer toxin forms ion channels in sensitive yeast spheroplasts and in artificial liposomes, *Proc. Natl. Acad. Sci. U.S.A.,* 87, 6228, 1990.
81. **Zhu, H. and Bussey, H.,** Mutational analysis of the functional domains of yeast K1 killer toxin, *Mol. Cell. Biol.,* 11, 175, 1991.
82. **Pfeiffer, P. and Radler, F.,** Comparison of the killer toxin of several yeasts and the partial purification of a toxin of type K_2, *Arch. Microbiol.,* 137, 357, 1984.
83. **Schmitt, M. and Radler, F.,** Molecular structure of the cell wall receptor for killer toxin KT28 in *Saccharomyces cerevisiae, J. Bacteriol.,* 170, 2192, 1988.
84. **Schmitt, M. and Radler, F.,** Blockage of cell wall receptors for yeast killer toxin KT28 with antimannoprotein antibodies, *Antimicrob. Agents Chemother.,* 34, 1615, 1990.
85. **van Vuuren, H. J. J. and Wingfield, B. D.,** Killer yeasts — cause of stuck fermentations in the wine cellar, *S. African J. Enol. Vitic.,* 7, 113, 1986.
86. **Tredoux, H. G., Tracey, R. P., and Tromp, A.,** Killer factor in wine yeasts and its effect on fermentation, *S. African J. Enol. Vitic.,* 7, 105, 1986.
87. **Jacobs, C. J. and van Vuuren, H. J. J.,** Effects of different killer yeasts on wine fermentations, *Am. J. Enol. Vitic.,* 42, 295, 1991.
88. **Heard, G. M. and Fleet, G. H.,** Occurrence and growth of killer yeasts during wine fermentation, *Appl. Environ. Microbiol.,* 53, 2171, 1987.
89. **Moore, K. J.,** Yeasts from Arkansas Wine Grapes: Sulfite Resistance and Killer Toxin Production as a Function of pH and Temperature, Ph.D. thesis, University of Arkansas, Little Rock, 1989.
90. **Rosini, G.,** The occurrence of killer character in yeasts, *Can. J. Microbiol.,* 29, 1462, 1983.
91. **Suzuki, C., Yamada, K., Okada, N., and Nikkuni, S.,** Isolation and characterization of halotolerant killer yeasts from fermented foods, *Agric. Biol. Chem.,* 53, 2593, 1989.

92. **Petering, J. E., Symons, M. R., Langridge, P., and Henschke, P. A.,** Determination of killer yeast activity in fermenting grape juice by using a marked *Saccharomyces* wine yeast strain, *Appl. Environ. Microbiol.,* 57, 3232, 1991.
93. **Shimizu, K., Adachi, T., Kitano, K., Shimazaki, T., Totsuka, A., Hara, S., and Dittrich, H. H.,** Killer properties of wine yeasts and characterization of killer wine yeasts, *J. Ferment. Technol.,* 63, 421, 1985.
94. **Young, T. W.,** The genetic manipulation of killer character into brewing yeast, *J. Inst. Brew.,* 87, 292, 1981.
95. **Young, T. W.,** The properties and brewing performance of brewing yeasts possessing killer character, *J. Am. Soc. Brew. Chem.,* 41, 1, 1983.
96. **Sasaki, T., Watari, J., Kohgo, M., Nishikawa, N., and Matsui, Y.,** Breeding of a brewer's yeast possessing anticontaminant properites, *J. Am. Soc. Brew. Chem.,* 42, 164, 1984.
97. **Hammond, J. R. M. and Eckersley, K. W.,** Fermentation properties of brewing yeast with killer character, *J. Inst. Brew.,* 90, 167, 1984.
98. **Hara, S., Iimura, Y., and Otsuka, K.,** Breeding of useful killer wine yeasts, *Am. J. Enol. Vitic.,* 31, 28, 1980.
99. **Seki, T., Choi, E.-H. and Ryu, D.,** Construction of killer wine yeast strain, *Appl. Environ. Microbiol.,* 49, 1211, 1985.
100. **Boone, C., Sdicu, A.-M., Wagner, J., Degré, R., Sanchez, C., and Bussey, H.,** Integration of the yeast K1 killer toxin gene into the genome of marked wine yeasts and its effect on vinification, *Am. J. Enol. Vitic.,* 41, 37, 1990.

INDEX

A

B

D

E

M

N

O

P

Q

R

S

Z